Aircraft Display Systems

Aircraft Display Systems

Malcolm Jukes

Professional Engineering Publishing

Professional Engineering Publishing Limited,
London and Bury St Edmunds, UK

This edition published 2004 by Professional Engineering Publishing, UK.
Published in USA by American Institute of Aeronautics and Astronautics, Inc.

ISBN 1 86058 406 3

A CIP catalogue record for this book is available from the British Library.

Printed by The Cromwell Press Ltd, UK

About the Author

Malcolm Jukes *BSc, FRAeS, FIEE* has over 35 years' experience in the aerospace industry, mostly working for the Smiths Group at Cheltenham, UK. Among his many responsibilities as Chief Engineer for Defence Systems Cheltenham, Malcolm managed the design and experimental flight trials of the first UK Electronic Flight Instrument System (EFIS) and subsequently the development and application of shadow-mask CRT technology for multi-function head-down displays on the F/A-18, AV8B, Eurofighter Typhoon and EH101 aircraft. Malcolm also managed the development of Head-Up Displays and mission computing for several aircraft types, including the Harrier, AV8-B, Sea Harrier, and Hawk lead-in fighter. In this role, and subsequently as Technology Director, he was responsible for product technical strategy and the acquisition of new technology for Smiths UK aerospace products. One of his most significant activities was the application of AMLCD technology to civil and military aerospace applications.

Malcolm is now an aerospace consultant operating in the areas of displays, display systems, and mission computing.

Related Titles

Title	*Author*	*ISBN*
Aircraft Systems – Second Edition	I Moir and A Seabridge	1 86058 289 3
Civil Avionics Systems	I Moir and A Seabridge	1 86058 342 3
Air Engines	T Finkelstein and A J Organ	1 86058 338 5
Aircraft Conceptual Design Synthesis	D Howe	1 86058 301 6
Optimizing the Super-Turbocharged Aeroengine	J Panting	1 86058 080 7
Aviation Ground Support Equipment	IMechE Seminar	1 68058 194 3
Aerospace Transmissions Technology	IMechE Seminar	1 86058 199 4
Journal of Aerospace Engineering	Proceedings of the IMechE, Part G	ISSN: 0954/4100

For the full range of titles published, by Professional Engineering Publishing (publishers to the Institution of Mechanical Engineers) contact:

Marketing Department
Professional Engineering Publishing Limited
Northgate Avenue
Bury St Edmunds
Suffolk IP32 6BW
UK
Tel: +44 (0) 1284 763277; Fax: +44 (0) 1284 704006
E-mail: marketing@pepublishing.com
www.pepublishing.com

Contents

Series Advisors' Foreword

The field of aerospace is wide ranging and covers a variety of disciplines and domains, not merely engineering but many related supporting activities, all combining to produce exciting and technologically challenging products.

The Aerospace Series is a practical and topical series of books aimed at engineering professionals, operators, and users in the aerospace industry. The range of topics is wide ranging, covering design and development, manufacture, operation and support of aircraft as well as topics such as infrastructure operations, and developments in research and technology. This series will provide a source of relevant information that will be of interest and benefit to all those people working in aerospace.

Aircraft Display Systems is written by a prominent practitioner in the field. Malcolm Jukes has spent most of his working life developing military and civil aircraft displays. The book addresses how the displays system performs a key part of the human–machine interface on modern aircraft and how technology has evolved to meet an increasingly demanding requirement. The topic is covered in a historical, contemporary, and forward-looking treatise that is both easy to read and very informative.

Ian Moir
and
Allan Seabridge

Author's Preface

Aircraft Display Systems is part of a series of books that together are intended to describe the complete set of systems which form an essential part of modern civil and military aircraft.

Aircraft Display Systems is a broad subject, encompassing operational requirements (what the crew needs in order to safely prosecute their task), sensors and data sources (such as flight, navigation, and targeting), and finally the display medium itself.

Aircraft Display Systems provides on overview of the historical development of the civil flight deck and the military cockpit (or crewstation) from the beginnings of manned flight to modern aircraft, highlighting – by specific aircraft case studies – the significant steps in the transitions from mechanical to electromechanical instruments to modern electro-optical displays. In these transitions the interplay between operational need, sensor developments, and display technologies interact. New sensors often drive new displays as much as new display technologies drive new displays, both to satisfy a new operational requirement. Now in the information age, future operational catalysts for change to crewstation displays include the Future Air Navigation Systems in the civil transport arena and Digital Battlespace in the military arena

Aircraft Display Systems provides an in-depth explanation of current display technologies applied to Head-Down, Head-Up, and Helmet-Mounted Displays and emerging display technologies candidate for use in next generation aircraft crewstations.

Becoming ever more complex and providing as they do the interface between the human and the machine, it is imperative that display systems be developed against the backdrop of proven approaches and techniques that assure safety and integrity. *Aircraft Display Systems* provides an insight into human vision and the perception of display information in the severe airborne lighting environment together with an overview of good design practice and the regulatory requirements.

Aircraft Display Systems is aimed at practitioners in the aerospace environment – researchers, engineers, designers and operators, and crew. It is aimed at a wider audience than engineers and will be of interest to people working in marketing, procurement, manufacturing, commercial, and legal departments. Furthermore, it is intended to complement undergraduate and postgraduate courses in aerospace systems to provide a path to an exciting career in aerospace.

The book will operate at a number of levels:

- provide a top-level overview of aircraft display systems with some historical background;
- provide more in-depth description of specific display equipments and technologies;
- provide references and suggestions for further reading for those who wish to develop their knowledge further.

I have tried to deal with this complex subject in a straightforward and descriptive manner, setting the context for new technologies into the background of historical display evolution. Though focussed on current display systems, technologies, and design practices I have included a comprehensive overview of potential new technologies and systems appropriate to next-generation aircraft display systems. To meet the challenge of next-generation aircraft display systems designers will require a thorough understanding of the operational needs and the capabilities and limitations of the display technologies available. My hope is that this book makes a contribution to that understanding.

Malcolm Jukes

Acknowledgements

Though I have spent most of my career working in the field of Aircraft Display Systems, the more knowledge and experience I have gained, the more I have realized there is to gain. I have been privileged through my employment at Smiths Aerospace, and through the contacts I have made with those in a similar fields in other companies and organizations worldwide, to experience first hand many of the advances made in the transition from electromechanical instruments to the modern 'glass' flight deck of today's civil and military aircraft. I have found and still, find the entire subject of matching technical implementation to operational need totally absorbing. However, the writing of this book would not have been possible without the encouragement, guidance, and generous support of many colleagues and specialists in those companies.

Particularly I should like to thank Ian Moir, a former colleague at Smiths Aerospace, Professor John Roulston, Technical Director of BAE SYSTEMS Avionics Group, and Robb Brockie, Leon Skorczewski, and my other friends and colleagues of the UK Industrial Avionics Working Group (the IAWG) without whose encouragement I would not have begun to write the book.

My thanks and appreciation also go especially to Christopher Bartlett, BAE SYSTEMS Rochester who, with David Craig and others, provided much of the material concerning BAE SYSTEMS display products and concepts.

I should also like to thank my friends and former colleagues at Smiths Aerospace, in particular David Fraser, Peter Allsopp, Andrew Hall, Andy Harwood, Chris Overton, Alison Starr, and Julie Williams who tirelessly supplied my seemingly endless requests for data. Also to Curt Casey, Kaiser Electronics (now part of Rockwell Collins), Brian Riddle, librarian of the Royal Aeronautical Society, Faye Lomax of the Boeing Company, and Jochim Kaiser, Darmstadt University now with DASA.

Invaluable support has been given by the following companies and organizations:

Airbus
BAE SYSTEMS
Brilliant Technology
Cambridge Display Technologies
Darmstadt University
DASA
Duxford Imperial War Museum
eMagin
Microvision Inc.
Planar Advance
QinetiQ
Rockwell Collins
Smiths Aerospace
Texas Instruments
Thales
The Boeing Company

Honeywell
Kentron
Korry Electronics
The Royal Aeronautical Society
VDO Luftfahrtgerate Werk

Finally I should like to thank my wife and family for their wholehearted support and encouragement to write this book.

Acronyms and Abbreviations

AC Alternating Current
ACFD Advanced Civil Flight Deck
ADF Automatic Direction Finding
ADI Attitude Direction Indicator
ADS Automatic Dependent Surveillance
AIMS Aircraft Management System
AMLCD Active Matrix liquid Crystal display
ANP Actual Navigation Performance
AOC Airline Operational Centre
ARINC Air Radio INC.
ASAR Arc Segment Attitude Reference
ASI Airspeed Indicator
ATC Air Traffic Control
ATF Advanced Tactical Fighter
ATI Air Transport Indicator
ATM Air Traffic Management

BAe British Aerospace
BARA British Aerospace Regional Aircraft
BFoV Binocular Field of View
BIT Built-In Test

CCIL Continuously Computed Impact Line
CCIP Continuously Computed Impact Point
CCRP Continuously Computed Release Point
CDU Control Display Unit
CFIT Controlled Flight into Terrain
cgpi Colour groups per inch
CJND Chrominance Just Noticeable Difference
CNI Communications, Navigation, Information
CNS Communication, Surveillance, Navigation
CNS/ATM Communications, Navigation, Surveillance/Air Traffic Management
COMED Combined Map and Electronic Display
COTS Commercial Off-The-Shelf
CPA Closest Point of Approach
CRT Cathode Ray Tube

DAS Defensive Aids System
DASS Defensive Aids Sub-System
DC Direct Current
DCDU Data Control Display Unit
DH Decision Height
DLP Digital Light Projector
DMC Display Management Computer
DMD Digital Micro-mirror Device
DME Distance Measuring Equipment
DoD Department of Defence (US)
DOHUD Diffractive Optics Head-Up Display (also sometimes DHUD)
DTI Department of Trade and Industry (UK)
DU Display Unit
DVI Direct Voice Input

EADI Electronic Attitude Direction Indicator
ECAM Engine Crew Alerting Monitor
ECM Electromagnetic Counter Measures
EFIS Electronic Flight Instrument System
EGNOS European Geo-stationary Navigation Overlay System
EHSI Electronic Horizontal Situation Indicator
EHVGS Enhanced Head-up Visual Guidance System
EICAS Engine Instruments and Crew Alerting System
EM Electro-mechanical
EO Electro-optica
ESM Electromagnetic Surveillance Measures
EVS Enhanced Vision System

FAA Federal Aviation Authority
FANS Future Air Navigation System
FAR Federal Aviation Requirements
FLIR Forward looking Infra Red
FMS Flight management System
FPT Flight Path Target
FPV Flight Path Vector
FSF Flight Safety Foundation

GNSS Global Navigation Satellite System

GPS Global Positioning System
GPWS Ground Proximity Warning System

HALE High Altitude Long Range Endurance
HDD Head-Down Display
HF High Frequency
HFDS Head-Up Flight Display System
HGS Head-Up Guidance System
HIDSS Helmet-mounted Integrated Display Sighting System
HMD Helmet Mounted Display
HMI Human–Machine Interface
HSI Horizontal Situation Indicator
HUD Head-Up Display
HVGS Head-Up Visual Guidance System

IC Integrated Circuit
IFALPA International Federation of Airline Pilots Associations
IFF Identification Friend or Foe
IFoV Instantaneous Field of View
IFR Instrument Flight Rules
IHADSS Integrated Helmet-And Display Sighting System
IIS Integrated Instrument System
IIT Image Intensifier Tube
ILS Instrument Landing System
IMC Instrument Meteorological Conditions
IN Inertial Navigation
IRS Inertial Reference System
IRST Infra Red Search and Track
ISIS Integrated Standby Instrument System
ITO Indium Tin Oxide

JAR Joint Aviation Requirements
JHMCS Joint Helmet Mounted Cueing System
JSF Joint Strike Fighter

LANTIRN Low Altitude Navigation and Targeting Infra Red for Night
LASS Local Area Augmentation System
LC Liquid Crystal
LCoS Liquid Crystal on Silicon
LED Light Emitting Diode
LEP Light Emitting Polymer
LIF Lead-In Fighter
LJND Luminance Just Noticeable Difference
LLTV Low Light TeleVision
LRMTS Laser Rangefinder/Marked Target Seeker
LRU Line Replaceable unit

MFD Multi-Function Display
MHDD Multifunction Head Down Display
MMI Man–Machine Interface
MoD Ministry of Defence (UK)
MPCD Multi-Purpose Colour Display
MPD Multi-Purpose Display
MTF Modulation Transfer Function
MWIR Millimetric Wave Imaging Radar

NASA National Aeronautical and Space Administration
ND Navigation Display
NDB Non-Directional Beacon
NVE Night Vision Equipment
NVG Night Vision Goggle
NVIS Night Vision Imaging System

OLED Organic Light Emitting Diode

PDU Pilots Display Unit
PFD Primary Flight Display
PJND Perceived Just Noticeable Difference
PMFD Primary Multi-Function Display
PPS Precise Position Service
PSR Primary Support Radar
PWB Printed Wiring Board

RA Resolution Advisory
RAE Royal Aircraft Establishment
RAF Royal Air Force
RAIM Receiver Autonomous Integrity Monitoring
R-LCD Reflective Liquid Crystal Display
RNP Required Navigation Performance
RTOS Real Time Operating System
RVSM Reduced Vertical Separation Minimum
RWR Radar Warning Receiver

SATCOM SATellite COMmunication
SG Symbol Generator
SMFD Secondary Multi-Function Display
SPS Standard Position Service
SSR Secondary Support Radar
SST Super Sonic Transport
SVS Synthetic Vision System

TA Traffic Advisory
TAB Tape Automated Bonding
TACAN TACtical Air Navigation system
TADS/PNVS Target Acquisition and Designation Sight/Pilots Night Vision System
TAWS Terrain Avoidance Warning System
TCAS Traffic Collision Avoidance System
TCV Terminal Configured Vehicle
TIR Total Internal Reflection
T-LCD Transmissive Liquid Crystal Display
TSO Technical Standard Order

UAV Unmanned Air Vehicle

UCAV Unmanned Combat Air Vehicle
UFCP Up Front Control Panel
UFD Up-Front Display
USAF United States Air Force
USMC United States Marine Corps

VDL VHF Data Link
VFR Visual Flight Rules
VGS Visual Guidance System
VMC Visual Meteorological Conditions
VOR Very high frequency Omni Range

VRD Virtual Retinal Scan
VSD Vertical Situation Display
VSI Vertical Speed Indicator

WASS Wide Area Augmentation System

CHAPTER 1

Introduction

This book, *Aircraft Display Systems*, is intended to introduce the reader to the means by which information is presented to the crew to enable the crew safely and successfully to prosecute their task. In the context of civil air transport that task is to carry passengers and/or freight from point of departure to destination in a complex air traffic environment on time and with optimum efficiency. In the context of military aviation that task will have a military mission objective, such as reconnaissance, transportation, or suppression of enemy air and surface defences.

Aircraft controls and displays technology has changed dramatically over the past 100 years. The advent of compact, high-power, rugged digital devices has facilitated on-aircraft, real-time processing of data. The digital impact has allowed a major shift from dedicated instruments to multifunction displays.

The history of aircraft displays can broadly be divided into three technology eras, the mechanical era, the electromechanical (EM) era, and the electro-optical (EO) era. Although the design boundaries are clear, the time boundaries are vague.

The mechanical era began at the outset of aviation history. The importance of instrumenting the information needed to fly an aircraft was recognized by the Wright brothers very early in their flying adventures.

The use of electromechanical instruments can be traced to the early days of flight, around the 1920s, but took a significant hold on aircraft design in the 1930s and gained impetus during World War II.

The catalyst for the electro-optical era was the dramatic increase in performance and capability of digital electronics in the late 1960s which led to an impetus to change the means of displaying information on the flight deck. Initially commencing with military aircraft, the technology migrated to civil aircraft in the 1980s. CRT technology was the first multifunction display medium, latterly superseded by AMLCD technology. Both provide a more flexible means for the display of information than had hitherto been possible with electro-mechanical instruments. Multifunction displays can show many formats on the same display surface and portray the same piece of information in a variety of different ways. Early CRT-based flight data formats emulated their electromechanical instrument predecessors. In civil transport aircraft the main motivation for the CRT was to improve horizontal situational awareness with a moving map. In civil aircraft a topographical map is seldom used. Map information generally comprises the planned route with navigational aids, destination airport, and possible diversionary airfields presented as a line drawing, overlaid with weather information. In military aircraft the moving map presentation is generally a topographical map overlaid with mission-related features such as preplanned route, potential targets, and threat areas to be avoided.

Chapter 2 traces the evolution of the civil and military transport flight deck from the very beginnings of manned flight through to the present day. Classic aircraft of each period are chosen to illustrate the transition from the mechanical to the electromechanical and finally to the electro-optical 'glass cockpit'. The transition of primary flight instruments from visual flying conditions to 'blind' instrument flying is illustrated to introduce the 'basic 6' and 'classic T' configurations of primary flight instruments and their subsequent transformation into today's Primary Flight (PFD) and Navigation (ND) displays.

Chapter 3 traces the evolution of the military fighter cockpit, or crewstation as it is now known, from its earliest beginnings in World War I to today's fast jet fighter. Again, classic aircraft of each period are chosen to illustrate the transition from the World War I simple fixed gun with a 'rifle' sight to the World War II electromechanical gyro gunsight and the radar 'scope' to current electro-optical Head-Down, Head-Up, and Helmet-Mounted Displays for advanced tactical and global situational awareness in today's digital battlespace.

Chapter 4 introduces the fundamental primary flight instruments of the angle of incidence meter, the airspeed indicator, the tachometer, the altimeter, barometric instruments, and the primary flight instrument set plus the Attitude Direction Indicator (ADI), Horizontal Situation Indicator (HSI), Airspeed Indicator (ASI), and Altimeter instruments found in 'classic' electromechanical cockpits of today.

Chapter 5 discusses the principles of operation and device features of the shadow mask CRT and the active matrix LCD Head-Down Display. Practical examples are given from civil and military aircraft to explain the role of the Head-Down Display to present primary flight, navigation, topographical and tactical map information, systems, sensor (radar and FLIR) images, and weapons information to the crew. The issues of obsolescence, standardization and the use of ruggedized COTS technologies (including AMLCD resizing) are discussed.

The Head-Up Display has been applied predominantly to military fast jet fighter aircraft. Chapter 6 discusses the optical principles of the Head-Up Display and, using specific aircraft examples, describes the role of the Head-Up Display and its principles of operation for air-to-air and air-to-ground weapon aiming. Chapter 7 discusses the emerging application of the Head-Up Display to civil and military transport applications.

The rotorcraft community were the first to apply Helmet-Mounted Display technology. Using specific application examples, Chapter 8 describes the operating principles and the role of the Helmet-Mounted Display to put the Head-Up Display on the pilot's head, thereby overcoming the field of view limitation of the HUD and greatly enhancing the crew's ability to acquire and designate targets and to release weapons off-boresight. Indeed, as currently planned for the joint strike fighter, it is entirely likely the Helmet-Mounted Display will become the primary flight instrument in future fast jet fighters, and the Head-Up Display will no longer be fitted.

Chapter 9 reviews the emerging and potential future display technologies of rear projection reflective LCD, transmissive LCD (Liquid Crystal on Silicon – LCoS), and Digital Micromirror Devices (DMD) and the means to apply these technologies to large-area megapixel Head-Down Displays, Head-Up Displays, and Helmet-Mounted Displays. Virtual Retinal-scan Display (VRD) technology is described with potential application to Helmet-Mounted Displays to offer wrap-around virtual crewstation capability. An overview is provided of other emerging technologies including Organic Light Emitting Diodes (OLEDs).

The introduction of Future Air Navigation System (FANS) concepts necessitated by the rapid increase in air traffic density together with the Terrain Avoidance Warning Systems (TAWS) and Traffic Collision Avoidance Systems (TCAS) has and will bring an information explosion onto the flight deck. Interaction with Air Traffic Control in the future FANS CNS/ATM environment will require displays of air traffic, planning, and presentation of datalink route clearances. These systems and initiatives drive the need for the display of information which reduces crew workload and provides improved situational awareness in an intuitive manner. Chapter 10 describes some of the work in this field using large-area Head-Down Displays to provide intuitive and predictive representations of the aircraft flight path in relation to ground features and other air traffic in order that the crew can make appropriate decisions in a timely manner aimed at reducing Controlled Flight into Terrain (CFIT) and near-miss incidents using 3D and 4D pathway-in-the-sky display formats. These enhanced displays will be fully integrated with the other aircraft systems to include, for example, programming the FMS through the Navigation Display, and providing pictorial presentations of the FMS-generated vertical descent profile as well as the traditional lateral profile.

Similarly, the military flight deck is experiencing an explosion of information resulting from the introduction of the digital battlespace. Military aircraft are now required to operate within a system of systems of cooperating similar and dissimilar manned and unmanned aircraft (UAV and UCAV).

Other aspects of the man–machine interface are discussed, including voice control, audio cueing, touch screen, and cursor control.

Display suite components and their assimilation into dumb, semi-smart and smart architectures are discussed in Chapter 11. Practical realizations are discussed in Electronic Flight Instrument Systems (EFIS) and Engine Instrumentation and Crew Alerting Systems (EICAS/ECAM) of the Airbus A330/340 and Boeing 777. The topics of availability, integrity, and implications for dispatch are introduced. The similarities and differences of federated and modular crewstation architectures are discussed using military fast jet and rotorcraft examples.

Visual performance aspects are discussed in Chapter 12. The performance of the human visual system drives the display visual requirements, and these are reviewed in the context of civil and military applications. The crewstation ambient lighting conditions in day, dusk/dawn, and night are described, together with the principal accepted industry visual performance metrics and optical test methods.

The requirements and means to achieve compatibility with night vision enhancement devices are discussed.

The book is intended to provide both an appreciation of the current information needs and the potential future information needs in civil and military flight decks and an appreciation of the technologies available to the display engineer to provide solutions to meet these needs.

Variously the reader will come across the terms 'cockpit', 'flight deck', and 'crewstation'. Often these terms are used synonymously. In early aircraft the term 'cockpit' was used, and that term has tended to remain in military parlance for a single-seat aircraft, but in more recent times it has been replaced by the term 'crewstation'. In civil aircraft the term 'cockpit' has for some time been replaced by the term 'flight deck'.

CHAPTER 2

Transport Aircraft Flight Deck

Introduction

This chapter chronicles the development of the transport aircraft flight deck from the developments of early flying instruments to today's 'all-glass' flight deck. The periods of development can be classified by their technology as:

- The mechanical era.
- The electromechanical era.
- The electro-optical era.

However, the implementations are blurred.

At the outset of manned flight, instruments developed ad hoc on a needs-must basis with little standardization. Within 30 years, passenger air transportation had become almost routine, and of necessity flight procedures were becoming standardized, as were the flight deck instruments to support that. World War II intervened, but much useful experience was gained in those years, and for that reason this chapter traces the evolution of the transport flight deck through the bombers of World War II. The jet age brought air transportation to the masses. Increasing air traffic density drove the need for automation and changed forever the role of the pilot from a 'seat-of-the pants barnstormer' to a 'manager of flight information and control systems'.

To include all the many flight deck iterations during this period would be overwhelming, so examples have been chosen to illustrate the salient points of the evolutionary process. The parallel evolutionary strands of the civil and the military flight deck are closely interwoven. The technological development in one often feeds the other, and so the chapter closes with a military transport example.

Figure 2.1 provides a timeline illustration of the evolution of the transport aircraft flight deck.

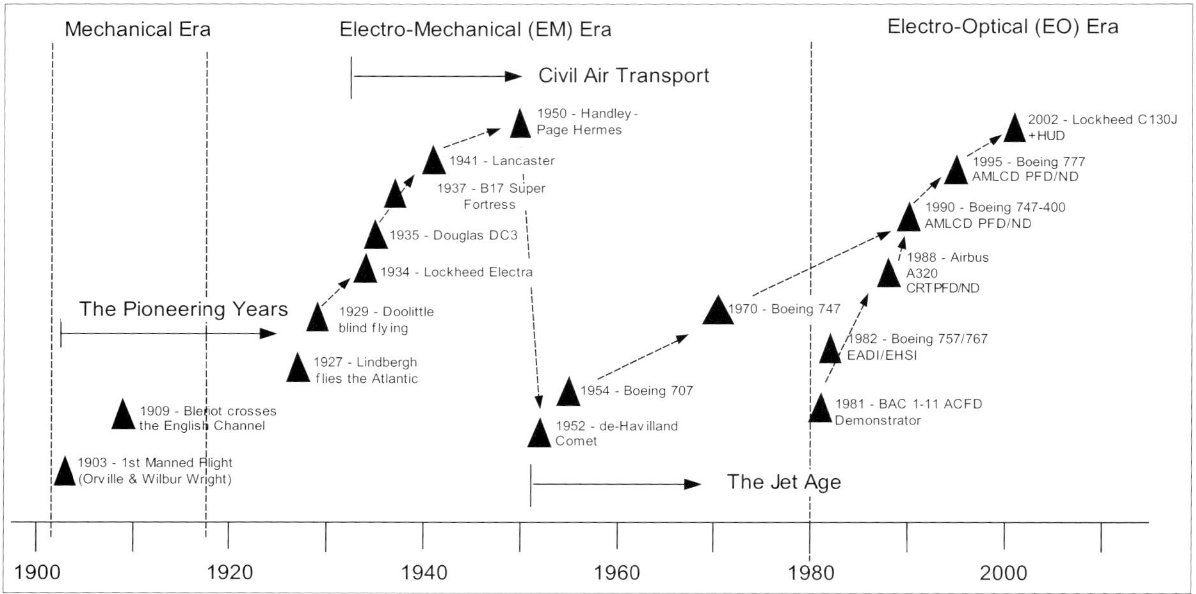

***Fig. 2.1** Transport aircraft flight deck timeline*

The mechanical era

The Wright brothers

The first flight of a heavier-than-air machine is generally recognized to have taken place in Kitty Hawk USA on the 17 December 1903. The aircraft known as the Wright Flyer was built by the Wright brothers and flown by Orville Wright. The aircraft was a biplane, powered by a single 12 hp engine that drove two 'pusher' propellers. The wingspan was 40 ft. The first flight was a mere 850 ft, at a height of 10 ft and a speed of 30 mile/h. The pilot lay in a prone position, facedown. Flying was by 'seat of the pants'. The flying controls warped the wings to achieve control of the aircraft rotational (roll) motion. Vertical (pitch) motion was controlled by a foreplane mounted in front of the wings. A twin rudder controlled the aircraft lateral (yaw) motion.

The importance of providing visual representation of the information needed to fly an aircraft was recognized by the Wright brothers early in their flying ventures, though initially the instruments were recorders of information rather than as pilot aids. It has been said that the first 'instruments' were string and chewing gum, and this is probably not far from the truth. The Wright brothers attached a piece of cloth to the strut between the two landing skids of their early aircraft to indicate critical information to maintain a safe turn to the landing field. If the cloth was streaming roughly parallel to the skids, then the slip angle of the airstream was satisfactory (**1**).

From early times the brothers installed an anemometer (a mechanical device to measure wind speed) a weather vane (to measure angle of incidence), and a rev-counter to indicate engine speed.

Fig. 2.2 Bleriot X1 monplane of 1909 (The Royal Aeronautical Society)

Bleriot

As aviators gained more flying experience, they recognized the need for additional pieces of information in the cockpit, which in turn meant there was a need for some kind of instrument. It did not take many engine failures before the need for data that would warn of an impending problem became obvious. In these early aircraft, engine thrust was proportional to the shaft speed of the propeller, and in the Bleriot X1 monoplane of 1909 a rev-counter was hastily lashed to the aircraft structure to indicate engine speed (see Fig. 2.2).

Armed with this rudimentary instrumentation and a map taped to the 'instrument panel', Bleriot established a new European endurance record of 36 min 55 s on 13 July 1909. At 4.41 a.m. on 25 July 1909 Bleriot set off from Les Baraques (near Calais) to cross the English Channel at a height of 330 ft and to land at Dover Castle at 5.17 a.m. For this he won the Daily Mail prize of £1000.

The barnstorming years

It was not until 1912 that a scientific paper was written on the design of instruments specifically for aircraft. Up to then, reliance had been placed on the direct use or adaptation of existing scientific instruments such as the aneroid barometer for indicating height. Even 10 years after the Wright brothers' pioneering first flight, the demand for flight instruments was not significant. Reasons for this include the preoccupation with aerodynamics, structures, and controls, as well as the reliability of the engines then available. There were more important things to worry about than instruments. If you got off the ground, the difference between cruising speed and take-off speed was small. Flights were limited to fine weather and light winds, and, with flight levels below the clouds, the pilot's instinctive sense of direction and of what was up and what was down was sufficient to keep control of his frail machine. The only other significant instrument to be added to the list of 'essential' instruments prior to the outset of World War I was the airspeed indicator.

World War I was to change totally the emphasis of aircraft and instrument development from an interest for the rich sportsman to front-line fighter pilot. The evolution of the military cockpit for that era is discussed in Chapter 3.

By the end of World War I manned aviation experience had multiplied many times over. At the end of the war there were thousands of aircraft that were superfluous to a peacetime economy. The embryo airlines of the immediate post-war era adapted military aircraft with few modifications, other than the removal of armament, to meet the needs of air transport. Only when it was realized that the key to successful air transport was adherence to a timetabled schedule was time and effort devoted to improving the design of the flight deck.

Lindbergh

On 22 May 1927 Lindbergh landed his Ryan monoplane, the Spirit of St Louis, at Le Bourget Paris after a 28 h flight across the Atlantic. Lindbergh had equipped his aircraft with the most advanced instruments available at the time, but the control interface was crude and his forward view extremely limited.

By 1927 there had been little or no study made of crewstation design and human factors. Instrumentation layout followed no particular scheme. The difficulties of instrument flying are described in detail in Lindberg's book.

The Spirit of St Louis was equipped with 11 instruments (see Fig. 2.3). There was only one gyroscope instrument – a turn and slip indicator. Dominating the instrument panel was a compass, the primary navigation aid. The altimeter, engine speed indicator, and airspeed indicator with the compass gave an approximately rectangular instrument grouping. To the top left of this group was a periscope by which Lindbergh could see along the fuselage and the engine. The lower centre of the instrument panel carried lateral and fore–aft inclinometers. The other instruments and controls were associated with the engine (**2**).

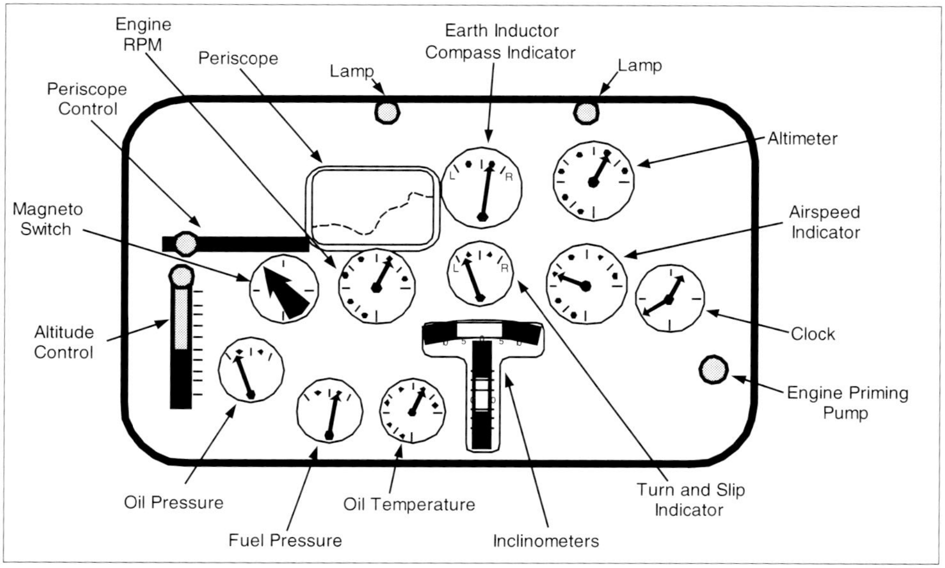

***Fig. 2.3** Flight deck of the 'Spirit of St Louis'*

Lieutenant James Doolittle

Lieutenant James Doolittle was not only a foremost competition flyer but was immensely interested in the whole subject of instrument flying and the loss of spatial orientation of pilots attempting to fly 'blind'. He was convinced that the recently developed gyroscopic instruments were the key to the problem. Working with Paul Kollsman, the developer of the sensitive altimeter, and with Elmer Sperry, the gyrosystems pioneer, Doolittle evolved techniques for 'blind' flying. After more than 100 proving flights, on September 1929 he flew his aircraft, a Consolidated NY-2 single-engined biplane, from take-off, round the airfield circuit, and back to landing with the cockpit hooded so that he was unable to see the horizon or the ground.

Doolittle followed the indications of a radio guidance system and used the recently developed flight instruments to maintain the correct attitude and to descend steadily and at a precise rate to land his aircraft successfully at Mitchell Field. For its time, the specially equipped NY-2 had more than the usual number of instruments. They included:

- A Sperry gyro artificial horizon.
- Direction finder.
- Compass.
- Turn and slip indicator.
- Airspeed.
- Vertical speed.
- A special 0–20 000 ft Kollsman altimeter.

The electromechanical era

The electromechanical era had begun. This period in aviation history saw the standardization of flight deck instrumentation using the then state-of-the-art electromechanical instruments. The initial configuration was known as the 'basic 6' and subsequently migrated to today's standard 'basic T' configuration of attitude (ADI), Airspeed (ASI), height, and compass/direction indicator (HSI) (see Fig. 2.12).

This period saw much rationalization and standardization in instrumentation and information/data formats. After extensive evaluation the basic ergonomic questions were answered such as which direction should a pointer move and how pointers and flags should move in relation to one another. In the end, formats and procedures emerged that were easier to follow, read, and cross-check than the old mechanical instruments. Although the display formats were simpler, the electromechanizations were significantly more complex and ingenious. The servomechanisms, with their complex mechanical gearings, required the skills of a watchmaker to make and maintain. Even so, the data were processed in a relatively simple manner within the constraints imposed by analogue processing of electrical signals and mechanical gear mechanisms.

Those instruments and formats have served the industry well and are still in use on many aircraft in service worldwide. However, the pure economics of maintaining those types of instrument is forcing the pace of upgrade to solid-state displays. Modern electronic 'glass' instrument displays replicate those instrument formats, although they offer extended and more intuitive display formats also.

The Lockheed Electra – 1930s

After the barnstorming years, the 1930s saw the advent of routine air travel for those who could afford it. The Lockheed Electra was one of the first aircraft to enter service with an all-electric flight deck. A two-man crew operated the aircraft. The neat and stylish flight deck, shown in Fig. 2.4, provided all the critical flight instruments, but the layout is (e.g. atitude, altitude, airspeed, and heading) unfamiliar by any of today's standards.

The Electra was a low-wing, all-metal monoplane transport aircraft providing accommodation for ten passengers. Powered by two Pratt and Whitney Wasp Junior radial piston engines, the aircraft had a maximum speed of 200 mile/h at an altitude of 5000 ft and a range of 810 miles. The first prototype flew in February 1934, followed by 143 production aircraft, the first of which entered service with Northwest Airlines late in 1934.

Fig. 2.4 Flight deck of the Lockheed Electra (The Royal Aeronautical Society)

During and after World War II this aircraft with the Douglas DC3 were to help found the rapid growth in civil airline traffic. US airline traffic increased by more than 600 per cent during the period 1936–1941. However, it soon became clear that the small capacity of the Electra made it uneconomic to operate compared with the Douglas DC-3. The DC-3 carried between 21 and 28 passengers at a speed of 207 mile/h, but with a longer range of over 2000 miles. The DC-3 first entered service with American Airlines in 1936. A total of 436 aircraft were built before the war; some are still operating today.

The Boeing B17 Super Fortress – 1930s

The USAAC issued a specification for an advanced bomber in 1934. The Boeing Super Fortress won the competition and the first prototype flew on 28 July 1935. A crew of nine, including the captain and first officer, operated the B-17. The flight deck is shown in Fig. 2.5. The flight instrument layout owes more to the conventions of civil transport aircraft like the Electra than to those of a military aircraft.

The first production aircraft flew on 2 December 1936 and entered service in mid-1937. The B-17 was powered by four 1200 hp Wright R-1820-97 radial engines, had a maximum speed of 287 mile/h at 25 000 ft, a range of 2000 miles, and a maximum payload of 17 600 lb. Some 8700 B-17s were built during World War II, plus another 5000 of other variants later.

Fig. 2.5 *Boeing B-17 (Flying Fortress) flight deck (The Royal Aeronautical Society)*

The Avro Lancaster – 1940s

Perhaps the most successful and certainly the most well-known bomber aircraft of World War II was the Avro Lancaster. The mission role was long-range bombing.

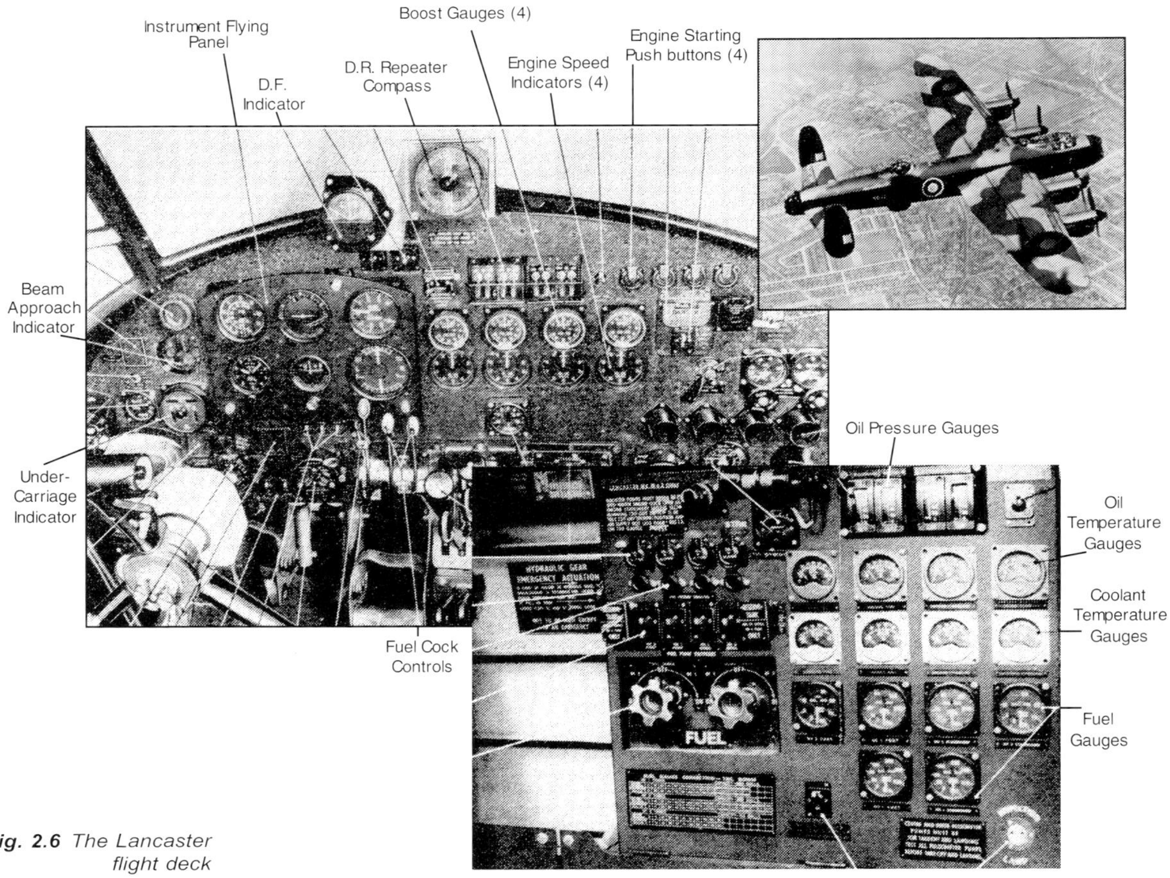

Fig. 2.6 *The Lancaster flight deck*

The flight deck was optimized for precision navigation to/from the target zone. Hence, in addition to conventional flight instruments, radio navigation/direction-finding instruments abound. A pilot (the captain) and a flight engineer flew the Lancaster. There was only one set of flying controls.

The Lancaster flight deck is shown in Fig. 2.6 (**3**). The basic flying instruments were in the configuration later to be known as the 'basic 6' (see in Fig. 2.7).

Cardholders were provided for important aide memoirs including compass deviation and airspeed correction factors. A direction finding indicator and a repeater compass are to be found in the centre of the glareshield (coaming). The direction finder gave bearing information to radio beacons so that the desired course could be flown and targets could be located by triangulation.

To the left of the main flight instruments is located a beam approach visual indicator which provided guidance to the runway at the return airfield. This was a precursor to the Instrument Landing System (ILS) refined after World War II. Also located here are the flaps indicator and the undercarriage indicator.

The four engine boost and speed indicators are mounted close to the centre of the instrument panel so as to be viewable by both the pilot and the flight engineer. The engine throttle and propeller speed controls were mounted in the centre

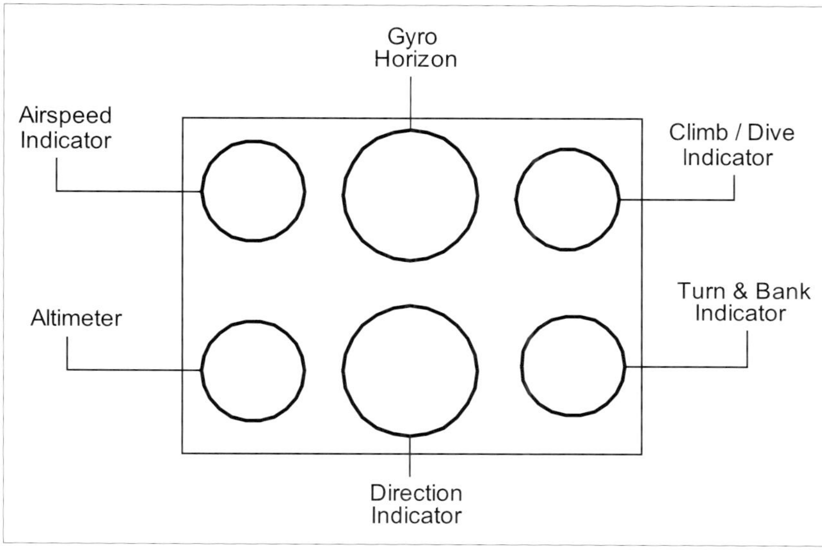

Fig. 2.7 *Basic '6' flight instrument configuration*

console and able to be operated by both crew members during take-off with a full weapons payload.

The controls for the flight engineer included engine start push buttons, two-speed supercharger controls, fuel cock controls, fuel mixer lever and propeller feathering switches. In a panel to his right side, the flight engineer also had engine oil pressure gauges, oil temperature gauges, and coolant temperature gauges. Fuel instrumentation and controls included fuel content gauges for each tank, fuel tank selector cocks, fuel booster pump switches, and warning lights.

The first production Lancaster was flown in October 1941, and the last was delivered on 2 February 1946. Four 1620 hp Merlin engines powered the Lancaster, which had a maximum speed of 287 mile/h and a range of 2530 miles. A total of 7377 machines were built of various mark numbers. The 59 Bomber Command squadrons that operated the Lancaster flew more than 156 000 sorties. The Lancaster dropped over 600 000 t of high-explosive bombs and over 50 million incendiaries. The bomb bay was progressively enlarged to carry larger and heavier bombs; eventually the Lancaster B.Mk1 (special) carried the enormous 12 000 lb 'Tallboy' and the 22 000 lb 'Grand Slam', the heaviest bomb carried by any aircraft in World War II.

The Handley Page Hermes – 1950s

World War II had provided significant experience about flying aircraft for long periods in adverse conditions by day and by night. The Douglas DC-3 had been the workhorse for troop transportation in wartime, and now this aircraft and its pilots were turning to peacetime occupations in the burgeoning civil air transport industry. The Avro Lancaster had showed how to navigate and fly precise missions under adverse conditions at night. Those lessons and procedures now migrated to civil transport aircraft.

The Handley Page Hermes, designed and built shortly after the end of World War II, though not a particularly successful aircraft as an aircraft type, encapsulated 'classic' flight deck design practice of the day. The flight deck layout and the distinct roles of the crew give a real insight into the post WWII flying experience.

The Hermes was basically a commercial version of the RAF Hastings military transport aircraft. The maiden flight of the first production aircraft took place on 5 September 1948. The aircraft entered service with BOAC on 6 August 1950. The aircraft remained in service with a number of operators until the mid-1960s.

Fig. 2.8 *Hermes flight deck – first officer's station*

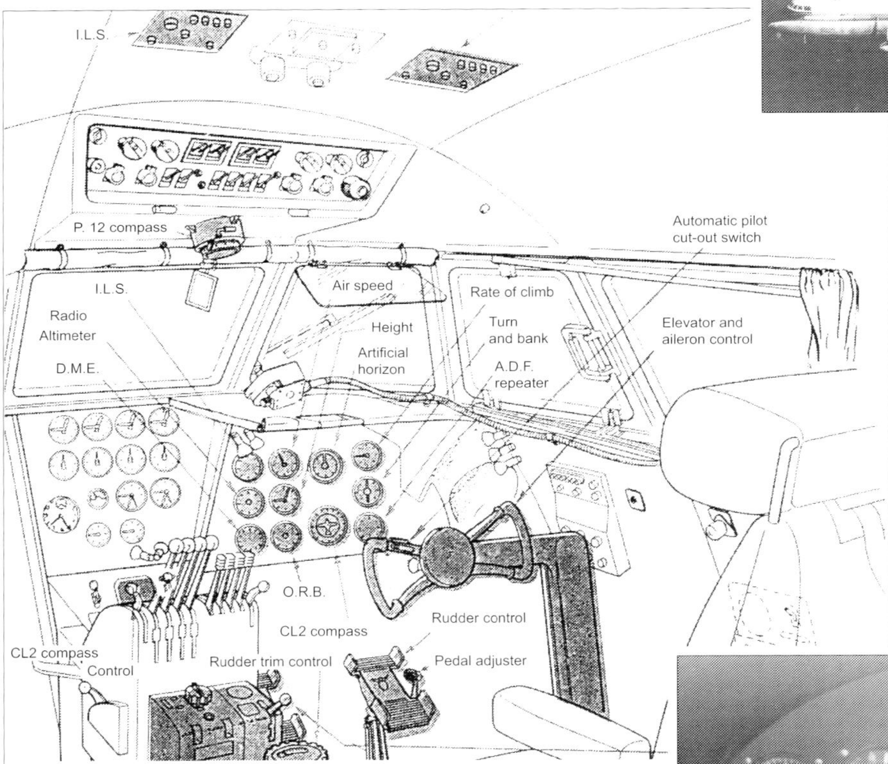

The Hermes was a luxury aircraft. It was one of the first aircraft to have a pressurized hull and therefore could fly above the weather and provide a smoother ride to its passengers who travelled in first-class luxury. A typical route was from London to Sydney. Four Bristol Hercules radial engines powered it, each rated at 2100 hp. The aircraft had a range of 2000 miles, a cruising speed of 275 mile/h, and a ceiling of 24 000 ft.

The aircraft had facilities for 40 passengers plus cabin staff. Five crew, the captain, the first officer, the navigator, the flight engineer, and the radio operator, operated it.

The flight deck of the Hermes is shown in Fig. 2.8. The primary flight instruments were arranged in the 'basic 6' configuration.

The barometric instruments (airspeed, altimeter, and rate of climb) are updated versions of those discussed in Chapter 4 and operate on the principles described directly from the pitot static probes. Two probes were used: one connected to the captain's instruments, the other to the first officer's instruments to provide independence and redundancy. The captain and first officer were required to cross-monitor each other's instruments to check integrity and detect and resolve any discrepancies.

The gyro artificial horizon has been introduced; an essential instrument for flying in Instrument Flying Rules (IFR) conditions. It comprises an integral two-axis gyroscope with mechanical pick-offs to drive the horizon bar to indicate aircraft pitch and roll angles with respect to the ground.

The compass was a repeater instrument driven from the navigator's station that could be slaved to one of two gyrocompass units indicating aircraft heading or operate as a direction indicator to fly to a selected course. A magnetic compass was installed in the roof. As an option, an Automatic Direction Finder (ADF) repeater instrument could be fitted to indicate direction to the selected radio beacon.

On the left of the 'basic 6' instrument cluster there was the option to fit an Instrument Landing System (ILS) instrument which would indicate lateral and vertical deviation from the ILS glideslope and localizer beams, a low-range radio altimeter to provide greater height accuracy during approach and landing, and a Distance Measuring Equipment (DME) instrument to indicate distance to the ILS transmitter (positioned at the runway threshold).

It was the job of the navigator to calculate present position and to plot the course to the destination. To do this he would use a number of on-board systems including radio navigation to triangulate and dead-reckon the aircraft position. If outside the range of radio stations, the navigator could resort to stellar navigation to augment his dead reckoning calculations using the sextant provided, sighting the stars through a periscope in the roof. The navigator's station is shown in Fig. 2.9.

There is at least one published incident of failure correctly to sight stars. On 26 May 1952, BOAC Hermes flight 251 took off from Tripoli in North Africa on an overnight leg across the Sahara desert, making for Accra in Ghana. The navigator inadvertently mis-set the magnetic deviation on the gyrocompass by some 30°. Out of range of radio communications, the crew chose to ignore the readings of the magnetic compass (which was correct) and rely on stellar navigation. The star sightings were misinterpreted to confirm their location to be in accordance with

Fig. 2.9 *Hermes flight deck – navigator's station*

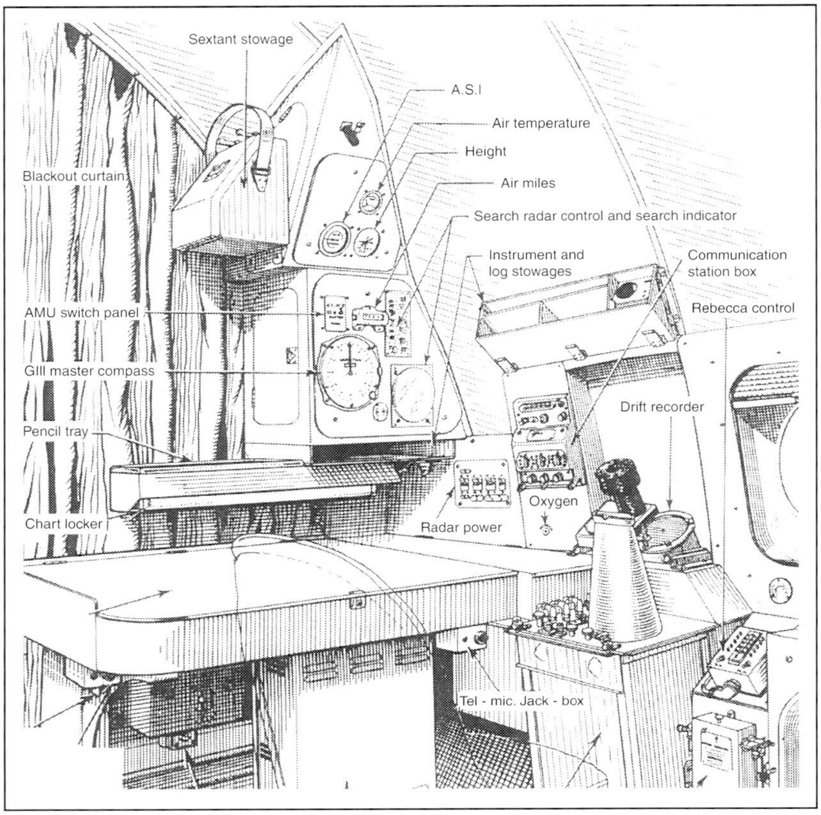

the dead-reckoned position from the incorrect gyrocompass. By daybreak it was clear they were lost. Eventually, the aircraft ran out of fuel and had to force land in the Sahara. The aircraft broke a wing, but all crew and passengers survived the landing. They were rescued by a local tribe of Bedouin and made it out of the desert with the help of the French Foreign Legion. Unfortunately, the first officer later died from an injury sustained during the crash landing and subsequent exposure to the rigours of the desert rescue.

The captain and the first officer operated the throttles of the four-engine aircraft and had repeater engine instruments on the centre console to indicate engine speed and boost.

The flight engineer had a complex and comprehensive set of instruments and controls to monitor and adjust the performance of the engines and of the other on-board systems. The flight engineer's station is shown in Fig. 2.10.

These instruments and controls included:

- Panel No. 1 – engines:
  - Torque, RPM, boost (supercharger),
  - Oil pressure, temperature and warning lights,
  - Fuel pressure, temperature and warning lights,
  - Airspeed and height repeaters,
  - Engine primer switches, magneto and starter switches and warning lamps,
  - Engine synchronization.

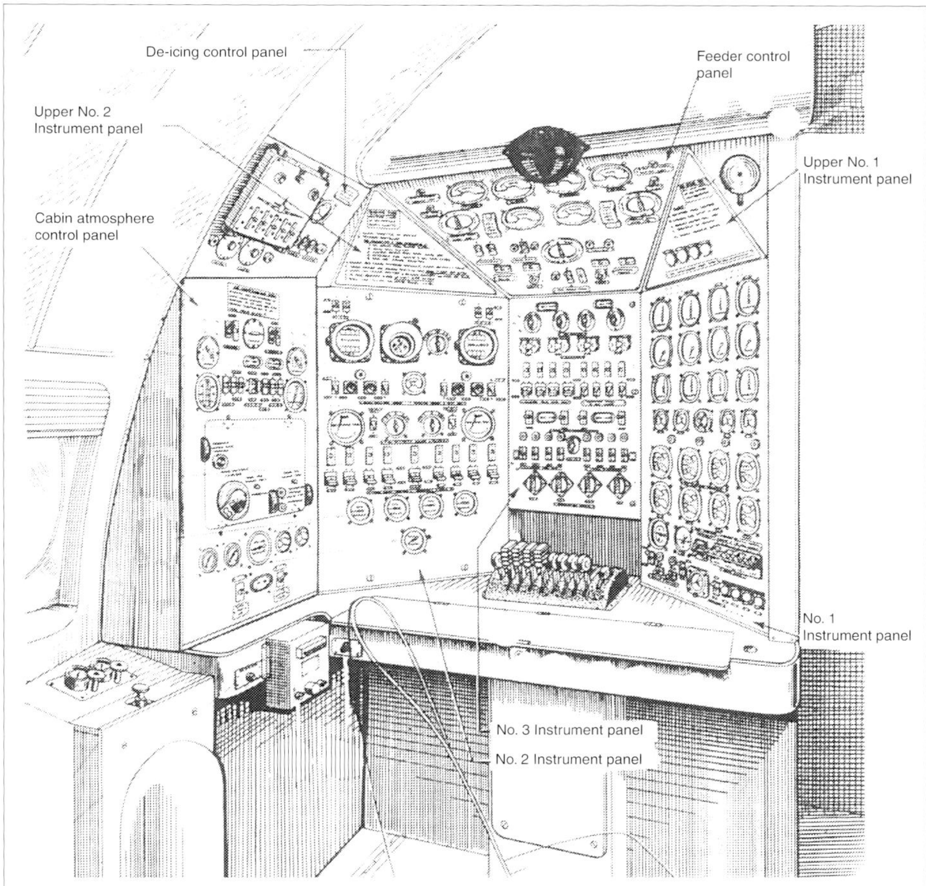

Fig. 2.10 Hermes flight deck – flight engineer's station

- Panel No. 2 – fuel system:
  - Fuel contents gauges and switches,
  - Fuel rate of flowmeter and selector,
  - Fuel pumps, valves, switches, and indicators,
  - Oil pumps, valves, and switches,
  - Hydraulics.
- Panel No. 3 – other systems:
  - Oil cooler controls,
  - Fire extinguishers,
  - Fuel cocks,
  - Cowl gill switches and indicators,
  - Air intake controls,
  - Supercharger controls,
  - Propeller feathering and de-icing controls.
- Upper panels – electrical systems:
  - Generator voltmeters and ammeters,
  - Battery accumulator charging controls and indicators.
- Cabin atmosphere controls.

Fig. 2.11 USAF integrated instrument panel

Finally, the fifth crew member, the radio officer, managed the communication and radio location systems including the ILS glideslope and localizer receivers.

Migration from 'Basic 6' to 'Basic T' instrument configuration

In the late 1950s the US Air Force formed a team of engineers, pilots, and human factor specialists to produce a complete instruments panel. The concept was known as the Integrated Instrument System (IIS). The changes were driven because aircraft were flying faster, were more manoeuvrable, and the weapon systems were becoming more complex. This complexity reduced the time available for the pilot to perform an instrument cross-check. The result was a further step in the presentation of flight parameters to the pilot.

The solution was to display all air data (that is, angle of attack, Mach, airspeed, altitude, rate of climb/descent) on vertical moving tapes rather than on round dials. The IIS panel is shown in Fig. 2.11 (**4**).

The vertical tape scales are read on a fixed horizontal lubber line that extends continuously across all of the tape displays and the Attitude Direction Indicator (ADI). Lateral information is read on a vertical reference line that traverses through the centre of the ADI and the Horizontal Situation Indicator (HSI). The two reference lines thus form the 'T'. Programmable or manually selectable command markers are positioned relative to the tape scales to provide easily noticeable deviations from desired value.

The civil air transport community in general adopted these recommendations, though since space was available on a civil flight deck, the circular counterpointer instruments were preferred to vertical tape scales.

The arrangement of the instrument cluster therefore transitioned from the 'basic 6' configuration to the 'basic T' now universally adopted on all

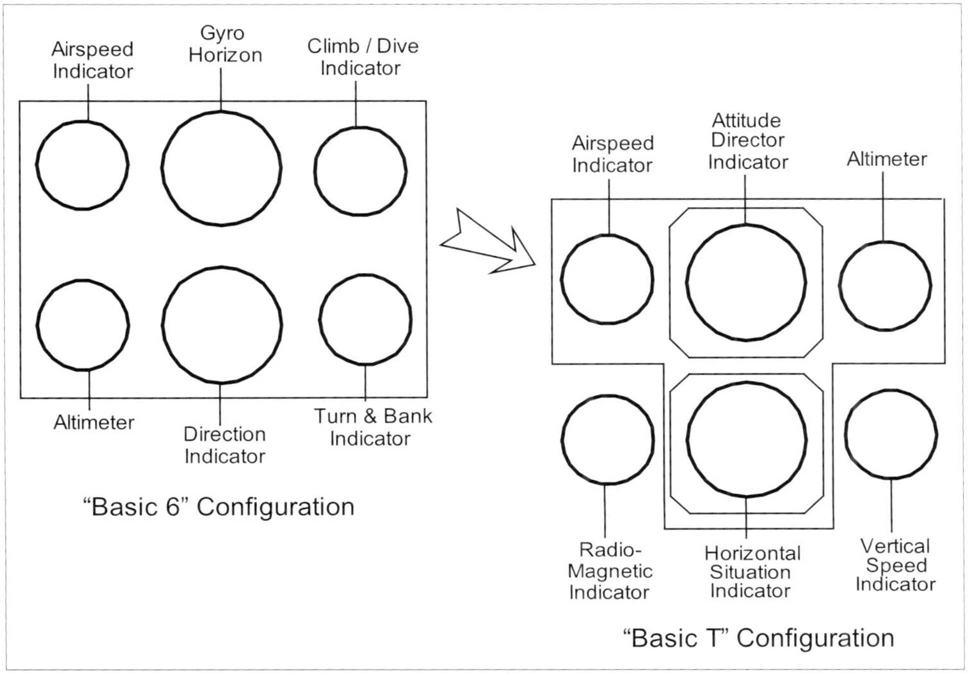

Fig. 2.12 *Migration from 'basic 6' to 'basic T' configuration*

electromechanically instrumented civil transport aircraft operating today (see Fig. 2.12).

The 'basic T' configuration comprises:

- Attitude Direction Indicator (ADI) – top centre.
- Horizontal Situation Indicator (HSI) – bottom centre.
- Airspeed Indicator (ASI) – top left.
- Altimeter – top right.

Ancillary instruments augmenting the 'basic T' primary flight instruments are:

- Radio Magnetic Indicator (RMI).
- Vertical Speed Indicator (VSI).

This transition required the amalgamation of several previous instruments into fewer instruments and used innovative, and sometimes complex, mechanical counters and pointers to make the presentations more intuitive. The altimeter represented the most significant change from a safety perspective. Previous presentations of altitude had used a combination of three rotating pointers to indicate height, each pointer representing a higher order of altitude:

- Outer pointer – 100s of feet.
- Middle pointer – 1000s of feet.
- Inner pointer – 10 000s of feet.

This had led to much confusion.

The new instrument indicated altitude as a single pointer, calibrated in 100s of feet (as the outer pointer of the earlier instrument), with the complete height value

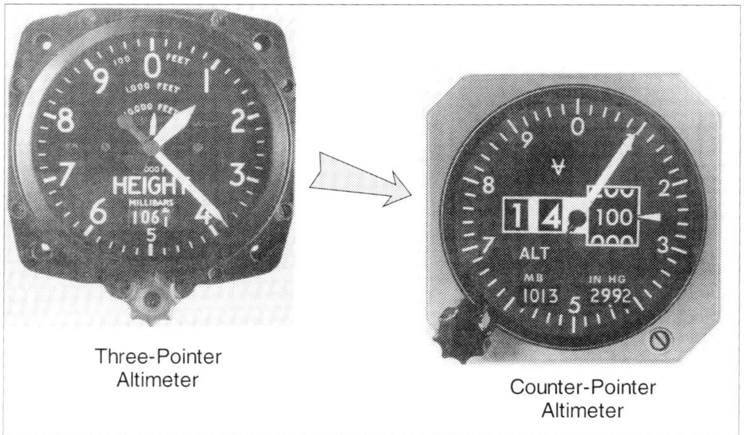

Fig. 2.13 *Transition from 3-pointer to counter-pointer altimeter (Smiths Aerospace)*

itself presented as a numeric readout in the centre of the instrument on a set of rotating drums, viewed through a window.

The layout of both instruments is shown in Fig. 2.13.

The ADI and HSI are commonly found in 5ATI format, that is, having a physical form factor in accordance with the Air Transport Indicator (ATI) standard, and having a 5 × 5 in square faceplate. The other instruments are commonly found in 3ATI format, that is, with a 3 × 3 in square faceplate.

The instruments are available from a number of manufacturers, have standardized functionality, and are certified to comply with Technical Standard Orders (TSOs) published by the Federal Aviation Authority (FAA).

The TSOs of relevance to these primary flight instruments are:

- TSO-C133: airborne multipurpose electronic displays.
- TSO-C3d: turn and slip instrument.
- TSO-C4c: bank and pitch instruments.
- TSO-C52b: flight director equipment.

In addition, a third set of primary flight instruments, known as standby instruments, provide fully independent information to aid the flight crew to resolve instrument discrepancies. The standby instruments (usually in 3ATI format) are totally segregated from the prime instruments and operate on separate electrical power, usually the DC essential bus.

The Boeing 747 – 1970s

The 1970s heralded the mass market for civil transport aircraft with the introduction of the Boeing 747 Jumbo Jet. The 747 first entered service with Pan American in January 1970. The aircraft has continuously been developed throughout its operating life and is still the workhorse of all the major long-haul airline operators. Typically, the aircraft carries over 500 passengers at a cruising altitude of 32 000 ft with a range in excess of 5000 miles at a speed of 480 knots.

At its introduction, the flight deck of the 747, shown in Fig. 2.14, had a set of conventional electromechanical instruments and was operated by a three-man crew, the captain, the first officer, and the flight engineer. Today, as we shall see

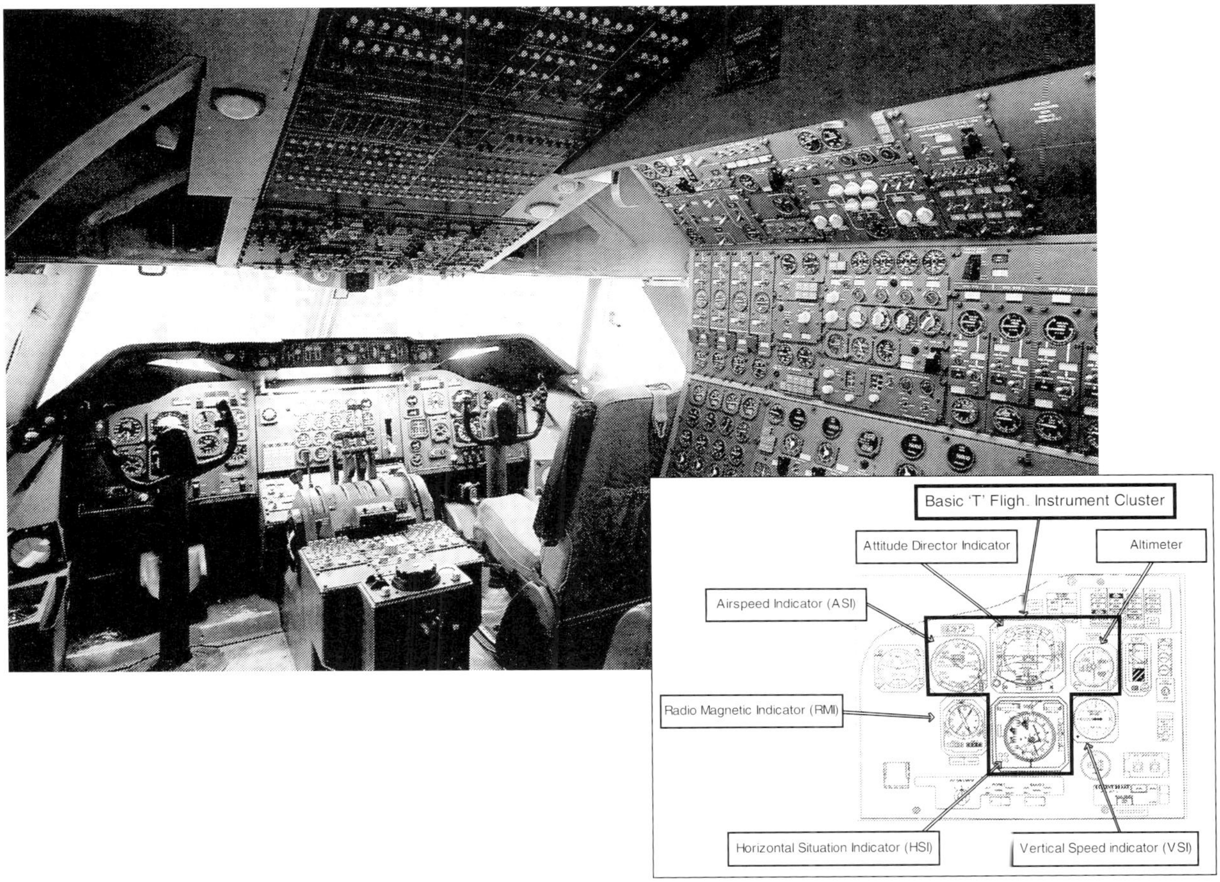

Fig. 2.14 Boeing 747 classic flight deck

later, the 747-400 aircraft has a fully integrated ‘glass cockpit’ with six AMLCD displays and is operated by a two-man crew.

By today’s standards, the cockpit of the early ‘classic’ 747 might be considered to be fairly basic, but the aircraft did have some early satellite communications equipment, an automatic landing system, and an Inertial Navigation (IN) system. The IN was a direct descendant of that used on the Apollo moon landing missions, and its inclusion in the 747 allowed the aircraft to be certificated to fly anywhere in the world without a specialist navigator on-board (**5**).

The Attitude Direction Indicator (ADI)

The ADI combines the functions of three earlier ‘basic 6’ generation instruments into one:

- Artificial horizon.
- ILS direction indicator.
- Turn and slip indicator.

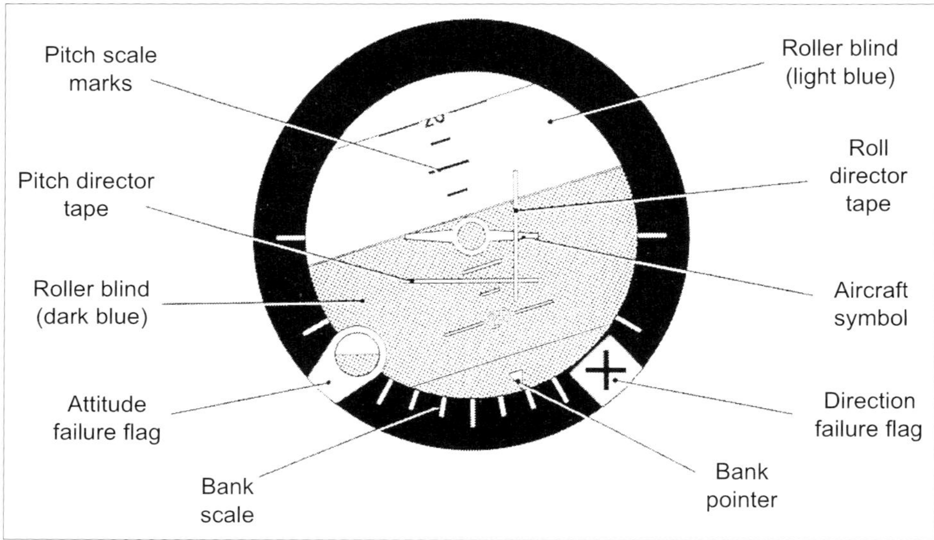

Fig. 2.15 Attitude Direction Indicator (ADI) (Smiths Aerospace)

The instrument shown in Fig. 2.15 comprises:

- *An attitude ball*, shaded blue for sky and brown for ground, which indicates aircraft pitch and bank angles against a fixed aircraft symbol in the centre of the instrument. The attitude ball is inscribed with pitch bar markings. Bank angle is indicated against a scale on the lower circumference of the instrument.
- *A flight director*, which consists of two bars indicating lateral and vertical steering guidance to capture and maintain a planned flight path. Different navigation aids are coupled to the flight director, depending on the phase of flight. These include radio aids (e.g. ADF, VOR, and TACAN), the flight management computer, and the autopilot itself as a monitor. During approach and landing, these bars are coupled to the ILS receiver to indicate glideslope and localizer deviation.
- *A turn and slip bubble*, which is attached to the faceplate of the instrument.
- *Warning flags*, to indicate warnings or cautions requiring crew action. The captain's and first officer's instruments are fed with data from different sources to provide both redundancy and independence. Electrical cross-monitoring triggers a warning flag if a discrepancy is found between the indications on the two instruments. The crew must resolve the discrepancy, for instance by using the standby instruments.

The Horizontal Situation Indicator (HSI)

The HSI combines the functions of three earlier 'basic 6' generation instruments into one:

- Compass (gyro and magnetic).
- Radio magnetic indicator.
- Distance measuring indicator.

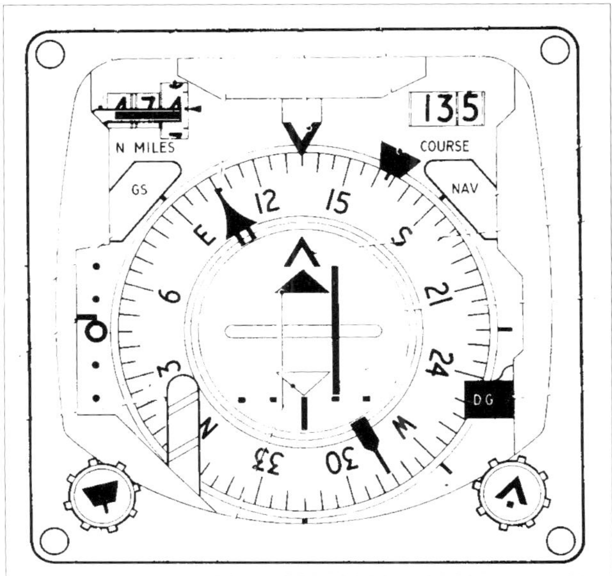

Fig. 2.16 *Horizontal situation indicator (HSI) (Smiths Aerospace)*

The instrument shown in Fig. 2.16 functions as a magnetic compass or as a directional gyro indicator, depending on the mode selected, and comprises:

- *Aircraft heading* is indicated by rotation of the compass card against a fixed lubber mark at the top centre of the instrument.
- *Aircraft course* is indicated by the course index, which registers against and rotates with the compass card. *Selected course* may be set relative to the compass card by means of a knob at the bottom righthand corner of the instrument. A drum counter display of selected course is also incorporated.
- *Selected heading* is indicated by the heading index, which registers against and rotates with the compass card. The index may be set relative to the compass card by means of the heading selection knob.
- *Relative bearing* information is displayed by the relative bearing pointer which registers against the compass card but is driven independently of the compass card from the radio navigation equipment.
- *Deviation in azimuth* from a selected relative bearing radial or ILS localizer beam is represented by the lateral displacement of the course deviation bar relative to the course deviation scale.
- *Deviation in pitch* from the ILS glidepath is represented by the vertical displacement of the glideslope deviation scale.
- *The DME counter* displays the distance of the aircraft from a beacon. A *to/from flag* indicates the aircraft direction relative to the beacon.

The Altimeter

The primary altimeter in current-generation aircraft (see Fig. 2.17) contains an electromechanical servomechanism driven by signals from the air data computer

Fig. 2.17 The Altimeter (Smiths Aerospace)

to indicate barometric altitude by means of a pointer and a counter (numerical readout). A knob with a numerical readout sets the barometric pressure datum.

The Airspeed Indicator (ASI)

As with the primary altimeter, the primary airspeed indicator in current-generation aircraft is driven by an electromechanical servomechanism slaved to the air data computer.

Fig. 2.18 The Airspeed Indicator (ASI) (Smiths Aerospace)

As shown in Fig. 2.18, in addition to airspeed it is not unusual also to find a numerical readout of Mach number. A warning bar indicates the maximum safe speed. Bugs may be set around the circumference of the instrument to indicate placard speeds.

Standby instruments

The availability and integrity of flight critical information are always augmented with standby instruments. The standby instruments must operate independently of the main display suite and the avionic systems that source data to them. The DC essential bus powers them independently. By these means, flight critical data are

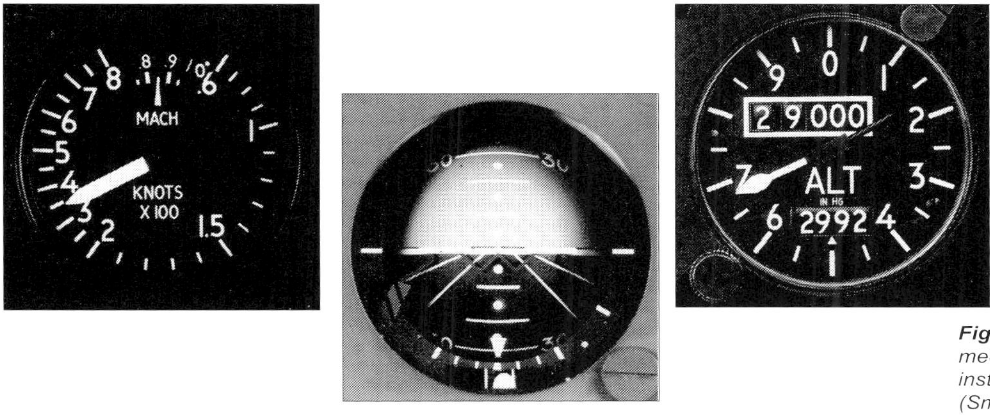

Fig. 2.19 Electro-mechanical standby instrument set (Smiths Aerospace)

still available on the flight deck even in the event of major avionic system failures and/or electrical power failure.

The standby instruments are typically a single set of miniature (2ATI and 3ATI) self-sensing instruments using integral transducers (see Fig. 2.19). They are the modern-day equivalents of the pitot static barometric instruments and gyro artificial horizon found in 1930s aircraft.

The electro-optical (glass) flight deck

By the 1970s, increase in air traffic, improved radio and inertial navigation aids, and on-board digital computing had driven the introduction of more efficient – yet more complex – route structures. The technology of the electromechanical instruments was limiting the means to convey data clearly and intuitively to the crew. The Cathode Ray Tube (CRT), previously a technology used primarily by the military, was to provide the solution. However, CRT brightness and contrast limitations had meant that early military CRT multifunction displays were monochrome (green). With a more benign environment, civil air transport led the use of the colour shadow mask CRT to provide colour electronic flight instrument systems.

Until recently an exclusive tool of the military fast jet, Head-Up Display technology is now migrating to civil and military air transport applications to aid approach and landing in poor visibility.

Advanced civil flight deck research

Research into the use of Cathode Ray Tube (CRT) technology to display crew information in civil transport aircraft began in the mid-1970s. In the United States, studies were being undertaken as part of the supersonic transport programme. When that was cancelled, the work transitioned to NASA for evaluation in a Boeing 737 as part of the Terminal Configured Vehicle (TCV) programme. That work was the natural starting point for the 757/767 flight deck. In the United Kingdom research was carried out at BAe Weybridge with contributions from GEC and Smiths Aerospace Industries. The Department of Trade and Industry (DTI) sponsored the programme, known as the Advanced Civil Flight Deck (ACFD).

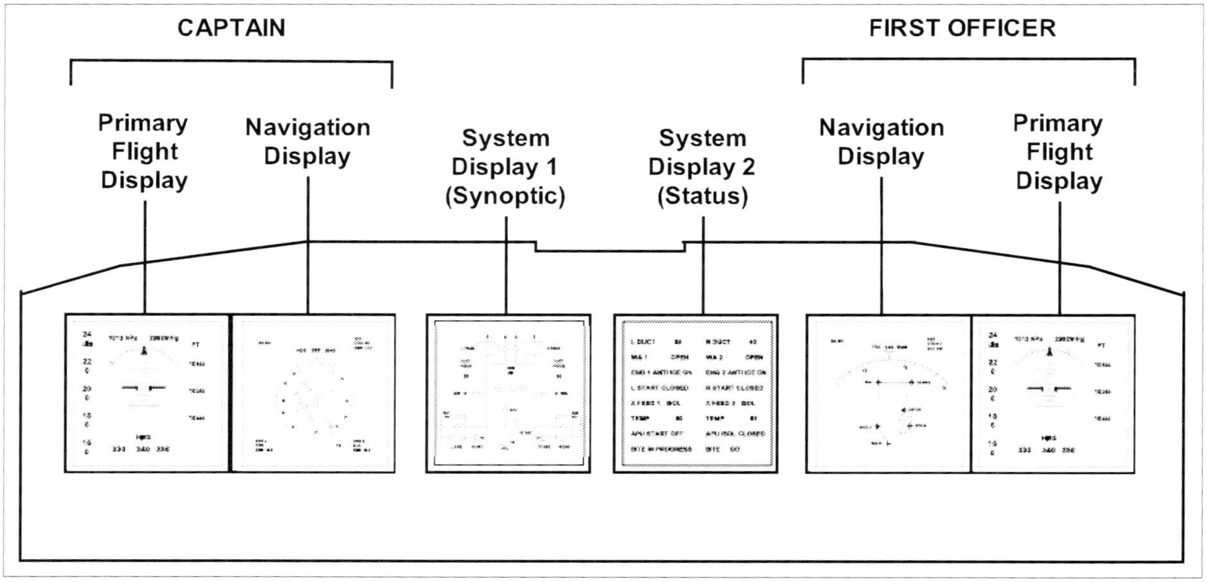

Fig. 2.20 The advanced civil flight deck

The ground-based simulator (shown in Fig. 2.20) had six CRTs each with a 6×4.5 in landscape (4:3 aspect ratio) usable screen area, arranged in landscape format in a side-by-side configuration across the flight deck.

The displays were monochrome (white on black) and the images were generated in 625 line 25:50 Hz interlaced video (TV standard).

The research activities included extensive human factor evaluation by pilots and demonstrated the viability of an 'all-glass' flight deck and the 'side-by-side' configuration of Primary Flight (PFD) and Navigation (ND) displays.

Although the advanced civil flight deck and the BAC 1-11 demonstration established the viability of a fully integrated flight deck with side-by-side PFD and ND displays, it would be some years before the aviation world was fully confident to adopt that format, though it has become the de facto standard today. It was necessary first to demonstrate with smaller display sizes that the CRT was sufficiently robust to operate reliably in the civil aircraft environment. In addition, and very rightly, the pilot community had also to gain experience with the medium.

BAC 1-11 technology demonstrator

In 1980 the ACFD research programme moved to a flight demonstration phase. Two displays (the primary flight and navigation displays) were installed into one crewstation of the BAC 1-11 civil transport aircraft operated by the Royal Aircraft Establishment (RAE) at Bedford. The programme was launched with monochrome CRTs (green on black), but, once it was established that colour shadow mask CRTs could be ruggedized to withstand the civil air transport environment, the programme transitioned to 6.25×6.25 in usable screen area square colour displays (see Fig. 2.21).

The display suite was fully integrated with the aircraft systems, and the aircraft could be flown from the left seat, the crew member in the right acting as the safety

Fig. 2.21 *BAC 1-11 advanced civil flight deck (April 1981) (The Royal Aircraft Establishment)*

pilot. The aircraft first flew in the spring of 1981 and from the autumn of 1981 made an extensive series of test and demonstration flights in Europe and the United States. During a comprehensive US tour, the aircraft visited nine sites, inviting guests from aircraft manufacturers, airlines, and research organizations to fly the aircraft. Some 34 sorties were flown (55 flight hours), most by guest pilots who were able to fly en-route and touch-and-go procedures to evaluate the concept.

Most pilots found the displays intuitive and easy to use; they adapted quickly to the side-by-side PFD/ND configuration.

The display formats used on the BAC 1-11 aircraft drew on the experience of the advanced civil flight deck ground-based programme. The formats mirrored closely the style of the electromechanical instruments they were replacing to facilitate transition to the new media.

The Primary Flight Display (PFD) shown in Fig. 2.22 preserved the 'basic T' configuration of airspeed, attitude, and altitude. Counterpointer presentations of speed and height were retained, but in addition digital readouts of speed and height were presented on the horizontal centre-line and adjacent to the attitude ball so the pilot could rapidly assess his primary flight data without having to scan the whole display.

The PFD also included autopilot mode selection annunciation and flight director guidance. A horizontal scale along the bottom of the display provided aircraft heading information.

Great care was taken to make digital readouts emulate the rotating drum counters of mechanical instruments to aid the ability to read the digits when changing and to preserve the perception of rate information.

The Navigation Display (ND) shown in Fig. 2.23 provided two format styles, the compass rose format and the map format. The compass rose format preserved the original electromechanical HSI compass rose style. The map format presented the planned route, updated with present position, in real time. This later format demonstrated the real freedom of the media to overcome the limitations of the electromechanical instrument and present for the first time in the cockpit a picture of the plan situation, a picture that hitherto the pilot had had to form in his head.

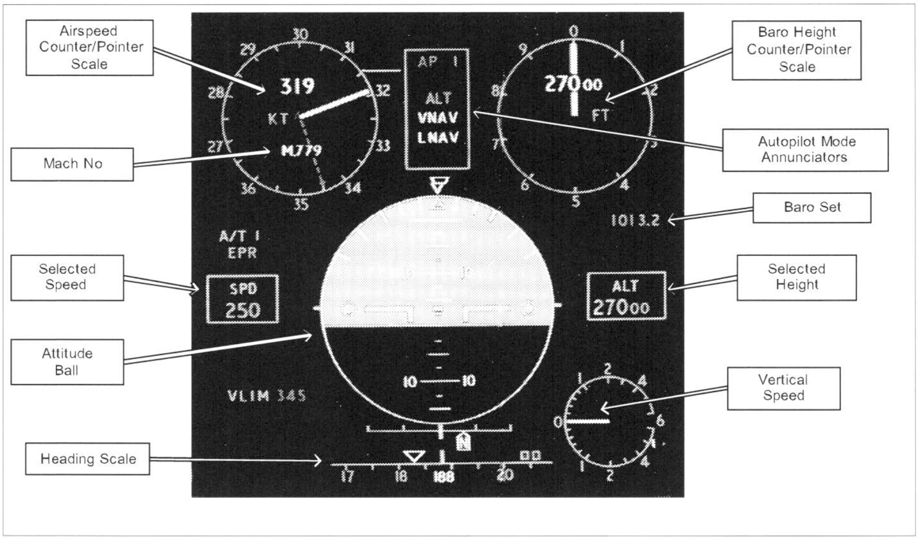

Fig. 2.22 BAC1-11 primary flight display (Smiths Aerospace)

The map could be oriented heading-up, track-up, or north-up, with the aircraft position portrayed either in the lower quadrant or in the centre of the display. Map range was selectable. Information on the flight plan included waypoints, navigation aids, and ground features of interest in the local area. Weather radar data could also be added to facilitate route replanning to avoid storm centres. On each side of the map was presented navigational information relative to the planned route, including waypoint identification, direction, distance, and time-to-go.

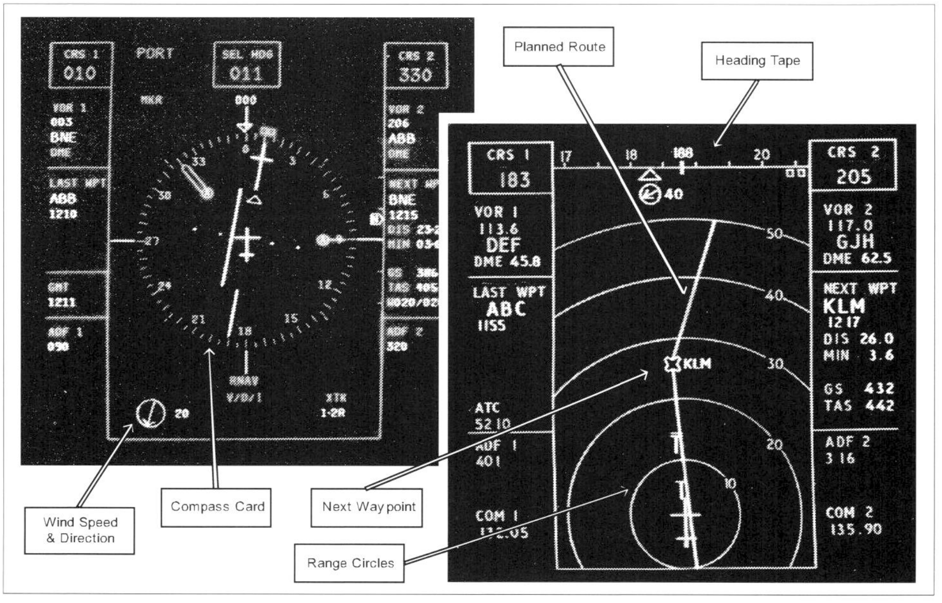

Fig. 2.23 BAC1-11 navigation display (Smiths Aerospace)

Fig. 2.24 Boeing 757 and 767 flight deck (The Boeing Company)

Boeing 757 and 767

The Boeing 757 and 767 aircraft were the first to enter service with a 'glass' flight deck in September 1982 (**5**).

In front of each pilot is a pair of colour CRT displays. The displays are vertically mounted with a conventional electromechanical airspeed indicator and altimeter to the left and right respectively, preserving the 'basic T' configuration (see Fig. 2.24).

The upper instrument, the EADI, supersedes the electromechanical artificial horizon (ADI) and provides information on roll, pitch, and slip angle. The EADI display format emulates the electromechanical instrument but with a more sophisticated flight director symbology afforded by the flexible CRT media. In addition, ground speed is indicated and autopilot status. It has a 5×4 in CRT display, in portrait mode.

The lower instrument, the EHSI, supersedes the electromechanical HSI with a 5×6 in CRT display, in landscape mode. The EHSI display format offers both the conventional compass and also the map format. The map indicates track, distance to go with estimated time of arrival, aircraft heading, and wind speed and direction. Additional modes that can be selected include VOR or ILS plus optional weather radar information overlay.

Located in the centre of the main instrument panel in front of the pilots is a pair of colour CRT displays arranged vertically one above the other which supersede the traditional electromechanical engine instruments and cautions. These displays are known as the Engine Indication and Crew Alerting System (EICAS) and provide information selected by the crew appropriate to the particular phase of the flight profile. The upper EICAS display shows warnings, cautions, and advisory information, and engine status primary information. The lower EICAS display is usually dedicated to engine secondary information, systems status, or maintenance data. The major advantages of the EICAS include centralized engine and alert displays, reduced pilot and maintenance crew workload, automatic and manual event recording, increased display and interface flexibility, reduced costs of ownership, and the capability of future system enhancements through software upgrades.

The Boeing 757 and 767 Electronic Flight Instrument System (EFIS) was founded upon research undertaken by NASA as part of the US Supersonic Transport (SST) programme. When the SST programme was cancelled, the work continued into an evaluation phase in a Boeing 737 for the programme Terminal Configured Vehicle (TCV). When the 767 programme (then called 7X7) committed to EFIS displays in 1978, the 5 years of TCV programme results were the natural starting point.

The upper EFIS display, the Electronic Attitude Director Indicator (EADI), started by containing the same information as a mechanical attitude director indicator. First, consideration was given to including additional information on this display. Those instruments located near the ADI on the captain's instrument panel were the likely candidates – radio altitude, baroaltimeter, airspeed indicator, and/or vertical speed indicators. Different combinations of these additional parameters were tried, and the flight test pilots flew a full-up display of several of them. In most cases, displays with all the additional data were considered more confusing than helpful. This was due in part to a different display format (vertical rather than round dial), partly because of combining moving scales and moving bugs to fit the display, and partly because a mixture of CRTs and conventional instruments was found to be less tiring to monitor than either one by itself. The final configuration developed included the normal radar altimeter display, autopilot mode display, and ground speed, which were added to a conventional ADI-type indication.

The Electronic Horizontal Situation Indicator (EHSI) allowed the use of a new display mode called the map mode. Rather than just a compass rose indication with VOR/ADF deviation, the pilot for the first time could see a visual map showing his true position in relationship to airports and radio navigation devices on the ground. In addition, he could select to have weather radar indications superimposed on the map. Besides the map mode, a VOR/ILS mode consisting of a partial compass rose with deviation indications was available to allow operation with displays similar in appearance to those existing in today's airplanes.

One of the major decisions that had to be made was the size of the map display. If the display size used by the TCV 737 airplanes were used it would require using a new unconventional control column wheel device in the flight deck. If the display size could be reduced to that of the ADI, some advantage from spare parts could take place. The studies showed that reducing the map display down to that of an ADI provides too large a compromise to crew operation in the terminal area. The requirements for good go-around navigation and weather radar look-ahead information, and the disappearance of navigation references during manoeuvring turns, were some of the major items that led to increasing the map display some 25 per cent larger than the ADI.

The larger map display area was not used for the ADI function since it would have spread apart the panel instruments, increasing the pilot's scanning time, and, because of its size, would force a compromise in its location on the panel.

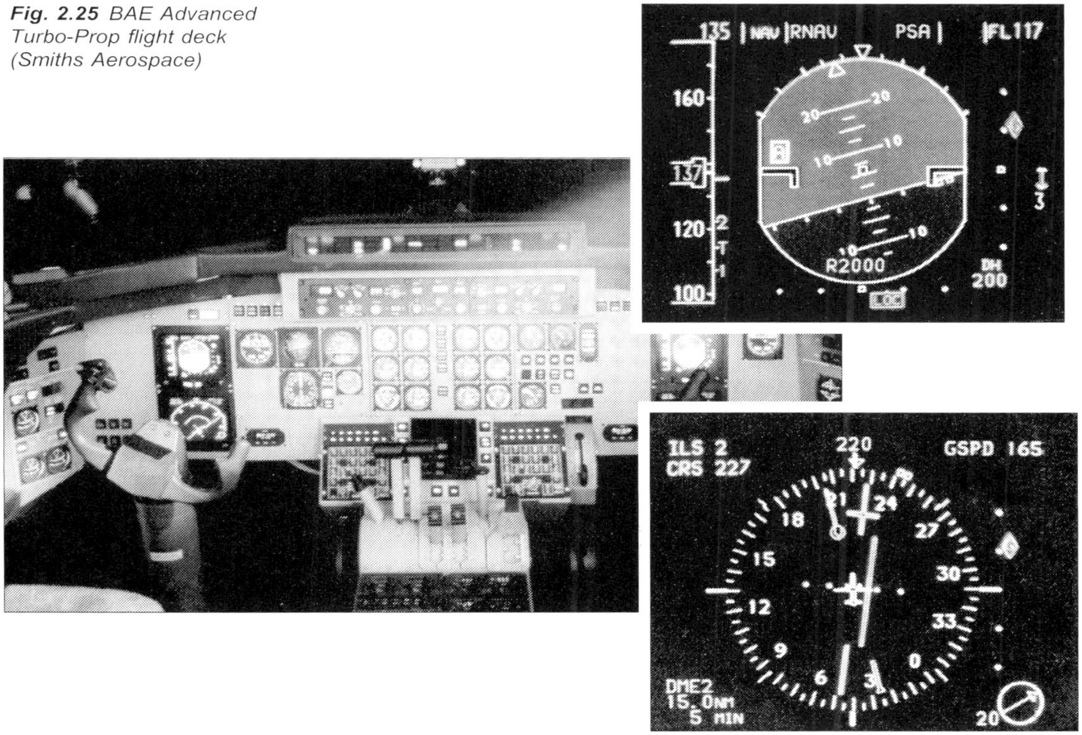

Fig. 2.25 *BAE Advanced Turbo-Prop flight deck (Smiths Aerospace)*

Therefore, each flight instrument was selected for its optimum size to provide the best information display with a minimum of compromises. The result was that the Boeing 757 and 767 ended up with six colour displays in the flight deck, utilizing two different sizes – one for the ADI and one for the HSI and EICAS (**6**).

British Aerospace Advanced Turbo-Prop

The British Aerospace Regional Aircraft (BARA) Advanced Turbo-Prop (ATP) is an upgraded version of the original Hawker Siddley 748. The flight deck design followed the precedent set by the Boeing 757/767 and placed the EADI and EHSI displays vertically above one another. It illustrates the marketing imperative now established for a 'glass cockpit'. The displays have a CRT with a usable screen area of 5×4 in, both in landscape orientation.

The original concept had been to include both speed and height onto the upper display as tape scales to the left and right respectively of the attitude 'ball'. However, it soon became apparent that the maximum display size that could be accommodated in the instrument panel would simply not permit the desired information to be written at a sufficiently large size to be read comfortably. Therefore, the final certificated configuration included only the speed scale on the EADI, with height presented on a conventional electromechanical altimeter to the right of the EADI (see Fig. 2.25). The EHSI presented conventional HSI formats plus a map, albeit constrained by the limited display area also.

Fig. 2.26 *Airbus A330 flight deck (March 1988) (Airbus)*

Airbus A320

The Airbus A320, which entered service in March 1988, was the first civil transport aircraft to adopt the original ACFD concept of side-by-side PFD/ND configuration with six 6.25×6.25 in CRT displays installed on the flight deck. The A330 has a similar configuration (see Fig. 2.26). For space reasons, the two system displays are installed one above the other in the centre of the flight deck.

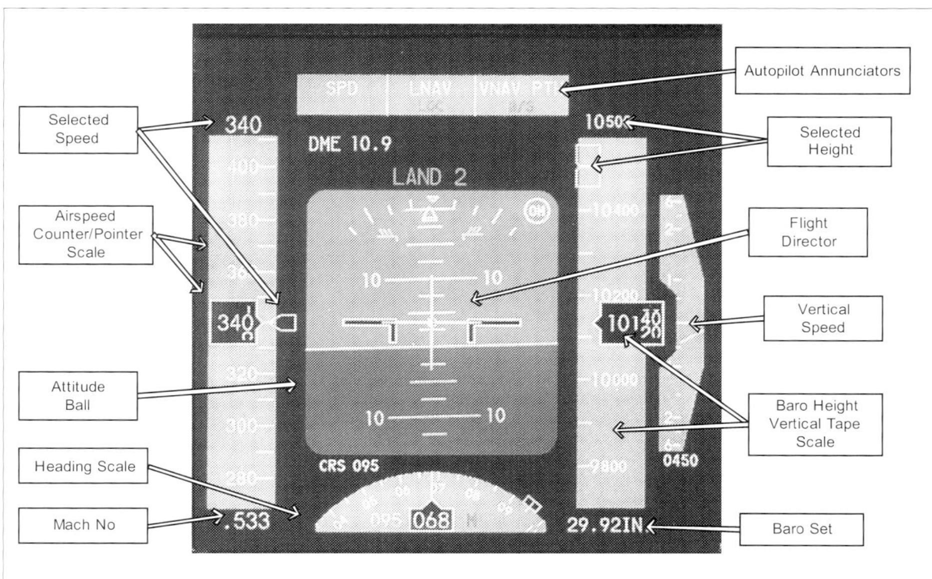

Fig. 2.27 *Airbus A320 primary flight display (Airbus)*

The A320 Primary Flight Display (PFD) shown in Fig. 2.27 introduces airspeed and altitude as two tape scales positioned either side of the attitude 'ball'. A digital readout is positioned as a window in the tape scales on the horizontal centre axis of the attitude 'ball'. This concept allows the pilot quickly to acquire the key flight parameters without having to scan the whole display surface.

The tape scales facilitate the acquisition of selected speed and height queues, which are presented on and attached to the moving scales. The moving scales also offer a representation of rate of change in the parameter, although it has to be said that this is recognized to be inferior to a circular scale and pointer. A vertical speed indicator therefore augments the height tape. At the bottom of the PFD format the upper segment of a compass rose provides heading and lateral guidance queues. Autopilot mode annunciators are incorporated along the top of the display.

Boeing 747-400

Boeing followed suit with the side-by-side, six-display configuration for its −400 variant of the 747 aircraft which entered service in June 1990 (see Fig. 2.28), and then subsequently for the Boeing 777 aircraft (**7**). As with Airbus, the two systems displays are positioned one above the other in the vertically extended centre portion of the instrument panel.

The 747-400 displays employ CRT technology. The display formats are similar to those of the Airbus A340, with strip speed and height scales positioned either side of the attitude 'ball'.

Fig. 2.28 *Boeing 747-400 flight deck (June 1990) (The Boeing Company)*

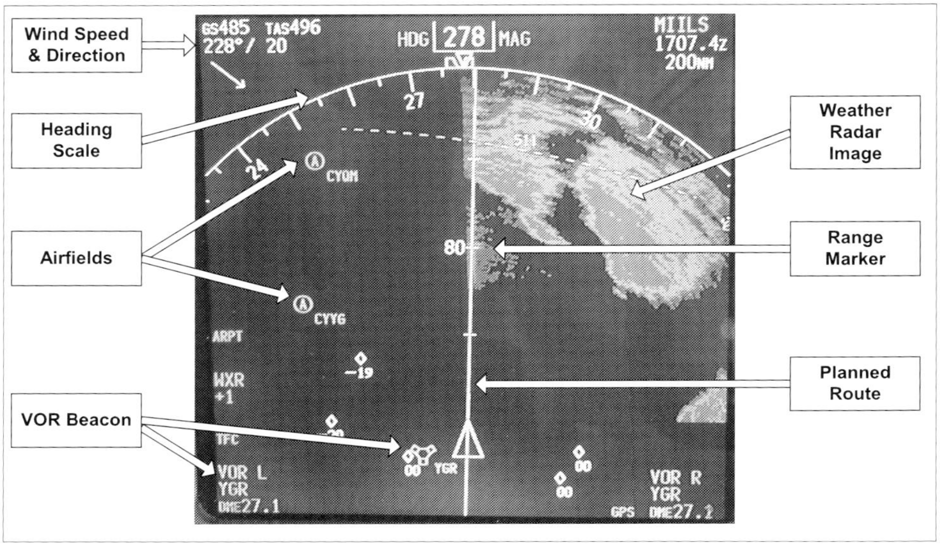

Fig. 2.29 Boeing 777 navigation display with weather radar (The Boeing Company)

It is worth noting that in this configuration the Boeing 747 (as with all 'glass cockpit' aircraft) operates with a two-man crew. The flight engineer's panel functions have been absorbed into the aircraft systems and presented and controlled through the two power/system displays.

The Boeing 777 (in service date June 1995) adopts the same cockpit configuration and display size but an integrated modular avionics architecture (see Chapter 11). The displays themselves were the first to utilize AMLCD technology in a wide-body civil transport aircraft and have a usable screen area of 6.7×6.7 in square.

The Boeing 777 Navigation Display (ND) shown in Fig. 2.29 provides planned and actual route together with lateral guidance command cues optionally overlaid with weather radar data.

Lockheed C130-J Hercules

The C130 is the West's most popular and widely used military transport aircraft, and has been in production longer than any other aircraft type in history. Very large numbers remain in service on every continent, and many operators have replaced earlier variants with newer versions. The first C-130 prototype made its maiden flight on 23 August 1954.

The development of the much improved, high-technology C130-J was launched in the late 1990s. The UK MoD was the launch customer. Four Rolls-Royce Allison AE2100D3 turboprop engines driving Dowty propellers each with six curved blades of composite construction power the aircraft. The aircraft has a cruising speed of 400 mile/h and an operating range of 3200 miles.

The flight deck, shown in Fig. 2.30, comprises the now accepted norm of six side-by-side displays. The screen area is increased to 6×8 in full colour AMLCD displays in portrait mode, providing enhanced primary flight, navigation, and tactical displays for the crew. Most significantly two head-up displays complete

Fig. 2.30 Hercules C-130J crew station (The Royal Aeronautical Society)

the crewstation to provide the pilot with primary flight real-world contact-analogue, or conformal symbology for approach, landing, and take-off at austere airfields in difficult terrain/weather conditions.

Flight deck upgrades

The power of the glass cockpit has been demonstrated, and now new air traffic procedures are being introduced to take advantage of the improved route planning and freedoms afforded by the glass cockpit and other aids. There is a demand to upgrade the flight deck of legacy (or classic) aircraft originally fitted with electromechanical instruments to a 'glass' flight deck so that these aircraft too can enjoy more efficient route structures and operating procedures.

To strip and replace the instrument panel and reconstruct the aircraft systems to support a fully integrated glass cockpit is neither practical nor affordable. Replacement of the existing electromechanical instruments with form, fit, but functionally enhanced 'glass' instruments is a viable option.

One such example, shown in Fig. 2.31, is the upgrade to many of the 747-Classic aircraft operated by KLM. Seven 5ATI 'glass' instruments replace the existing 5ATI ADI and HSI instruments, together with some of the engine

Fig. 2.31 KLM 747-Classic upgraded flight deck (Smiths Aerospace)

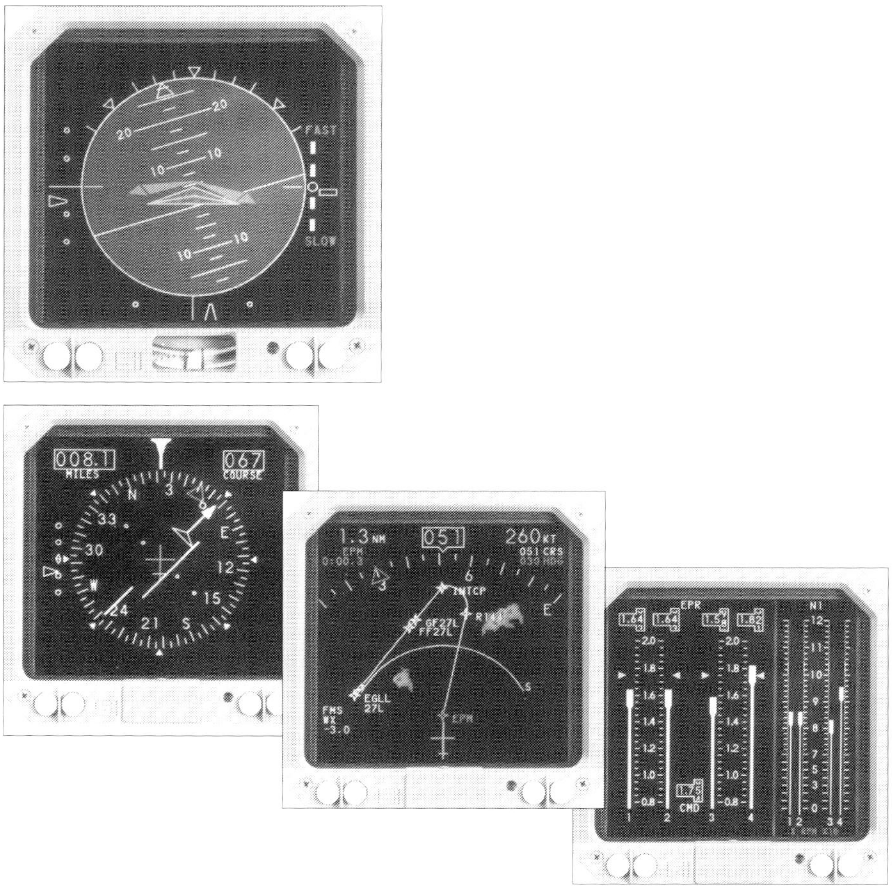

Fig. 2.32 KLM 747-classic upgrade – EHSI and engine formats (Smith Aerospace)

Fig. 2.33 *KLM 747-classic upgrade – combined display formats (Smiths Aerospace)*

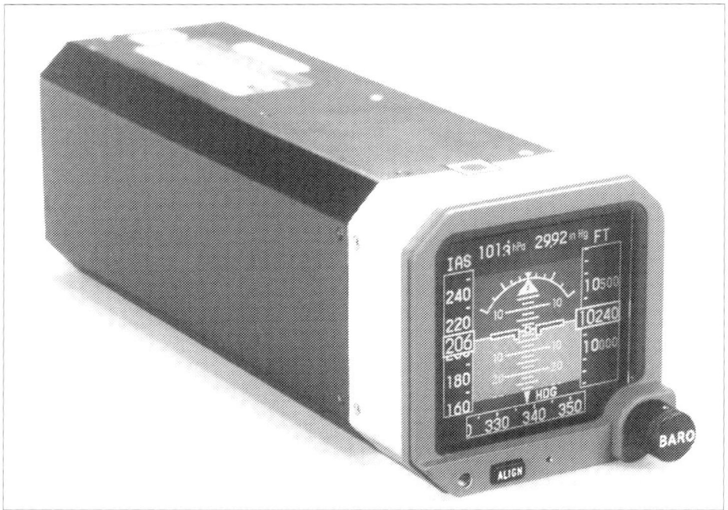

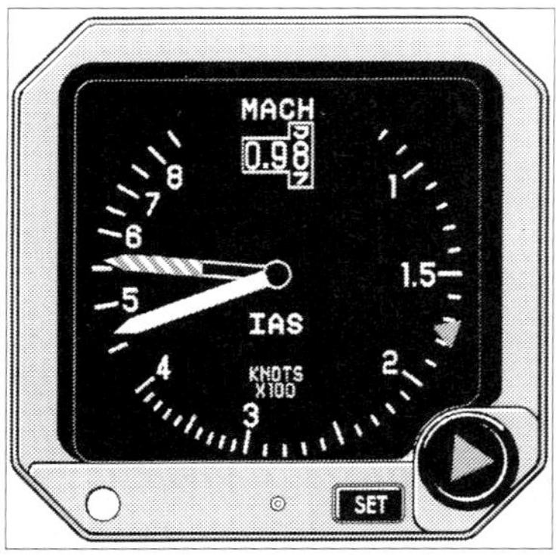

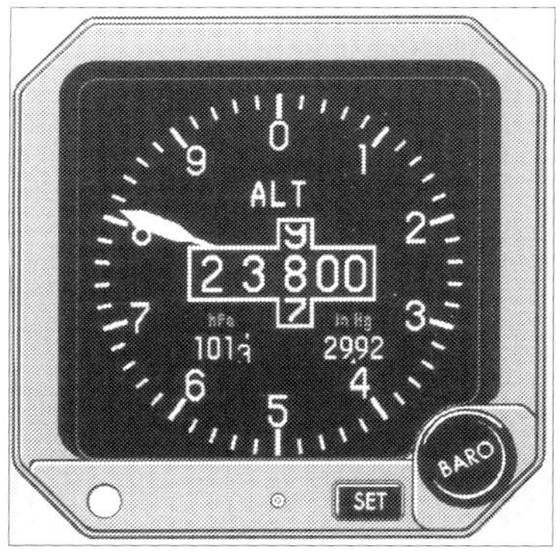

Fig. 2.34 *3ATI ISIS with display formats (Smiths Aerospace)*

instruments. These instruments use AMLCD technology. All the instruments are identical, their function being contained in software that configures itself when the instruments are installed, sensing their location in the aircraft by programme pins.

The display formats are shown in Figs 2.32 and 2.33.

- The EADI emulates the pre-existing electromechanical ADI, but with improved flight director presentation.
- The EHSI emulates the pre-existing electromechanical HSI, but with the addition of the map format optionally overlaying weather radar.
- Two instruments combined provide engine data in tape scale format.

- A combined EHSI and EADI format is provided as a standby instrument in the event of failure of either or both of the prime EADI or EHSI displays.

Glass standby instruments

As described earlier, the availability and integrity of flight critical information is always augmented with standby instruments. This remains true in a 'glass flight deck'.

In early glass flight decks the standbys are the traditional set of miniature (2ATI and 3ATI) dedicated electromechanical instruments. More recently these have been superseded by electronic displays using AMLCD technology with integral solid-state sensors.

Piezoresistive pressure sensors have replaced the aneroid capsules. Micromachined rate sensors and accelerometers have replaced the rotating gyroscope. A single instrument can replace all the dedicated instruments, pin programmed to provide the desired display format. This instrument, shown in Fig. 2.34, is known as the Integrated Standby Instrument System (ISIS).

The ISIS provides a display of:

- Attitude.
- Indicated airspeed and Mach number.
- Barocorrected altitude and baroset.
- Slip/skid.
- Heading.

The display formats are designed to be compatible with the primary EFIS formats to reduce pilot adaptation time and workload.

On power-up the ISIS is automatically aligned. Fast manual align is also provided. A manual baroset control is provided on the front panel, but the instrument can be slaved to a central baroset also to synchronize the altitude display to the other cockpit instruments. ARINC 429 data bus inputs and outputs are provided for auxiliary data not self-sensed (e.g. heading) and to output data for cross-monitoring and maintenance purposes.

The ISIS shown is in 3ATI form factor and uses a high-resolution full colour AMLCD with a usable screen area of 2.4×2.4 in square. It operates from the 28 V DC essential bus and consumes 15 W at full brightness.

References

(1) **Chorley, R.A.** (1976) Seventy Years of Instruments and Displays, 3rd Folland Lecture, Royal Aeronautical Society, February.
(2) **Coombs, L.F.E.** (1990) *The Aircraft Cockpit*, Patrick Stephens Limited.
(3) **Tanner, J.** (1977) *The Official Air Publication for the Lancaster Mk I and III 1942–1945*, Airs and Armour Press.
(4) **Garland, D. J., Wise, J. A.**, and **Hopkin, V. D.** *Handbook of Aviation Human Factors*.
(5) **Gilchrist, P.** *Boeing Classic 747*, Airlife Publishing Limited.
(6) **Carr, G.P.** and **Montemerlo, M.D.** (1984) *Aerospace Crew Station Design*, Elsevier Science Publishers.
(7) **Birtles, P.** *Boeing 757*, Airlife Publishing Limited.

CHAPTER 3

Military Fighters and Rotorcraft

This chapter discusses the evolution of the cockpit/crewstation of military aircraft from the World War I biplane to the present day as typified by the F-22 Raptor fast jet fighter and the AH-64 Longbow Apache attack helicopter.

The parallel strands of civil and military flight deck evolution are closely interwoven. The technological development in one often feeds the other.

Undeniably, the first use of electro-optical (EO) devices in the cockpit lies with the military. The first step was airborne intercept radar installing the Cathode Ray Tube (CRT) into the cockpit at the end of World War II. Soon, the CRT was used in the gyro gunsight to produce a collimated (focused at infinity) aiming reticle. Very quickly the CRT gyro gunsight developed into a sophisticated projection device through which the operator could correlate the position or vector of the aircraft or weapon with the outside world. The Head-Up Display (HUD) had been invented.

Head-up display technology is now migrating to civil transport applications, while in the military fast jet the HUD is being superseded by the Helmet-Mounted Display (HMD).

As CRT technology improved, it became possible to augment the role of the radar to provide flight information also, and the embryo multifunction Head-Down Display (HDD) had been invented. However, CRT brightness and contrast technology limitations meant that early CRT multifunction displays were monochrome (green) and information presentation was limited to character/symbolic images. In these early 'glass' cockpits, the moving map display was produced by optical rear projection of a photographically produced 35 mm filmstrip topographical map. Civil air transport pioneered the use of the colour shadow mask CRT in the airborne environment. However, it was not long before further advances in colour CRT technology made it possible to present a direct view, full colour topographical map image in the severe lighting conditions and environment of a military fighter cockpit.

Figure 3.1 traces the evolutionary timeline of the military crewstation.

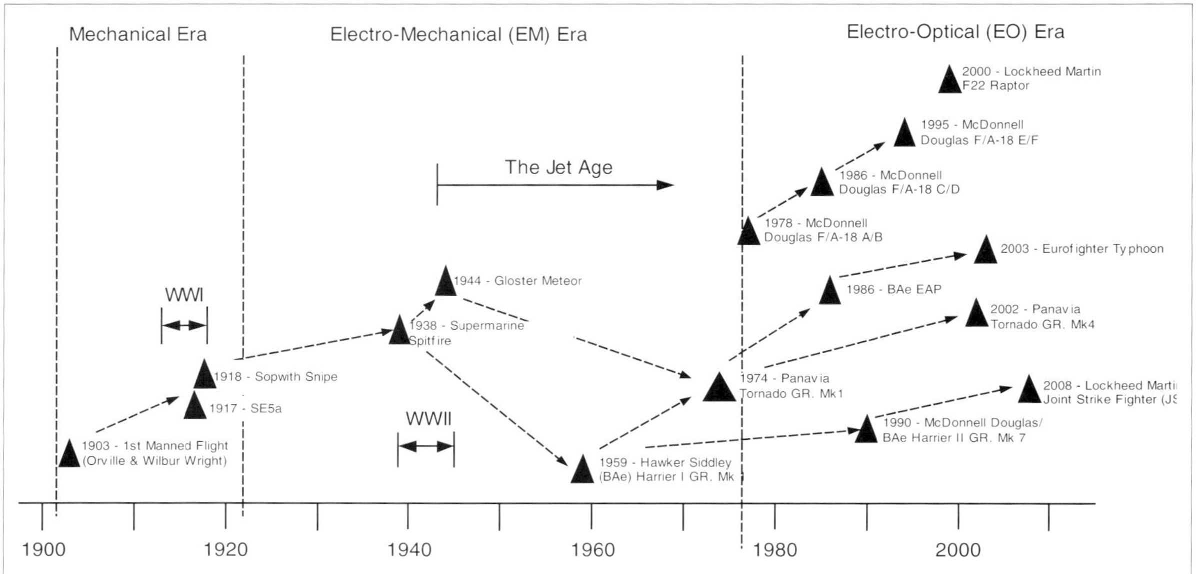

Fig. 3.1 *Military fast jet crew station timeline*

The mechanical era

Towards the end of World War I the art of flying had advanced considerably. Figure 3.2 shows an illustration from an aircrew 'manual' at the time called 'The Aeroplane Speaks', a semi-humorous publication of 1916 in which the features of the aircraft take on human voice attributes. By all accounts the book was well received by the trainee pilots of the day.

Fig. 3.2 *Extract from 'The Aeroplane Speaks'*

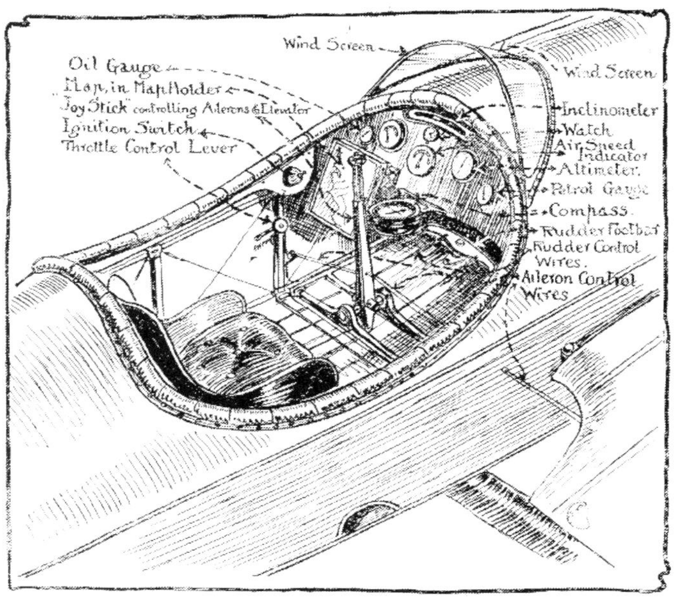

The Pilot's Cock-pit.

This manual identifies the cockpit/crewstation instrumentation of the time typically to include:

- Inclinometer
- Turn and Slip Indicator,
- Magnetic compass
- Airspeed Indicator
- Altimeter
- Watch

The Royal Aircraft Factory S.E.5a – 1917

The S.E.5a, shown in Fig. 3.3, was without doubt the most successful World War I aircraft to emanate from the Royal Aircraft Factory. The single-seat biplane designed by the team headed by H.P. Folland was simple to fly owing primarily to the 200 hp Hispano-Suiza V8 static piston engine. The first aircraft entered service in mid-1917. The aircraft was armed with one forward-firing Vickers synchronized machine gun and one Lewis gun mounted over the centre section of the upper wing, plus up to four 25 lb bombs. In addition to the flight instruments note the telescopic aiming sight for the fixed gun was mounted on the top of the fuselage.

Fig. 3.3 *Royal Aircraft Factory S.E.5a (BAE SYSTEMS)*

The Sopwith Snipe – 1918

Designed around a new 130 hp Bentley rotary piston engine, the Sopwith Snipe entered service towards the end of the war in the summer of 1918. The aircraft had a surprisingly comprehensive set of flight instruments (see Fig. 3.4) arranged as needs must around the two Vickers forward-firing synchronized machine guns.

A total of 497 aircraft were built, and the Snipe remained in first-line squadron service until 1926 and remained in use with training schools for some time after that. In its brief operational career, the Snipe proved an exceptional fighter.

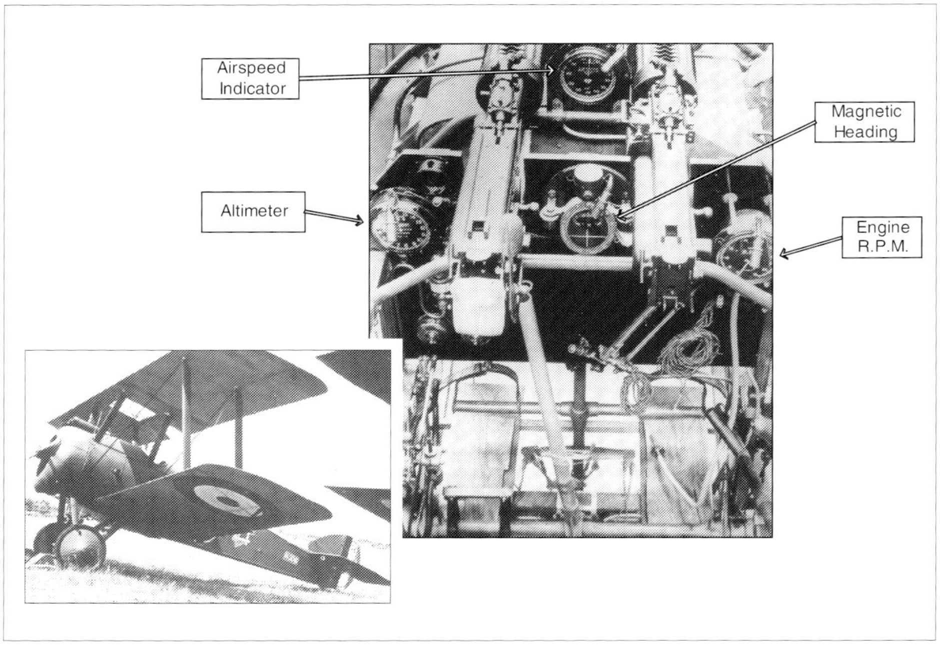

Fig. 3.4 The Sopwith Snipe

The electromechanical era

The Supermarine Spitfire – 1938

Without doubt, the best-known British fighter aircraft of World War II is the Supermarine Spitfire. Designed by R.J. Mitchell, the aircraft had a distinctive cantilever elliptical low-wing which housed eight machine guns and two cannon, all of them firing outside the propeller disc.

The Spitfire cockpit is shown in Fig. 3.5. The instrument panel contains the basic primary flight instruments required for daytime flying conditions, plus rudimentary engine instruments, namely:

- Attitude – top centre.
- Airspeed indicator – top left.
- Rate of climb/dive – top right.
- Altimeter – bottom left.
- Turn and slip indicator – bottom right.
- Compass – bottom centre.

The most notable additional instrument is the gunsight in the centre of the instrument panel. Initially the gunsight was a simple optical device projecting a collimated (focused at infinity) aiming reticle into the pilot's line of sight to aid gun aiming. Adjustments could be made to depress the reticle in elevation according to the weapon type and anticipated firing range.

The Spitfire was powered by a Rolls-Royce 1175 hp Merlin engine. Subsequent marks were fitted with progressively increasing performance engines, culminating with the 2375 hp Griffon.

Fig. 3.5 *Spitfire cockpit*

The prototype aircraft first flew on 5 March 1936. However, planned mass production was slow to start, and the first production aircraft did not enter service until July 1938 with No. 19 Squadron at Duxford. By the outbreak of World War II on 3 September 1939, the RAF had nine operational Spitfire squadrons, and on 16 October 1939 a Spitfire of No. 603 Squadron claimed the first German aircraft to be destroyed over the United Kingdom in World War II.

In total, 20 334 Spitfires were built, plus 2556 related Seafires, the naval fighter version.

The Gloster Meteor – 1944

The Gloster Meteor was the only Allied turbojet-powered aircraft to see action during World War II.

The cockpit of the Meteor was very cramped. The instrument panel layout of the F.Mk8 is shown in Fig. 3.6. The flying instruments are once again arranged in the 'basic 6' configuration in the centre of the instrument panel. The gyro gunsight is installed above the basic instruments.

Now incorporating gyroscopes, the gunsight senses aircraft body rates, and mechanical analogue computing provides lead angle compensation to the aiming reticle to allow more accurate aiming of the aircraft weapons while manoeuvring. The gunsight controls are presented to the pilot on the rear face of the instrument in the location now known as the up-front control panel. These controls allow for setting the weapon type, in this case the four cannon, rocket projectiles, or freefall bombs. Additional controls then allow for the adjustment of preset values to be entered for the particular weapon of each type to optimize the computing for the type of attack to be prosecuted.

Fig. 3.6 Gloster Meteor cockpit

The Meteor's twin-engined layout was necessitated by the low thrust produced by the turbojet engines then available. Later, two Rolls-Royce Welland turbojet engines each producing 1799 lb thrust powered the aircraft. In that configuration the aircraft had a maximum speed of 415 mile/h at 10 000 ft and a service ceiling of 40 000 ft. It could climb to 30 000 ft in 15 min. The Meteor's maiden flight is often said to have taken place on 5 March 1943 at Cranwell, but in fact the aircraft first flew a few days earlier at Hucclecote, the Gloster Aircraft test site. The first production aircraft flew on 12 January 1944. The first operational sorties were flown on 27 July 1944, and on 4 August near Tonbridge the aircraft destroyed the first V1 flying bomb using the Meteor's wing tip to roll the bomb into a spin after the aircraft's 20 mm cannon jammed. A second V1 was also shot down that day near Tenterden (**1**).

Over 3000 Meteors were produced in 20 marks, some conversions of earlier marks. The most numerous were the F.Mk8; the first of a total of 1183 flew on 12 October 1948. The F.Mk8 had a lengthened fuselage, a redesigned tail unit, an additional fuel tank, and a bubble canopy. Two 3600 lb Derwent 8 engines to provide a top speed of 600 mile/h powered the aircraft.

Hawker Siddley (BAE SYSTEMS) Harrier GR.Mk1 and Mk3 (RAF) and AV-8A (USMC) – 1970s

In the United Kingdom, the first aircraft to enter service with a 'designed-in' Head-Up Display was the vertical take-off Hawker Siddley (BAE SYSTEMS) Harrier.

The cockpit of the Harrier GR.Mk3 is shown in Fig. 3.7. The flight instruments are of the conventional electromechanical counterpointer type. In the centre of the instrument panel is the moving map. This large and complex instrument projects the image of a filmstrip topographical map onto a rear projection screen. The film transport mechanism is driven by aircraft inertial reference coordinates and heading to indicate aircraft current position and course/track with reference to ground topography.

Above the moving map is the head-up display. The HUD was small by today's standards. It was a refractive design with a 4 in exit lens and provided a modest instantaneous field of view of about 14°. The HUD itself contained only the CRT and the collimation optics. The CRT high-voltage supplies and the CRT beam deflection electronics were remote from the HUD to minimize space and weight in the cockpit. HUD symbology was produced by a remote waveform generator in the equipment bay which also contained the CRT drive electronics. The waveform generator was originally implemented as an analogue computer but was replaced shortly after entry into service with a digital design using then state-of-the art diode-transistor logic (DTL) integrated circuits.

The HUD operates in cursive (stroke) mode only to provide a daylight viewable symbolic image.

Display modes included:

- Navigation.
- Approach and landing.
- Precise local fix (IN update).
- Air–air attack (guns, rockets, and missiles).
- Air–ground attack (freefall and retarded bombs, CCIP manual and CCRP automatic release).

The HUD controls are located on the glareshield to the right of the HUD and select air–air or air–ground display mode, air data display options, baroset, and target aircraft wingspan.

Development of the vertical-lift turbofan engine began in 1959. The engine (later named the Pegasus) employed four nozzles, in fore–aft pairs, which pivoted to vector engine thrust between just forward of the vertical to directly backwards. The first prototype, the P1127, hovered for the first time on 21 October 1960, and by September 1961 complete transitions were being made between vertical and horizontal flight. Reaction control jets mounted in the nose, tail, and each wing tip provided stability in the hover and during low-speed manoeuvre. The aircraft achieved a maximum speed of 737 mile/h at low level, an initial climb rate of 29 000 ft/min, a service ceiling of 51 000 ft and a range of 550 miles.

With the concept proven, nine preproduction aircraft were built, designated the Kestrel. The first of 61 production version GR.Mk1 aircraft entered service with the RAF on 1 April 1969. These initial models were powered by the Pegasus

Fig. 3.7 Harrier GR.Mk3 cockpit (The Royal Aeronautical Society)

Mk101 turbofan rated at 19 000 lb. Subsequently, these aircraft, plus 40 new-build aircraft, were upgraded to GR.Mk3 standard with the uprated 21 500 lb Mk103 Pegasus engine. These aircraft equipped four operational squadrons, one in the United Kingdom and four in Germany. The aircraft equipment included an Inertial Navigation System (INS), a Head-Up Display (HUD), a Laser Rangefinder/Marked-Target Seeker (LRMTS), and a Radar Warning Receiver (RWR).

The US Marine Corps (USMC) took delivery of 110 aircraft, designated the AV-8A. The aircraft could be armed with 5000 lb of disposable stores carried under the fuselage and wing. The stores generally included two Aden cannon pods, freefall or retarded bombs, multiple rocket launchers, and AIM-9 Sidewinder missiles for the USMC.

Panavia Tornado GR.Mk4 – late 1970s

The multinational (United Kingdom, Germany, and Italian), multirole Tornado remains one of Europe's most important combat aircraft. The cockpit demonstrates the transition between electromechanical and true electro-optical design.

The fore and aft cockpits of the Tornado are shown in Fig. 3.8. The front cockpit (pilot) incorporates conventional electromechanical counterpointer

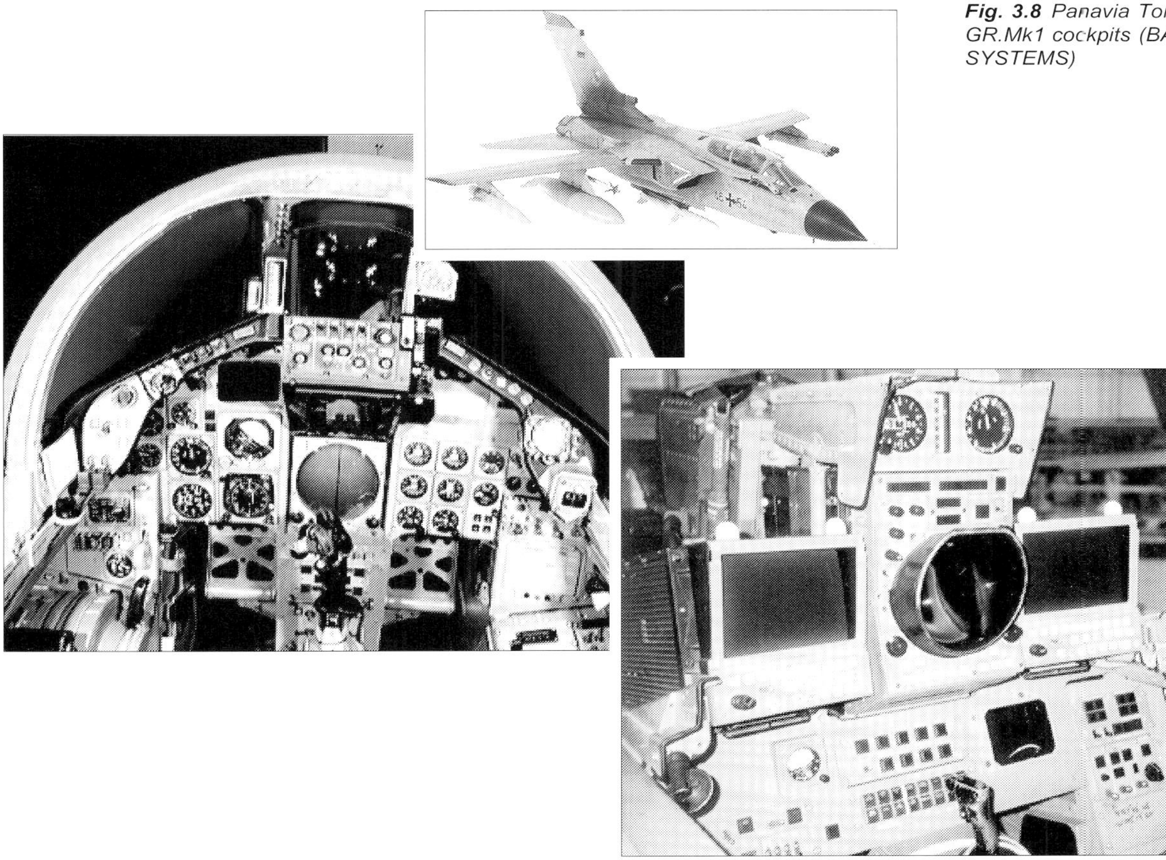

Fig. 3.8 Panavia Tornado GR.Mk1 cockpits (BAE SYSTEMS)

primary flight instruments. In the centre of the instrument panel is a moving map display. In early GR.Mk1 aircraft the map display was implemented using filmstrip rear projection technology. This was replaced with a 5×5 in usable screen area, high-brightness, high-resolution, full colour shadow mask CRT multi-function display in later GR.Mk1 and all GR.Mk4 aircraft. The map information itself was then generated remotely from the display by a digital map generator from data stored on a CD-ROM.

The head-up display is a refractive dual-combiner design having a 5 in exit lens. It provides an instantaneous field of view of around 16° horizontally by 20° vertically. The HUD operates in both cursive (stroke) mode for daytime use and raster plus stroke overlay for night-time use. At night the HUD can display real-world video from the on-board Forward-Looking Infrared (FLIR) camera.

HUD modes (or formats) include:

- Navigation.
- Air-to-surface weapon aiming (CCIP and CCRP).
- Air-to-air guns and rockets (CCIL).

Figure 3.9 shows typical HUD air-to-surface weapon aiming symbology. Typical HUD air-to-air symbology is shown in Fig. 3.10.

The rear cockpit incorporates three displays to facilitate the tasks performed by the navigator. In the centre of the instrument panel is the CRT-based tactical display which displays sensor video imagery. To the left and right of the moving map display and high on the coaming are two other displays known as the TVTabs. These displays were originally implemented as high-brightness, high-resolution monochrome (green) CRTs with a usable screen area of 5×3.75 in and 4:3 aspect ratio. In the GR.Mk4 these displays are replaced with colour AMLCD technology. The TVTab displays present tactical information to the navigator regarding threats, weapon status, and imagery.

Development of the Tornado began in 1968. The first prototype flew in 1974 and the aircraft entered service with the RAF in its GR.Mk1 configuration in 1979. In 2001 the RAF began using fully operational aircraft upgraded to GR.Mk4 standard with new cockpit displays, full compatibility with the Thermal Imager And Laser Designator (TIALD) pod, and integration of Night Vision Goggles (NVGs) with an upgraded Forward-Looking Infrared (FLIR) vision enhancer and an enhanced Defensive Aids Suite (DAS).

The aircraft is a two-seat tandem configuration; pilot in the front seat, navigator in the rear seat. Two Turbo-Union RB 199 Mk103 turbofans provide 8650 lb thrust each. The aircraft achieves a maximum speed of 1453 mile/h or Mach 2.2 at 36 000 ft and can climb to 30 000 ft in under 2 min with a service ceiling of 50 000 ft and a range of 863 miles. The aircraft is armed with two 27 mm IWKA Mauser cannon and can carry 20 000 lb of ordnance including

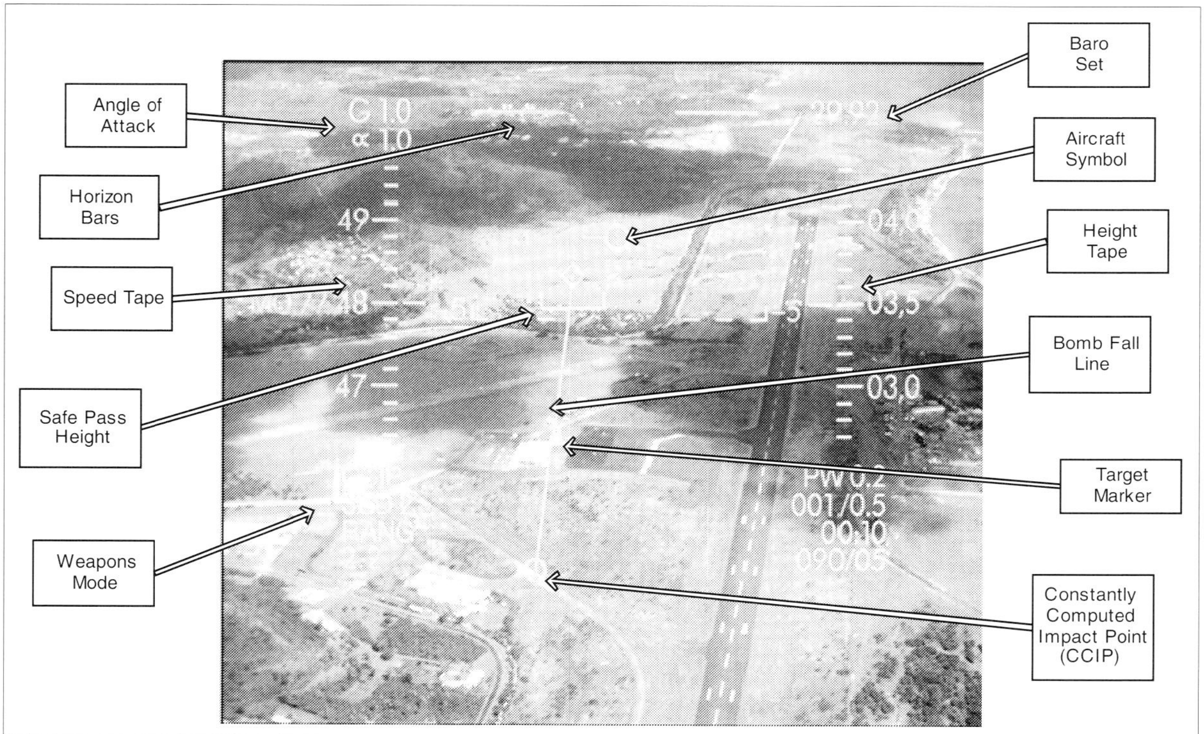

Fig. 3.9 *Typical HUD air-to-surface format (Smiths Aerospace)*

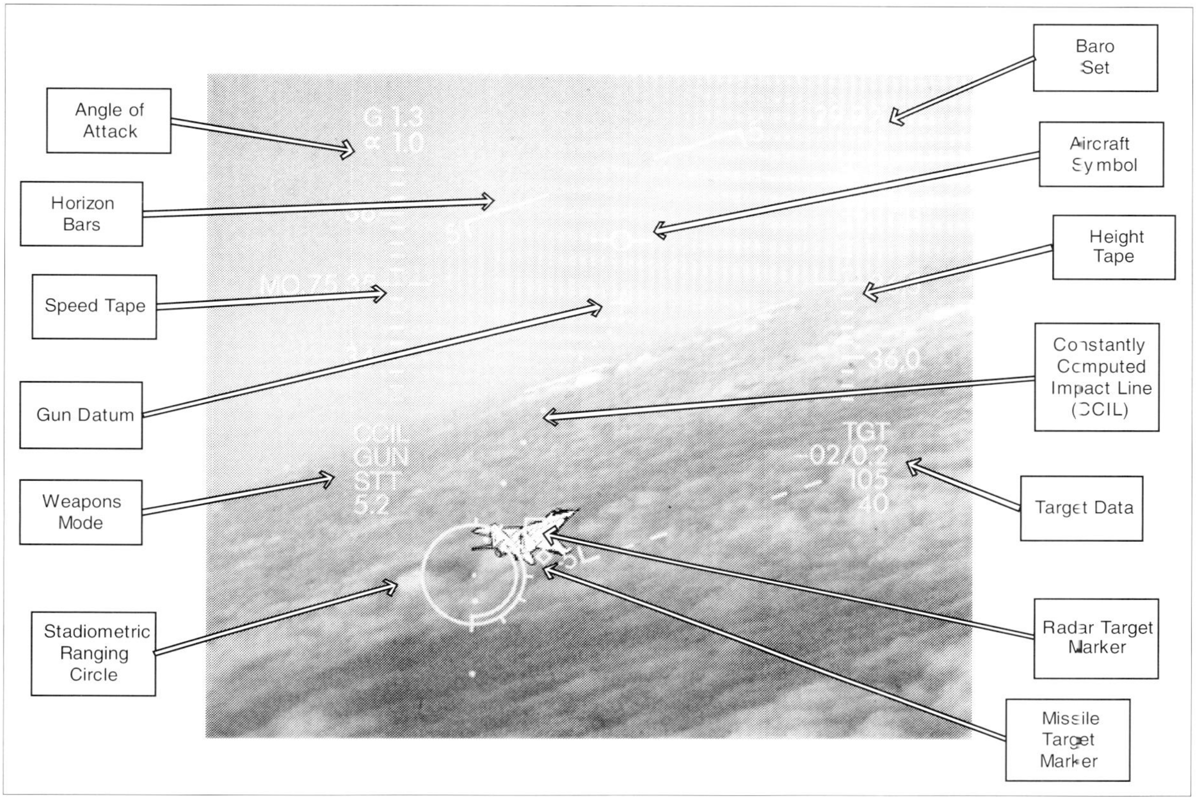

Fig. 3.10 Typical HUD air-to-air format (Smiths Aerospace)

freefall and retarded bombs, Paveway II laser-guided bombs, and Storm Shadow missiles.

The electro-optical era

McDonnell Douglas F/A-18 Hornet – early 1980s

The F/A-18 Hornet is generally recognized as the watershed in cockpit display design and can be considered the true beginning of the electro-optical (EO) era. Improved radar technology able to acquire and track multiple targets, together with thermal imaging technology and smart weapons, drove the need to be able to present tactical data to the crew in a flexible, graphical, and pictorial manner.

The cockpit displays of this era are composed largely of CRTs presenting data that have been digitally processed by the aircraft's on-board computer systems. The real impact of this digital processing technology is the design flexibility of the displays, and the ability to alter the display content according to the information required or needed by the user for the current aspect of the mission. Because of this characteristic, the displays are generally known as Multifunction Displays (MFDs). The ability to show large amounts of information on a relatively limited display surface shifted the emphasis of cockpit design from packaging of fixed-profile instruments to packaging of information. Specifically, the task is how best to format the displays and how best to structure the control menus so that the operator does not drown in an excess of data or get lost in a complex menu structure (**2**).

The cockpit of the F/A-18 is shown in Fig. 3.11. The initial F/A-18A/B was fitted with a refractive cursive (stroke) daytime only HUD. The left and right MFDs were high-brightness monochrome (green) CRTs with a 5×5 in square format usable screen area capable of operating in cursive (stroke) mode for high-quality graphic instrumentation formats and in raster mode with stroke overlay for weapon imagery. The central moving map used rear projected filmstrip technology.

In the F/A-18C/D night attack version, the HUD incorporates a raster mode for night-time use, presenting a collimated outside-world image from the Forward-Looking InfraRed (FLIR) pod overlaid with conventional stroke symbology (see Fig. 3.12). The left and right MFDs are upgraded to present limited colour formats using time-sequential liquid crystal shutter technology. This technology adds the additional dimension of red and orange symbology to green instrumentation formats and red and orange overlays to the already high-resolution, high-brightness green raster weapon imagery. The projected map display is replaced with a multipurpose high-brightness, high-resolution full colour stroke and raster shadow mask CRT display (MPCD) with a 5×5 in square format usable screen area. The map image is generated remote from the display in a digital map computer using CD-ROM technology. The cockpit lighting is fully compatible with night vision goggles and at night the pilot may aid his night vision with the use of 'cat's eye' goggles attached to the helmet.

The F/A-18 Hornet was the successful contender for the US Navy Air Combat Fighter requirement in the mid-1970s. It was first intended that the Hornet be produced in two versions, the F-18 fighter and the A-18 strike/attack warplane. Eventually it was realized that the match of the right airframe with a choice of two mission software packages would allow the procurement of a single aircraft type capable of undertaking both missions, hence the dual-role designation F/A-18. Both single-seat and twin-seat versions were produced.

The first prototype aircraft flew on 18 November 1978. A total of 371 aircraft were delivered in the initial F/A-18A (single) and F/A-18B (twin-seat) configuration. The aircraft were declared operational in January 1983.

The F/A-18A and B was superseded by the F/A-18C and D which made its maiden flight on 3 September 1986. After 137 aircraft had been delivered, production switched to a version capable of the night attack role. Two General Electric 17 700 lb F404-GE-402 turbofan engines power the F/A-18C. The aircraft achieves a maximum speed of 1190 mile/h or March 1.8 with a combat ceiling of 50 000 ft and a range of 460 miles (fighter) or 662 miles (attack), considerably more with drop tanks.

The aircraft can be armed with a formidable array of weapons which include AIM-120 AMRAAM, AIM-7 Sparrow, and AIM-9 Sidewinder in addition to a Vulcan 20 mm nose-mounted cannon for the air–air role and AGM-65 Maverick and AGM-84E stand-off land attack missiles plus a variety of freefall, retarded, and laser-guided bombs. The sensor suite includes the Hughes (now Raytheon) APG-65 radar and in the night attack role an FLIR targeting pod.

In the early 1990s, McDonnell Douglas commenced development of the F/A-18E and F Super Hornet. The aircraft made its maiden flight in November 1995. The first production aircraft was delivered in 1998, and initial operational capability was achieved in 2001. The US Navy had originally intended to procure

Fig. 3.11 *McDonnell Douglas F/A-18C cockpit*

1000 F/A-18E and F aircraft, but in 1997 this was reduced to 548 in the light of the capabilities offered by the Joint Strike Fighter (JSF) programme.

Major alterations in the F/A-18E/F are upgraded avionics, a larger airframe with changes to reduce the radar cross-section, increased fuel capacities, and greater power from the engines. The avionics upgrade is centred around the APG-73 radar. A new defensive aids suite enhances RF countermeasures and a revised digital quadruplex fly-by-wire system eliminates some of the previous mechanical

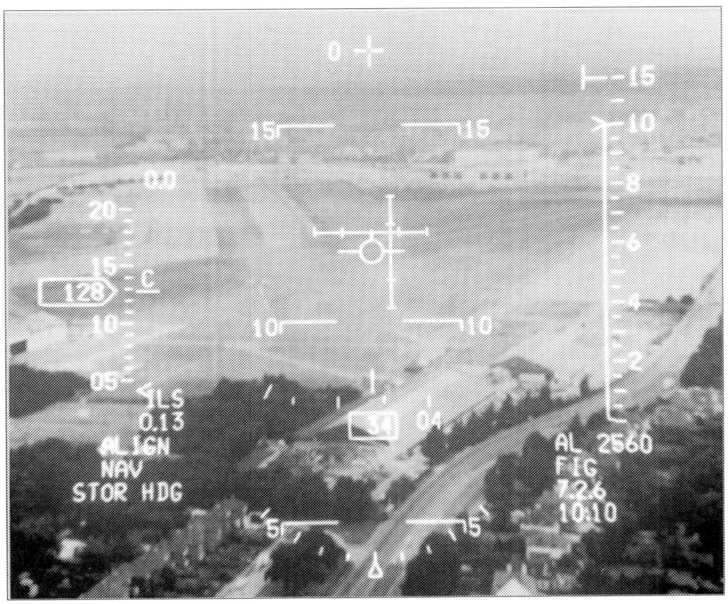

Fig. 3.12 *HUD FLIR sensor video plus stroke symbology overlay (BAE SYSTEMS)*

back-up systems to save weight. The cockpit of the Super Hornet is similar to the F/A-18C, with the exception of the central colour display which is replaced with a larger area (6×6 in square format display) employing AMLCD technology to provide enhanced brightness/contrast and resolution.

The F/A-18 cockpit truly broke new ground, but its introduction represented only the tip of a technological iceberg in terms of the challenge for the electronic cockpit designer. Although the MFD gave a degree of freedom over what could be displayed, the then CRT technology posed some serious limitations on the positioning of the display owing to its size, weight, and power consumption and on the information content owing to brightness/contrast and colour (or lack of it).

The problem to be solved by the crewstation designer is how to show to the crew the massive amount of data now made available by digital processing without them becoming swamped by data overload. The answer is to present only the data required for the current phase of the mission and to configure the display format accordingly, which in turn requires the displays to be changed, or controlled, during the course of the mission. Initially, this change was performed by the operator who decided what to display and when. Unfortunately, excessive operator involvement was found to be counterproductive in terms of reducing workload. The response to this problem is continually to develop more sophisticated decision aids to predict the information requirements of the user and then configure the display formats accordingly.

As will be seen in subsequent cockpit designs, there is a continuous struggle to reduce the bulk of the display device itself while increasing the display surface area and flexibility of the information content. The ultimate aim, possibly to be achieved in the Joint Strike Fighter (JSF), is to provide the operator with one contiguous, controllable display surface.

McDonnell Douglas/British Aerospace Harrier II AV-8B and GR.Mk7

The Harrier II came from the same stable and is of the same genre as the F/A-18C, and the reason for including it here is for completeness to trace the evolution of the VSTOL cockpit from the original Harrier AV-8A/GR.Mk1 through the Harrier II AV-8B/GR.Mk7 to the Joint Strike Fighter (JSF). The Harrier II cockpit is shown in Fig. 3.13.

The first production aircraft was delivered to the USMC during 1983. From the 167th aircraft, all AV-8Bs and GR.Mk7s were fully equipped for the night attack roll with the installation of a GEC FLIR and Smiths Aerospace improved wide field of view HUD, two Smiths Aerospace Head-Down Multipurpose Colour Displays (MPCDs) and a digital colour moving map. The first GR.Mk7 was delivered to the RAF in May 1990.

The cockpit is fully night vision compatible, allowing the pilot to use 'Cat's Eye' NVGs attached to his helmet. The two Head-Down MPCDs each have a 5×5 in usable screen area square format and high-brightness, high-resolution shadow mask CRTs to provide full colour cursive (stroke) symbology and raster (plus stroke overlay) for full colour map and high-resolution monochrome (green) FLIR imagery. The Head-Up Display is a refractive dual-combiner design with optics truncated in the fore/aft direction to provide a wide field of view of

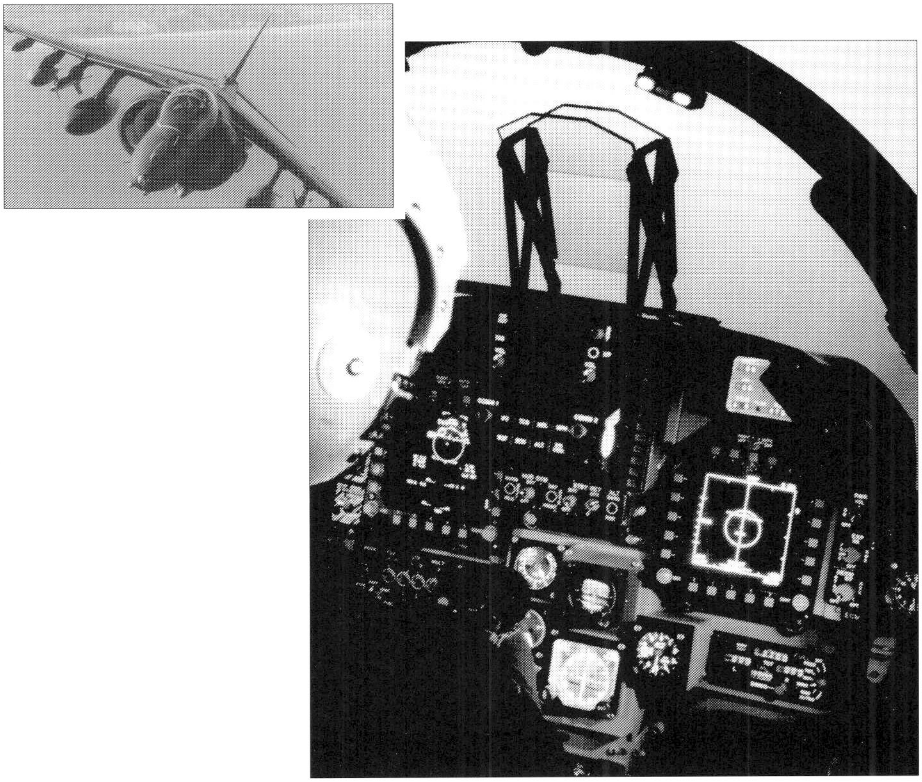

Fig. 3.13 McDonnell Douglas/British Aerospace Harrier II cockpit (The Royal Aeronautical Society)

approximately 20×18°. The HUD provides stroke mode for daytime use and raster plus stroke overlay for night-time use, portraying on the HUD a view of the outside world from the FLIR sensor.

In a typical operational scenario, the pilot will use the HUD as the primary flight instrument, the right-hand MPCD as the colour moving map and the left-hand MPCD as the systems/weapons display, selecting the FLIR image when appropriate to the phase of the mission.

The Harrier II was developed to provide for the carriage of a larger weapons load and to provide better range/endurance characteristics than the original Harrier I. The new design is based around a new and larger wing with a supercritical aerofoil section, the use of composite carbon fibre materials, and a completely revised cockpit with Hands-On-Throttle-And-Stick (HOTAS) controls.

The aircraft is powered by an enhanced Rolls-Royce F402-RR-408 turbofan engine delivering 23 800 lb thrust. Maximum speed at sea level is 662 mile/h with a service ceiling of more than 50 000 ft and a range of 684 miles. Armament includes one 25 mm GE fixed forward-firing cannon plus 13 235 lb of disposable munitions.

Eurofighter Typhoon – 1990s

The Eurofighter consortium was formed in June 1986 by the same three countries that developed the Tornado, namely the United Kingdom, Germany and Italy, and was shortly joined by Spain. France joined briefly, but then dropped out to proceed with a totally national effort, the Dassault Rafale.

Much experience was gained with the main concepts for the Eurofighter Typhoon from the British Aerospace Experimental Aircraft Programme (EAP). One EAP technology demonstrator aircraft was built. The aircraft first flew on 8th August 1986, reaching Mach 1.1 at 30 000 ft on that flight. The EAP programme ran until May 1991, and in that time a total of 259 flights generated a mass of data for the EFA. The EAP cockpit is shown in Fig. 3.14.

The man–machine interface used a Hands-on-Throttle-and-Stick (HOTAS) control concept augmented by direct voice control. The head-up display was a wide field of view diffractive pupil-forming holographic type with an integral up-front control panel. The instrument panel incorporated three 6×4.5 in, 4:3 aspect ratio, full colour, shadow mask, multifunction displays with soft-function keys which provided flight, navigation, and systems information together with weapons-related data. The HUD plus the three MFDs provided a fully integrated display suite. The HUD was the primary flight instrument.

The display formats shown in Fig. 3.14 include the control surface page, the attitude display, and the engine page. Other pages included the fuel quantity page which indicates the disposition of fuel in the many fuel tanks of the EAP.

The EAP validated a number of concepts to be used in Eurofighter Typhoon, including extensive use of carbon fibre composite materials and aluminium–lithium alloys and active fly-by-wire flight control technology. Fly-by-wire flight control is essential for an inherently unstable aircraft with a low set dihedral wing,

Fig. 3.14 The EAP technology demonstration cockpit (Smiths Aerospace)

elevons on the leading and trailing edges, close-coupled anhedral foreplane canards, and a high-fin inset rudder.

As finalized in September 1987, the Eurofighter Air Staff Target called for a relatively light but technically sophisticated single-seat fighter with a low radar cross-section, high supersonic performance, considerable agility, and carefree handling; capable of operating from short and austere airstrips and optimized for primary Beyond-Visual-Range (BVR) and close air combat roles with a secondary air–surface capability.

The development programme was launched in November 1998 and the first two prototype aircraft undertook their maiden flights on 27 March and 6 April 1994. Eurofighter has passed through a turbulent period of financial and technical problems, the latter mainly on the flight control software, but is now beginning to reveal considerable potential as an advanced warplane.

The Eurofighter Typhoon cockpit is shown in Fig. 3.15. The main instrument panel contains three colour Multifunction Head-Down Displays (MHDDs). In the prototype aircraft these displays used 6×6 in square format shadow mask CRTs to provide daylight-viewable, full colour, high-brightness, high-resolution images in both cursive (stroke) and hybrid (stroke + raster) modes. In production, the CRTs are superseded with high-resolution active matrix liquid crystal displays in 6.25×6.25 in square format. The MHDDs incorporate 18 multifunction keys around the bottom, left, and right edges of the display. Each key contains a daylight-viewable LED matrix of two rows of four 7.5 characters plus underline.

The HUD uses holographic technology to achieve an ultrawide 30×25° Field of View (FoV). The HUD provides stroke (cursive) operation for daytime use and raster plus stroke for night-time use with outside-world sensor video. The HUD incorporates a sophisticated up-front control panel with a 4×3 in daylight-viewable LED matrix display. The HUD is the primary flight instrument.

The Eurofighter Typhoon also incorporates the option for a 40° binocular Helmet-Mounted Display (HMD). The HMD can be configured into two variants. The daytime variant provides a symbology for the targeting and release of off-boresight weapons. The night-time variant adds Night Vision Goggles (NVGs) to the helmet to provide the pilot with enhanced night vision. The NVG image is electrically mixed with the CRT symbology image to provide a comprehensive night-time capability.

To either side of the HUD the left and right glareshield panels provide essential controls and warnings. The right-hand panel incorporates the standby attitude display. In the prototype aircraft this was an electromechanical instrument, in production it is superseded by an AMLCD. The farthermost part of the right-hand glareshield flips open to reveal a set of standby get-u-home instruments in the unlikely event that there is a major power failure.

The Eurofighter Typhoon provides Direct Voice Input (DVI) command control to avoid taking hands off throttle and stick for non-mission-critical functions such as communications equipment. The DVI speech recognizer has a vocabulary of about 100 words. The DVI system will function under all operational conditions including hi-*g* manoeuvres and low speed passes with significant wind buffet. Having been trained by the individual user, it will recognize speech of male and female voices of crew from any of the four Eurofighter nations with a 98 per cent

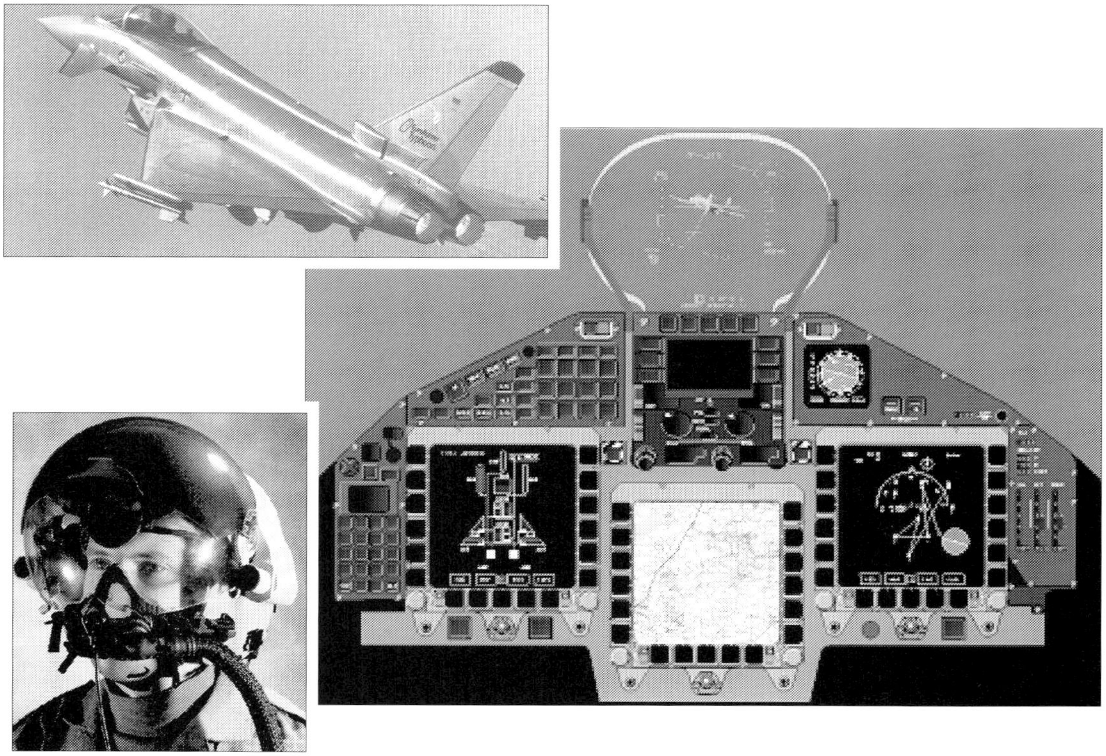

Fig. 3.15 *The Eurofighter Typhoon cockpit (BAE SYSTEMS)*

recognition probability. DVI commands are repeated back to the crew via the Head-Up Display for confirmation before execution (**3**).

The aircraft is powered by two Eurojet EJ200 turbofans rated at 13 940 lb thrust plus afterburner. The maximum speed is 1320 mile/h or Mach 2.0 at 36 000 ft. The service ceiling is 47 570 ft with a range of 864 miles. After much debate, the aircraft is equipped with one 27 mm Mauser fixed forward-firing cannon in the starboard side of the fuselage. Up to 17 673 lb of disposable stores can be carried on four semi-recessed missile stations under the fuselage and on 13 hard points under the fuselage and wing. Weapons include medium-range and short-range air–air missiles, optically or laser-guided bombs, antiship missiles, antiradar missiles, stand-off guided bombs and missiles, and freefall and retarded bombs.

The Eurofighter Typhoon incorporates the new ECR90 multimode pulse Doppler 'look-up/look-down' fire control radar with multitarget search and detect capability. The ECR90 is the fighter's primary active sensor, but the aircraft also features a new and entirely passive InfraRed Search and Track (IRST) sensor. A highly advanced Defensive Aids SubSystem (DASS) provides an integrated package of missile approach, laser and radar warning elements together with tip-mounted ESM and ECM pods and fuselage-mounted chaff/flare dispensers.

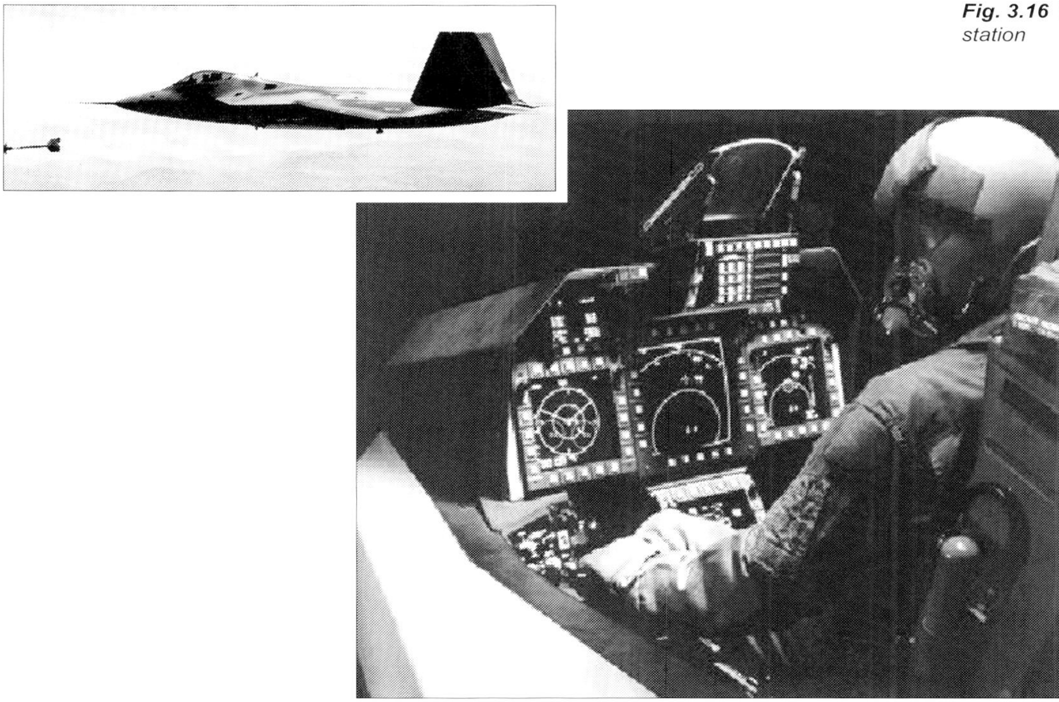

Fig. 3.16 *F-22 Raptor crew station*

Lockheed Martin F-22 Raptor – late 1990s

In April 1991 the USAF announced that it had selected the Lockheed Martin F-22 Raptor to meet its Advanced Tactical Fighter (ATF) requirement for a successor to the F-15 Eagle. The requirement called for a fighter combining low-observability, supersonic cruise, long range, and a very high level of agility.

Key features of the F-22 design are an angular but clean stealthy shape, an advanced composite structure, internal weapon bays, and an advanced modular avionics suite linked by fibre-optic data buses. The F-22 is powered by two Pratt and Whitney F119 P-100 turbofans and has an estimated maximum speed of 1157 mile/h or Mach 1.7 at 30 000 ft.

The F-22 advanced avionics system provides the pilot with fused situational awareness in the battlefield environment. The core of the offensive avionics is provided by the APG-77 multimode electronically scanned radar, and as a supplement to the radar an InfraRed Search and Track (IRST) system. The pilot controls the Raptor by means of a side-stick controller and a throttle. The advanced crewstation layout is shown in Fig. 3.16 and includes a number of large-format full colour AMLCD multifunction head-down displays, numerous general- and special-purpose knobs and switches, a wide-angle Head-Up Display, and a Helmet-Mounted Display. The displays contain integral video processing and graphics generation (**4**).

The four multifunction displays are arranged as a 'crowned T' configuration with a central single 8×8 in primary multifunction display flanked by three 6×6 in secondary multifunction displays. Two 3×4 in up-front displays are arranged either side of the wide field of view HUD which is at the top centre of the panel. Below the HUD is an integrated control panel housing some dedicated switches and alphanumeric text readouts. Bezel Option Selection Buttons (OSBs) surround each of the displays. The OSBs are used for menu navigation, paging, and function select with appropriate legends being shown on the display adjacent to the OSB.

The prototype aircraft also included touch-sensitive screens, voice recognition, and 3D audio, but these technologies were deemed to be immature and were not included in the production design.

The Primary Multifunction Display (PMFD) is used to display the Situation Display (SD) or the Attitude Director Indicator (ADI) display. The SD is a tactical format showing the entire track file icons as well as navigation data, with multiple levels of OSB menus for feature control. SD symbology is centred around ownship. The ADI is intended to back up use of the HUD and is a careful copy of HUD symbology and mechanization, yet with selectable colour or monochrome background fill for the earth–sky components of the pitch ladder.

The three Secondary Multifunction Displays (SMFDs) are used to display:

- Attack Display (AD).
- Defense Display (DD).
- Expand Display (EXD).
- Situation Display – Secondary (SD-S).
- Stores Management Display (SMD).
- Fuel display, the engine display.
- Mission Data Edit (MDE) display family.
- Backup integrated control panel display.
- Flight Test Display (FTD).
- Surface Position Display (SPD).
- Electronic Checklist (ECL).

The AD, DD, SD-S, and EXD are tactical formats. The AD and DD show a more narrowly defined set of track file icons and navigation data along with specialized data for weapon and countermeasure deployment, centred about ownship. The EXD format is centred on a pilot selectable icon. The remaining formats are self-explanatory.

The UFDs are used for either Communication, Navigation, and Identification (CNI) display or Standby Flight Group (SFG) display. Additionally, the Integrated Caution, Advisory, and Warning (ICAW) data (if any) are shown in the centre column of the CNI display. The CNI display shows green or white text depicting status information for radios, navigation steerpoint, Identification Friend or Foe (IFF), etc. The ICAW data are presented as red text for warnings, yellow for cautions, and cyan for advisories. The SFG display shows a simplified small version of the ADI and HUD attitude data with digital readouts for altitude, speed, and heading.

The Head-Up Display (HUD) provides a 24° horizontal×20° vertical binocular wide field of view monochrome (green) image. The HUD is the Primary Flight Display (PFD) and provides attitude, flight path, navigation, and weapons deployment symbology in general conformance to MIL-STD-1787 (**5**). The HUD has a built-in colour video camera and extensive user-editable controls to declutter symbology.

The Helmet-Mounted Display is the US standard Joint Helmet-Mounted Cueing System (JHMCS) with a 20° field of view. The HMD is monochrome (green) monocular (right eye only), and is used to show symbology similar to the HUD with some special symbology for depicting sensor volume limits. Like the HUD, extensive user-editable controls are provided to declutter symbology.

Rotorcraft

Boeing (McDonnell Douglas/Hughes) AH-64D Longbow Apache

The AH-64D Longbow Apache best illustrates the attack helicopter electro-optical 'glass' crewstation.

Formulated in the 1970s, the US Army's requirement for an advanced attack helicopter visualized a machine suitable for a day/night antiarmour role in all weather conditions. Bell and Hughes built two prototypes for competitive evaluation which flew in late 1975. Hughes was declared the winner and development proceeded. The US Army finally awarded a contract for an initial batch of vehicles in March 1982, and the first production AH-64A was delivered in January 1984.

The Apache is a tandem, two-seat helicopter with advanced crew protection systems, avionics, and electro-optics plus a weapon control system that includes the nose-mounted Target Acquisition and Designation Sight/Pilots Night Vision Sensor (TADS/PNVS). The copilot/gunner sits in the front seat, with the pilot in the higher rear cockpit. Both use sophisticated sensors and systems for the detection and attack of targets, including the Integrated Helmet and Display Sight System (IHADSS) which provides a monocular Helmet-Mounted designator sight. The copilot/gunner has the primary responsibility for firing both the gun and the missiles, but can be overridden by the pilot in the back seat.

The AH-64D crewstations, shown in Fig. 3.17, have two 6×6 in full colour, high-resolution, Active Matrix Liquid Crystal (AMLCD) displays replacing the monochrome CRT displays of earlier versions. The IHADSS is retained.

The Apache's principal payload consists of up to 16 Hellfire long-range, laser-guided antitank missiles carried on four hard points under the stub wing. The standard fixed armament is the 30 mm chain cannon. The other main disposable weapon is a pod carrying Hydra air–surface guided rockets. Air–air weapons can also be carried, including the AIM-9L Sidewinder.

The AH-64D adds the Longbow target acquisition and fire control system. It is easily identifiable by the mast-mounted antenna for the APG-78 radar. The new radar allows AGM-114L Hellfire missiles to be launched in an autonomous fire and forget mode. The radar can detect, classify, prioritize, and track 12 targets simultaneously.

Westland Lynx

The Westland Lynx is a battlefield light helicopter and is included here as an example of progressive upgrade of a cockpit from an early EO configuration to a

Fig. 3.17 *AH-64D Longbow Apache cockpit (The Royal Aeronautical Society)*

modern 'glass' crewstation. The Lynx was launched as a joint Anglo-French agreement in 1967. The design is wholly of Westland origin, but the manufacture is shared between Westland and Eurocopter. The Lynx has advanced digital flight controls plus all-weather avionics.

The crewstation of the current Lynx Mk8, in service with the Royal Navy, is shown in Fig. 3.18. The pilot flies the vehicle from the right-hand seat using conventional electromechanical instruments. The other crew member operates the weapons systems using the CRT display mounted on the instrument panel. The CRT is an early example with poor performance under high-brightness conditions, and hence the hood.

The Lynx has been upgraded as an export version to include an all-glass cockpit with an integrated display suite of four 6×6 in high-resolution, high-brightness AMLCD multifunction displays which provide primary flight and tactical information to the crew (see Fig. 3.19). Two 5ATI AMLCD displays mounted in the centre of the instrument panel provide systems and engine status

Fig. 3.18 *Lynx MK8 helicopter cockpit*

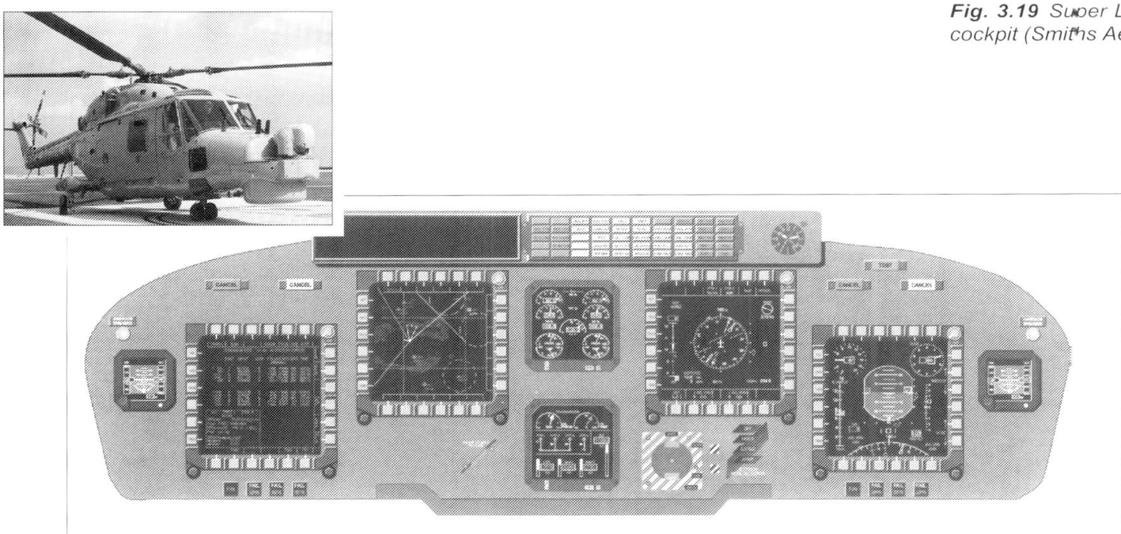

Fig. 3.19 *Super Lynx cockpit (Smiths Aerospace)*

information. 3ATI instruments at the extreme left and right hand of the instrument panel provide standby primary flight data in the event of a major systems failure. The Lynx is a small aircraft with very limited space in the equipment bay for electronic boxes. Therefore, all these displays are fully

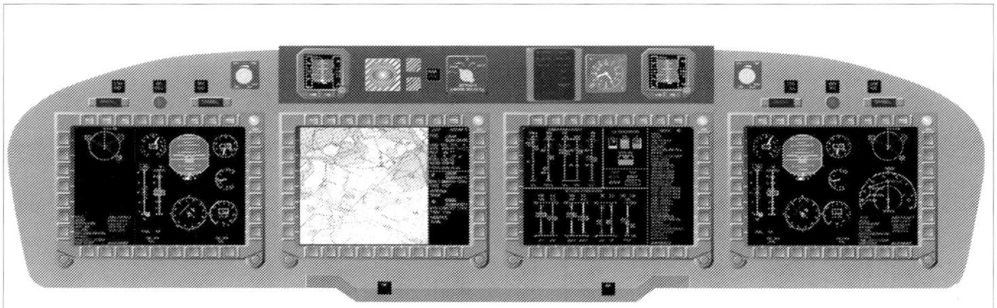

Fig. 3.20 UK Battlefield light utility helicopter (proposed) (Smiths Aerospace)

integrated, that is, they contain all the electronics to acquire data and process and generate the display formats. There are no additional boxes in the equipment bay to support the display suite.

A further iteration of the Lynx cockpit design is proposed for the UK Battlefield light utility helicopter which takes the level of integration one step further and utilizes newer and larger 10×8 in Commercial Off-The-Shelf (COTS) AMLCDs. This cockpit is shown in Fig. 3.20.

For the proposed Battlefield light utility helicopter cockpit, the primary flight and navigation information has been combined into a single display format presented to the crew on the two outer 15 in diagonal AMLCD displays. The two centre displays present a combination of tactical and systems information. Standby instruments are mounted in the centre glareshield.

References

(1) **Jones, B.** *Gloster Meteor. Crowood Aviation Series*, The Crowood Press.
(2) **Garland, D. J.**, **Wise, J. A.**, and **Hopkin, V. D.** *Handbook of Aviation Human Factors.*
(3) **Birch, S.** Eurofighter ergonomics. *Aerospace Engineering*, June 2001.
(4) **Greeley, K.** and **Schwartz, R.** *F-22 Cockpit Avionics – A Systems Integration Success Story*, Lockheed Martin, Marietta, Georgia.
(5) MIL-STD-1784, Head-Up Display Symbology, US Department of Defense.

CHAPTER 4

Electromechanical Instruments

The importance of instrumenting the information needed to fly an aircraft was recognized at the outset of manned flight. Early instrumentation was purely mechanical, measuring and displaying the parameter of interest at source. As instrumentation needs grew, technology in the guise of electrical signalling provided a route to separate parameter measurement from information display, and electromechanical instruments replaced their earlier mechanical counterparts. Eventually the cathode ray tube would revolutionize the flight deck, now no longer constrained by the (albeit ingenious) mechanical and electromechanical mechanisms of earlier times. The filmstrip moving map display provided a bridge between EM instrument and EO displays.

Mechanical instruments

Chapter 2 discussed the evolution of the transport aircraft flight deck from the beginnings of manned flight to today. Much of the information displayed on the 'glass' flight deck of today has its origins in the instruments developed during the first 20 years of powered flight. It is interesting, therefore, to review some of the (sometimes bizarre) instruments used in those pioneering days.

The requirement to display most pieces of information in a cockpit can be traced to the need to identify or solve a problem. Therefore, the process during most of the mechanical era was to invent a device or improvise from something that already existed. Any testing was generally done in-flight. During this era it was not unusual for a potential solution to progress from concept to a flight trial in a matter of weeks, as opposed to the years it takes today. However, it would be entirely wrong to imply that the mechanical era was simply one of simple-minded ad hoc evolution. On the contrary, many of the procedures still adopted today, for instance for instrument flying, were developed during this era.

Fig. 4.1 Engine rev-counter/tachometer (Smiths Aerospace)

Early instruments were sometimes just pieces of string to indicate sideslip, but the most important 'instrument' was the feel of the slipstream on the pilot's cheeks. The 'wind on the cheek' became a vital factor that remained of value for over two decades. It also contributed to the retention of the open cockpit. Completely enclosed, a pilot was deprived of this useful indication of sideslipping either inwards or outwards during a turn. Not until the turn and slip indicator was introduced in the early 1930s could the pilot dispense with this natural aid to precise flying (**1**).

The Rev-Counter (Tachometer)

In general, engine instruments came before flight instruments. In the early years, engines were temperamental and unreliable, and so needed continuous observation of their state of health and continuous adjustment. Airspeed was what you could coax out of an engine that had a small speed range; it was either going flat out or it had stopped. Figure 4.1 shows the rev-counter/tachometer in the Blackburn Monoplane of 1912. To the right of the rev-counter is a glass bulb 'pulsometer' which provided the pilot with a visual indication of oil pressure.

The Incidence Vane

Critical to the performance and to the safety of the Wright aircraft was accurate knowledge of the airflow angle of incidence over the flying surfaces. Too low an angle and there would be insufficient lift, too high an angle and the wing would stall. The incidence vane shown in Fig. 4.2 was designed and made by the Wright brothers in 1913.

The principle of operation of the Wright incidence vane is extremely simple. A vane is mounted in the airstream. It operates like a weathercock. Attached to the vane is a pointer which indicates angle of incidence against a calibrated scale. The

Fig. 4.2 *The Wright incidence vane*

instrument was placed on one of the vertical wing struts in direct line of sight of the pilot who could then control the flying surfaces so that the airflow angle of incidence was appropriate to achieve the degree of lift required for the aircraft to continue in safe flight (**2**).

Angle of incidence is still measured in today's aircraft using the same method. The only difference is that today the sensor is mounted on the aircraft skin outside the pilot's line of sight. The angle of incidence is measured by a transducer and signalled electrically to the aircraft systems. The parameter is no longer usually directly displayed on the flight deck, though it remains a key parameter in high-performance fighter aircraft.

The Airspeed Indicator

In addition to engine speed and angle of incidence, airspeed is crucial to obtain and maintain lift during all phases of the flight, particularly during take-off and landing.

The Wright brothers applied meteorological ball and cup anemometer technology which they attached to a wing strut. Captain Eteve (a French aviator) used the principle of air pressure in his airspeed indicator of 1911, shown in Fig. 4.3.

Fig. 4.3 *The Eteve airspeed indicator*

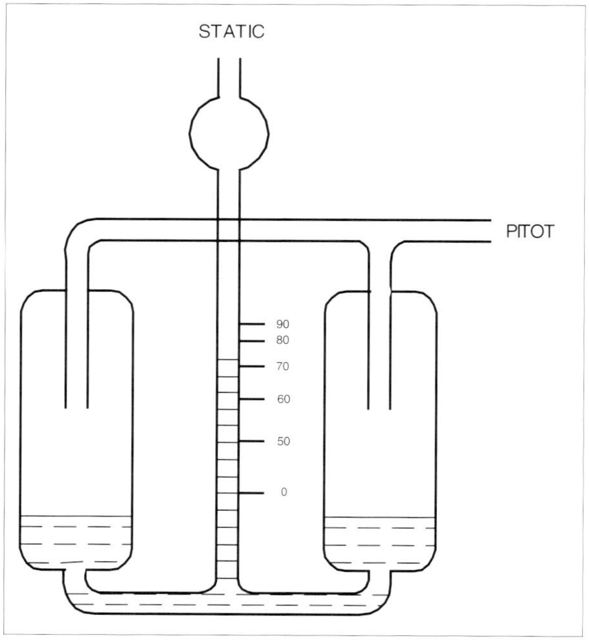

Fig. 4.4 Early manometric airspeed indicator

The principle of operation of the Eteve airspeed indicator is once again simple. The instrument uses air pressure to calculate airspeed. Forward air pressure on the vertical paddle works through a lever against a spring. An arm is attached to the paddle/lever mechanism and indicates airspeed against a calibrated scale. Like the incidence meter, the airspeed indicator was mounted in the airstream on a wing strut in direct view of the pilot. At low speeds and low height, this simple instrument would have been adequate to assist the flyer to obtain and maintain safe control of his flying machine.

An alternative means to deduce airspeed was (and still is) to measure the difference between forward air pressure and static air pressure. Figure 4.4 shows diagrammatically the arrangement of a typical apparatus installed on early aircraft. It owes more to the doctor's surgery than to aircraft engineering, using blown glass tubes, rubber hose, and coloured water.

The Altimeter

Early instruments were often the direct use or adaptation of scientific instruments allied with technology from the non-aviation world. The pocket watch altimeter, shown in Fig. 4.5, is such an example.

The instrument shows a marked similarity to a gentleman's fob watch. Indeed it was probably adapted from the same, the clock mechanism being replaced by an aneroid barometer. The basis of an aneroid barometer is an evacuated capsule. Air pressure works on the outside of the capsule and its centre deflection is proportional to the instantaneous static air pressure. Attached via a linkage, the instrument pointer indicates air pressure. Setting the instrument to read 0 ft before take-off, the instrument scale can be calibrated to indicate the height above ground at the local airfield for the current barometric pressure.

Fig. 4.5 The pocket watch altimeter (Ellictt Bros)

The Primary Flight Instrument Set

At least one company combined all the flight instruments into a single instrument panel as shown in Fig. 4.6. It can be said to be the first example of an integrated primary flight display, a concept to reappear later in the context of the 'glass cockpit'.

The integrated panel comprised:

- Airspeed (bottom right).
- Altimeter (top left).
- Vertical speed (centre vertical instrument).
- Engine tachometer (bottom left).
- Stopwatch (top right).
- Slip indicator (bottom curved horizontal instrument).

The slip indicator is a spirit level with an accentuated curve. The bubble will indicate the axis through which g is acting. To illustrate this, imagine the aircraft is banked with one wing down, but still flying in a straight line (not a realizable manoeuvre). The bubble will find the highest point in the tube and in this scenario indicate bank angle. In practice, of course, the aircraft banks into a turn. Without further action on the part of the pilot, the aircraft would 'slip' into the turn. In practice, the pilot will aim to compensate the control of aileron with elevator to produce a coordinated turn with no slip, in which case the g axis acts through the perpendicular to the aircraft wings and the bubble will return to the central position.

A magnetic compass would also be provided, though not included in this integrated panel but elsewhere in the cockpit (often the roof) to avoid being disturbed by metal structures. The pilot would have used the compass, together with airspeed and the stopwatch, to calculate present position by dead reckoning. He would then have used the compass to fly a course to the desired destination. Since most flying in this era was done in visual flying conditions, the pilot would have regularly correlated his calculated position with ground features such as church towers, rivers, roads, and railways.

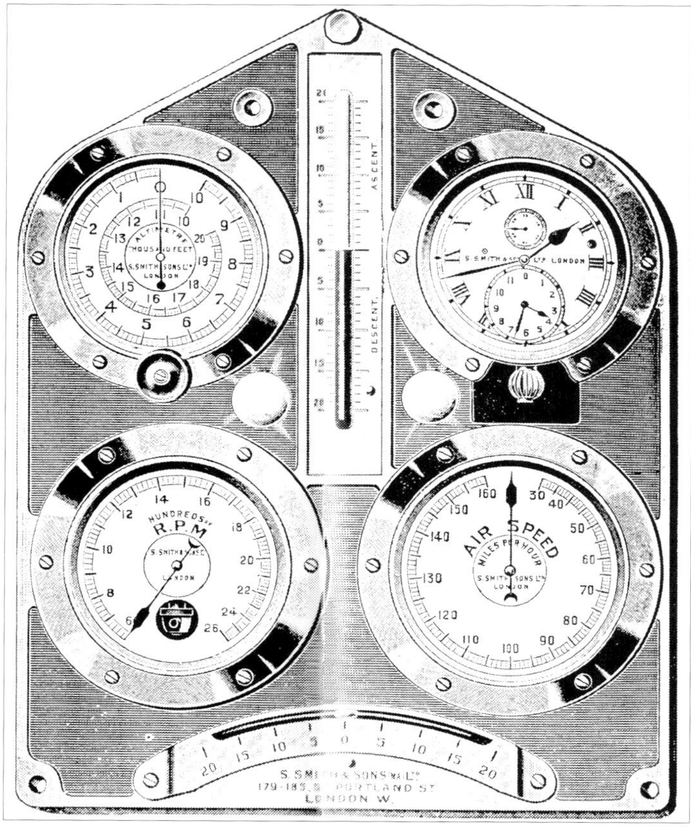

Fig. 4.6 The integrated flight instrument panel (Smiths Aerospace)

It is interesting to note the inconsistent direction of rotation of these early instruments.

On the S. Smith and Sons integrated instrument panel:

- The airspeed pointer rotates *clockwise* for increasing airspeed.
- The altimeter rotates *anticlockwise* for increasing height.
- The tachometer rotates *clockwise* for increasing speed.

On the Elliott pocket watch altimeter, the altimeter rotates *anti-clockwise* for increasing height.

Attitude (pitch and roll)

Early aircraft had no indication of pitch and roll. It was unnecessary. All flying was done on clear days with direct view of the visible earth. The pilot used aircraft structure, the bracing wires, tops of the engine cowling, and other features to provide cues for take-off, landing, and other manoeuvres.

Compass

Aircraft heading was indicated with a magnetic compass. The compass was free to maintain its orientation with respect to the ground and located so as not to be influenced by any local metal structures.

Fig. 4.7 *The magnetic compass*

The cockpit of the 1916 Bristol Scout Type 21 F in Fig. 4.7 shows a typical installation. The cockpit also included:

- Turn and slip indicator.
- Magnetic compass.
- Airspeed indicator.
- Altimeter.
- Clock.

Electromechanical instruments

Chapter 2 discussed the evolution of the transport flight deck and the migration of the flight instrument cluster from the 'Basic 6' to the 'Basic T'. The 'basic T' configuration is shown in Fig. 4.8.

The ADI and HSI are commonly found in 5ATI format, that is, having a physical form factor in accordance with the Air Transport Indicator (ATI) standard, and having a 5×5 in square faceplate. The other instruments are commonly found in 3ATI format; that is, with a 3×3 in square faceplate.

The instruments are available from a number of manufacturers, have standardized functionality, and are certified to comply with Technical Standard Orders (TSOs) published by the Federal Aviation Authority (FAA).

The TSOs of relevance to these primary flight instruments are (**3–7**):

- TSO-C133 – airborne multipurpose electronic displays.
- TSO-C3d – turn and slip instrument.
- TSO-C4c – bank and pitch instruments.
- TSO-C52b – flight director equipment.
- TSO-C6d – direction instruments, gyrostabilized.

In addition, a third set of primary flight instruments, known as standby instruments, provide fully independent information to aid the flight crew to

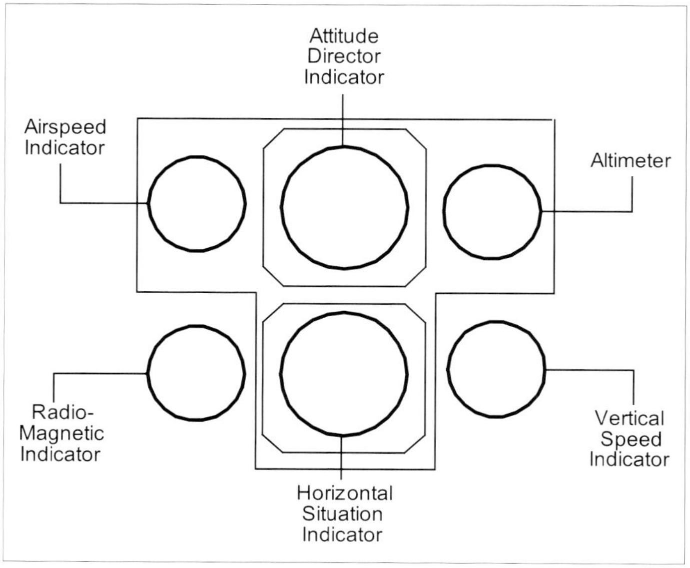

Fig. 4.8 *'Basic T' flight instrument cluster*

resolve instrument discrepancies. The standby instruments (usually in 3ATI format) are totally segregated from the prime instruments and operate on separate electrical power, usually the DC essential bus.

The Attitude Direction Indicator (ADI)

The ADI, shown in Fig. 4.9, provides:

- An attitude ball.
- A flight director.
- A turn and slip bubble.
- Warning flags.

The pitch and roll attitudes of the aircraft are represented by the position of a roller blind relative to the aircraft symbol. The blind, shown unrolled in Fig. 4.10, consists of a two-tone band, light-blue representing the sky and brown representing the ground. A white line separating the two zones represents the horizon. The aircraft symbol, in the form of a yellow coloured motif, is painted on the rear glass of the instrument.

Movement of the roller blind over rollers attached to a carriage assembly depicts a change in aircraft pitch attitude, while rotation of the roller blind with the carriage assembly depicts a change in aircraft roll attitude. Pitch angle is given by pitch scale marks on the roller blind. Roll or bank angle is given by a pointer, which rotates with the roller blind, against a scale painted on the instrument bezel.

Motion of the roller blind carriage is controlled by two electromechanical servomechanisms. The operating principle of each mechanism is similar. An error signal is produced in the control transformer between the aircraft actual pitch or roll attitude (received from the aircraft attitude reference system in synchro form) and the displayed attitude. The error signal is amplified and drives the servomotor which drives the blind through a series of gears until the error signal reduces to

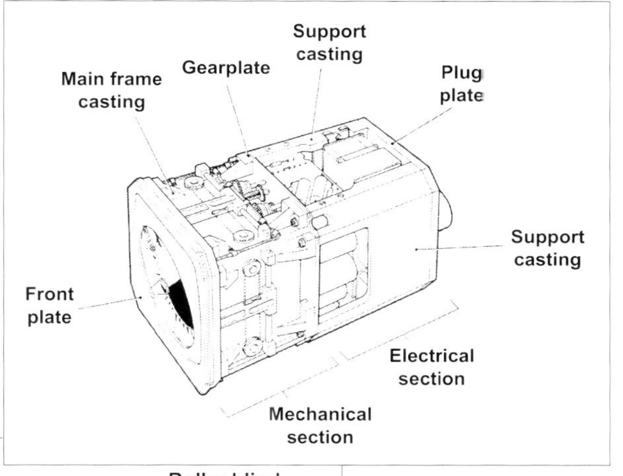

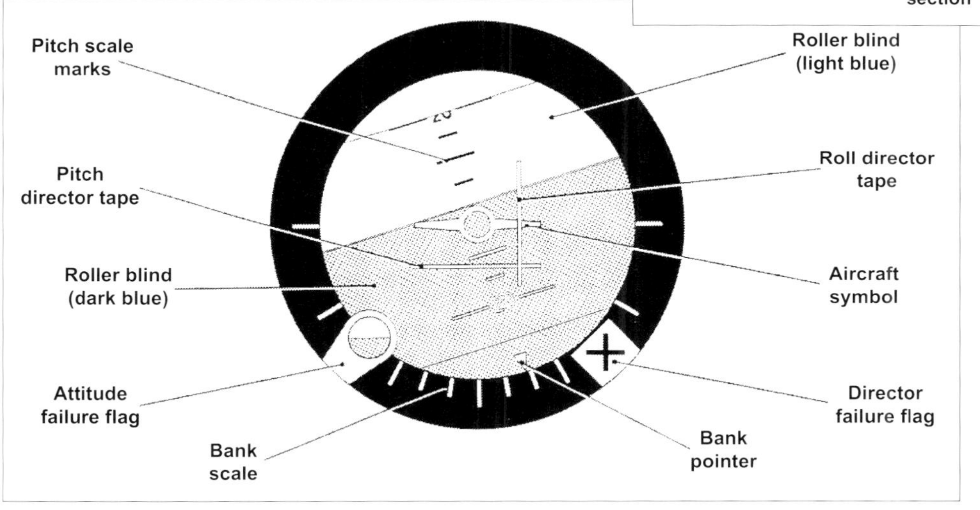

Fig. 4.9 Attitude direction indicator (ADI) (Smiths Aerospace)

zero, at which time the servomotor ceases to drive and the instrument displays the new aircraft attitude.

A requirement to manoeuvre the aircraft in pitch to achieve a desired flight path is represented by the vertical displacement of a horizontal yellow bar (or tape) from its datum position in the centre of the aircraft symbol.

Similarly, a requirement to manoeuvre the aircraft in azimuth to achieve a desired flight path is represented by the lateral displacement of a vertical yellow bar (or tape) from its datum position in the centre of the aircraft symbol.

Motion of the director bars is again achieved by two electromechanical servomechanisms, as shown in Fig. 4.11.

Failure of either the attitude system or the director system is indicated by the presence of failure flags. The flags disappear when the systems are functioning correctly.

Fig. 4.10 ADI pitch and roll servomechanisms (Smiths Aerospace)

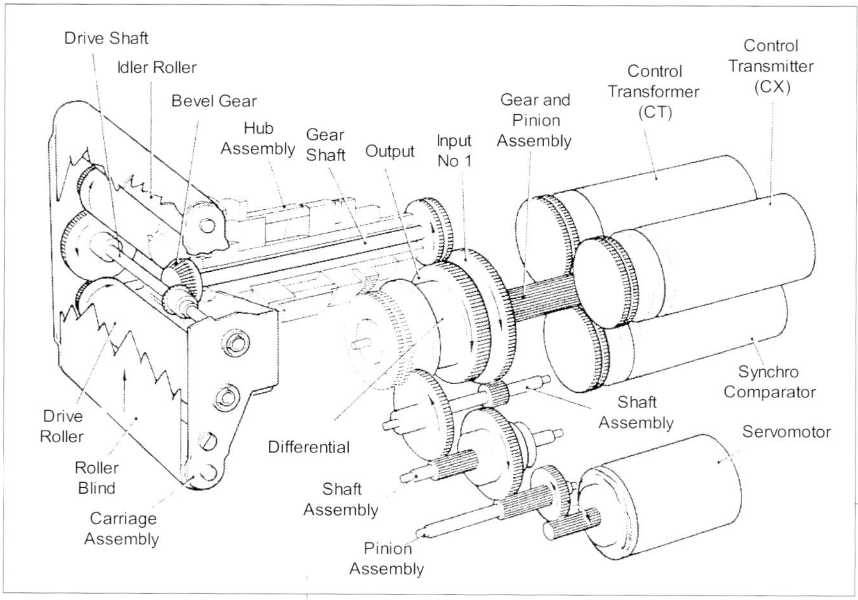

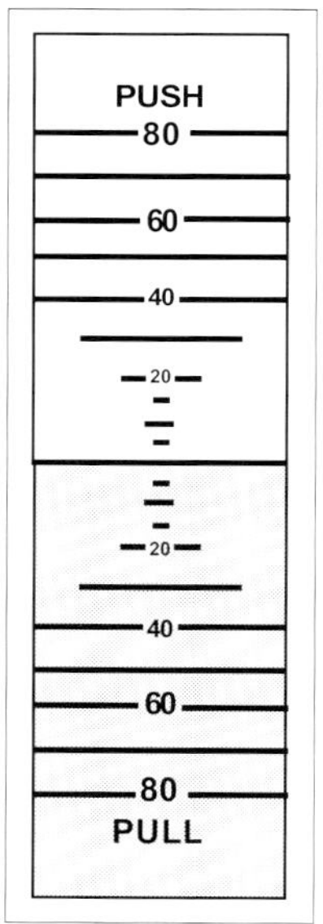

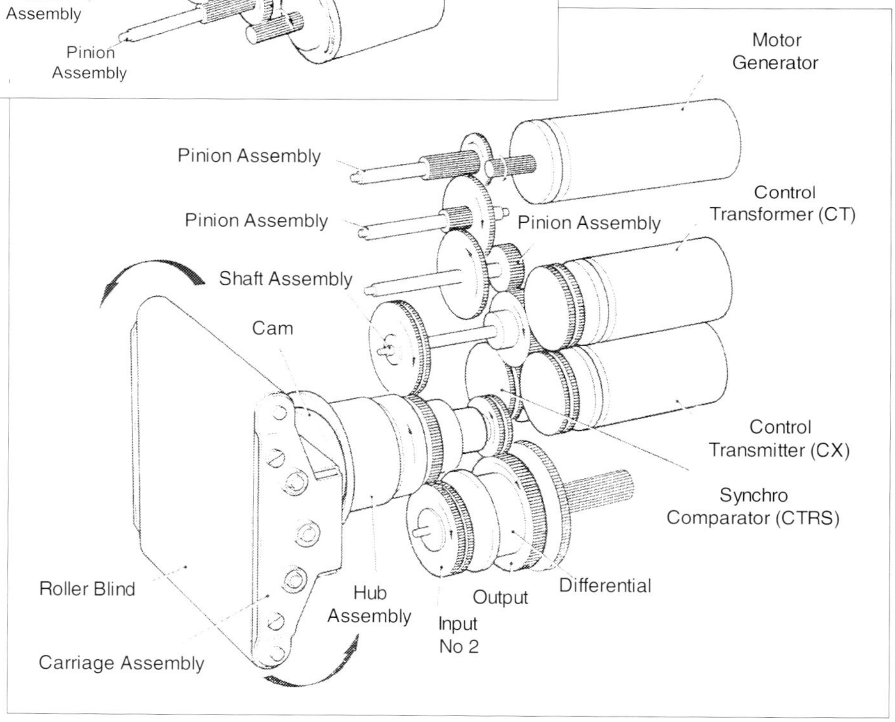

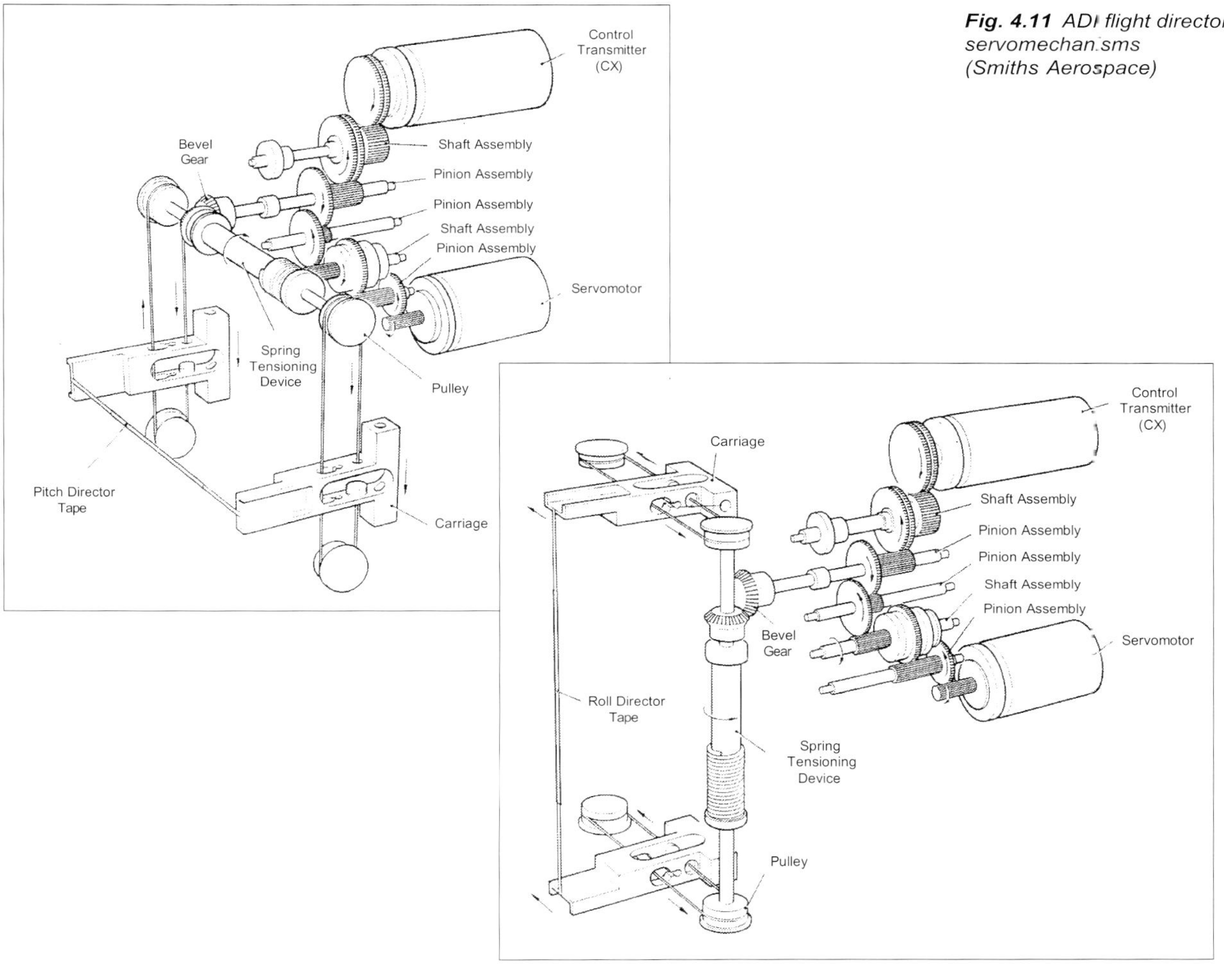

Fig. 4.11 *ADI flight director servomechanisms (Smiths Aerospace)*

Barometric instruments

The Altimeter, Airspeed Indicator (ASI) and a third instrument, the Vertical Speed Indicator (VSI), operate on the principle of measuring air pressures and are therefore often referred to as barometric or manometric instruments.

In the mechanical era, measurement of these parameters was achieved by direct connection of the pitot static tubes to the cockpit instruments themselves. In the electromagnetic aircraft and aircraft of today the air pressure sensors and computational corrections are performed in an air data computer and transmitted electrically to the flight deck instruments. Standby instruments on the latest generation of aircraft with glass cockpits must still operate in this way to provide independence and isolation from the other aircraft systems.

Fig. 4.12 Pitot static probe

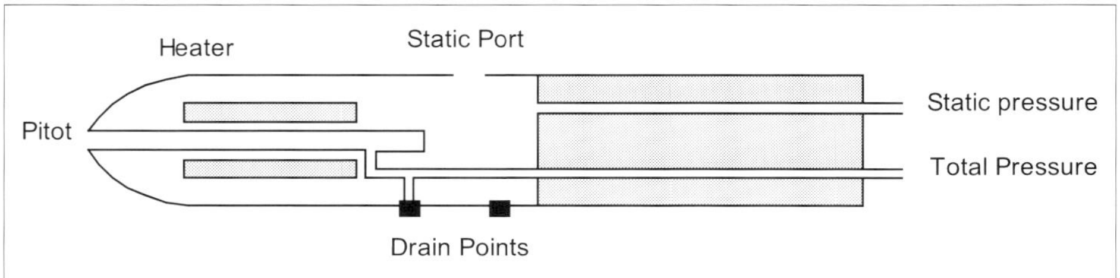

The air data system senses:

- Static pressure, P_s, the pressure of the surrounding atmosphere.
- Total pressure, P_t, a combination of dynamic pressure, P_d, created by the forward motion of the aircraft and the static pressure.

From the static pressure and the total pressure, the dynamic pressure can be found, and from these three pressures it is possible to determine:

- Airspeed; from dynamic pressure.
- Altitude; from static pressure.
- Vertical speed; from rate of change in static pressure.

The sensor is a pitot static probe that samples both the total and the static pressure (see Fig. 4.12). The probe is extended from the skin of the aircraft to be outside the boundary layer and also any turbulence created by the aircraft. Sometimes, the pitot and static ports are separated.

The functioning of a basic set of mechanical barometric instruments is shown in Fig. 4.13.

Fig. 4.13 Barometric instrument set

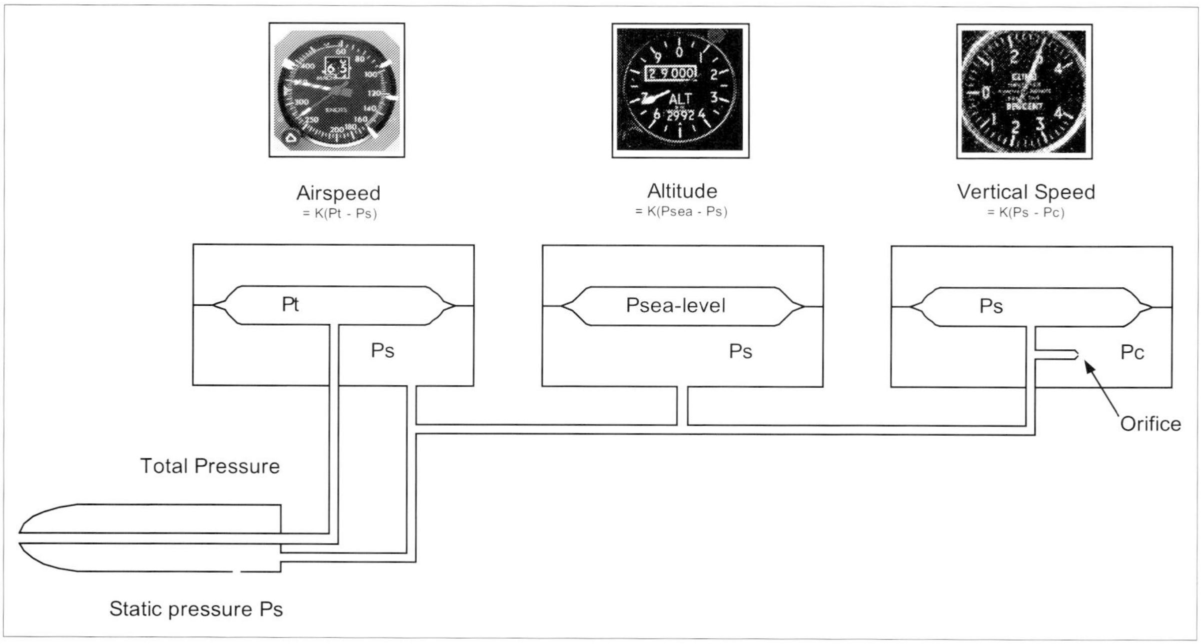

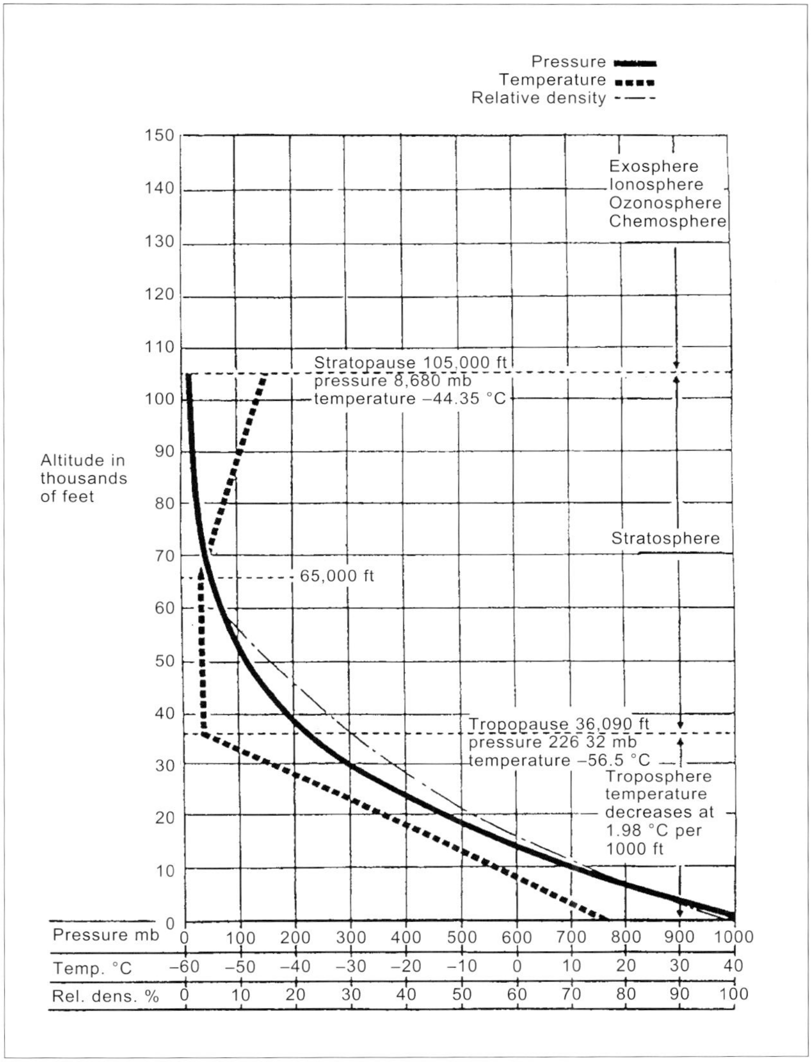

Fig. 4.14 *Standard atmosphere chart*

Standard pressure tables, shown in Fig. 4.14, model the change in atmospheric pressure with altitude. Universally adopted operating procedures are used to set the instrument pressure datum (or baro-set) to the air pressure at the take-off or landing airfield in order to indicate height relative to those locations. When at cruising altitude, the pressure datum is normalized to a standard pressure so all aircraft in the vicinity compute height to the same datum and therefore can deduce height relative to each other. The instrument and the procedures are of sufficient accuracy safely to permit air traffic with a vertical separation distance between aircraft of 1000 ft below 29 000 ft and 2000 feet above. Modern air data computers have improved accuracy, and changes are being made to the route structures to introduce Reduced Vertical Separation Minima (RVSM) to 1000 ft between 29 000–41 000 ft.

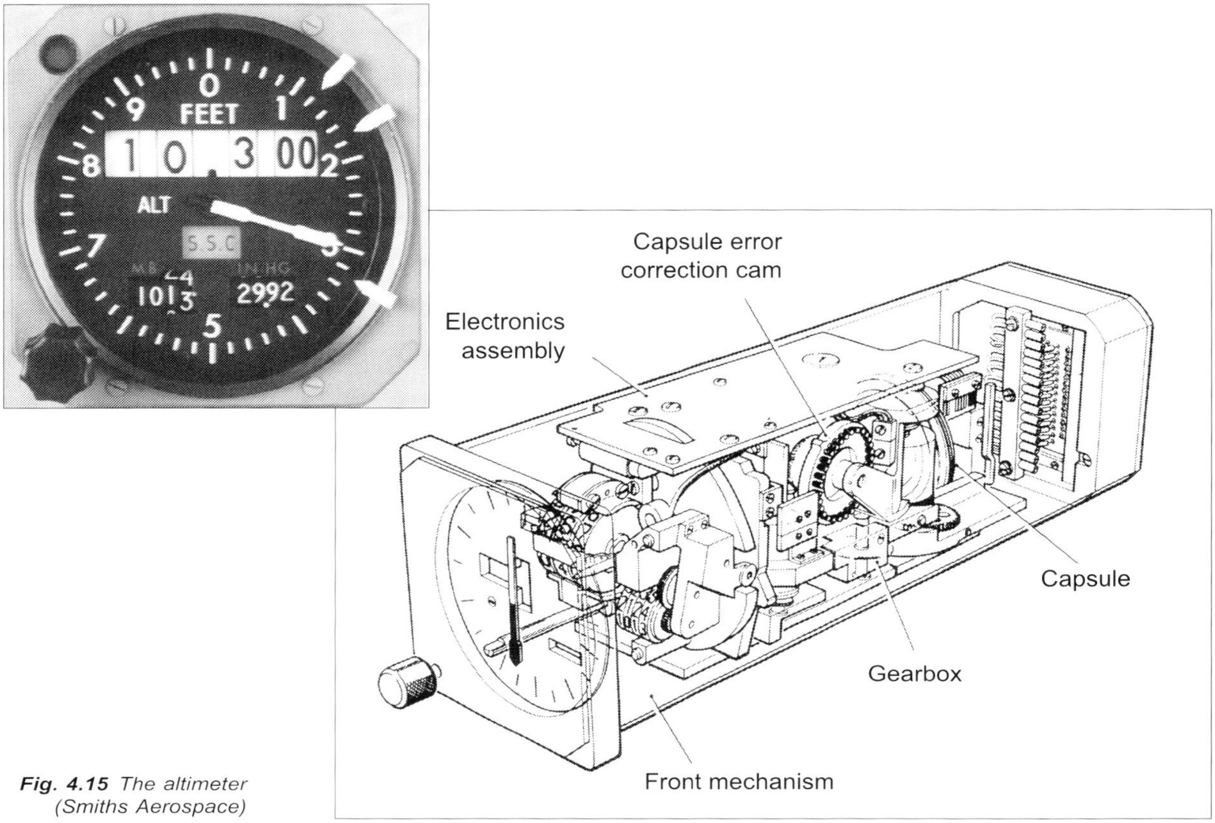

Fig. 4.15 The altimeter (Smiths Aerospace)

The Altimeter

The primary altimeter in current-generation aircraft contains an electromechanical servomechanism driven by signals from the air data computer to indicate barometric altitude by means of a pointer and a counter (numerical readout). A knob with a numerical readout sets the barometric pressure datum.

Early instruments and standby instruments operate directly from air pressure. By means of gears, levers, and cams shown in Fig. 4.15, the deflection of an aneroid (sealed) operating against a static pressure outer atmosphere is translated into motion of counters and pointers against a scale indicating height (in feet). The capsule operates against aircraft outer atmosphere static pressure (derived from the pitot static probe). The cams model the standard atmosphere chart. The pointer completes one revolution for every 1000 ft, rotating clockwise for increasing height. A bimetal strip provides temperature compensation.

The Airspeed Indicator (ASI)

As with the primary altimeter, the primary airspeed indicator in current-generation aircraft is driven by an electromechanical servomechanism slaved to the air data computer.

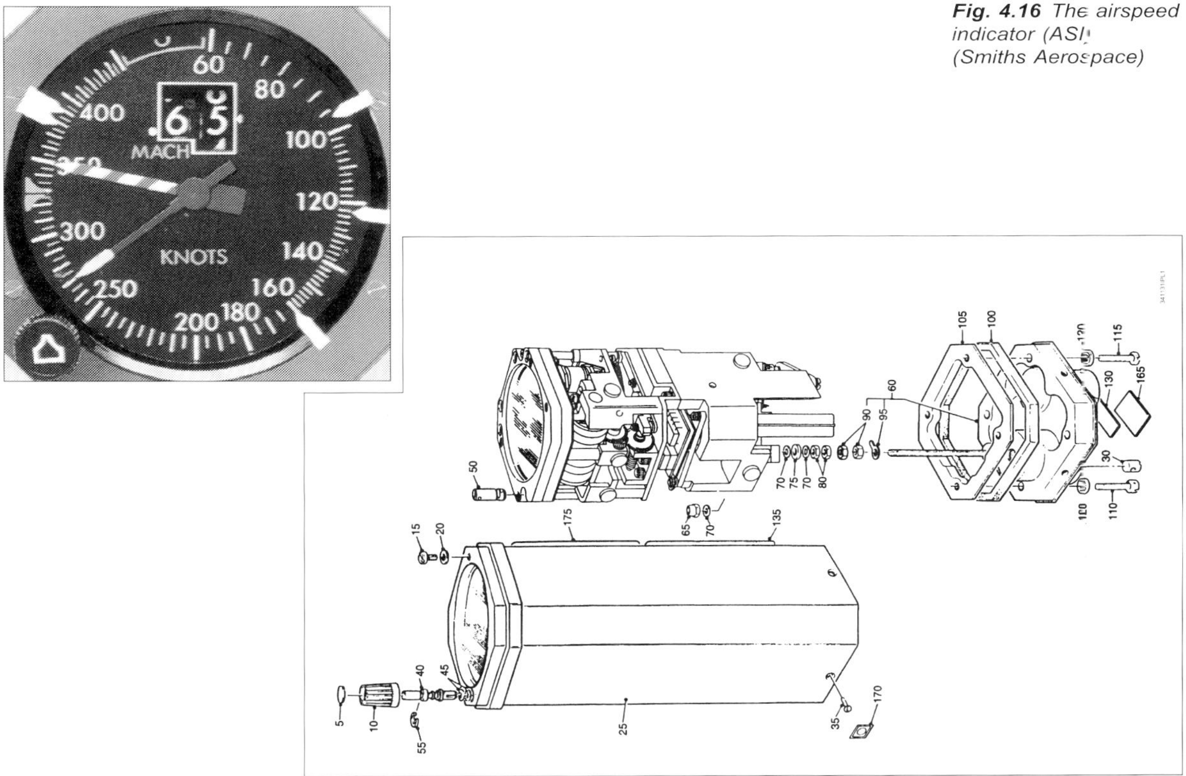

Fig. 4.16 *The airspeed indicator (ASI) (Smiths Aerospace)*

Standby instruments, and early primary instruments, operate directly from air pressure. The deflection of a capsule whose inner atmosphere is total pressure operating against a static pressure outer atmosphere is translated into motion of counters and pointers against a scale indicating airspeed in knots. In the ASI the capsule inner atmosphere is airstream total pressure operating against outer atmosphere static pressure. Strictly speaking, knowledge of the air density is also required, but the simple mechanical instrument cannot compute this quantity from the information to hand and assumes a standard air density. The instrument also assumes that air is incompressible (which is true at airspeeds below Mach 1). The parameter computed with these assumptions is called calibrated airspeed, K_{cas}.

As shown in Fig. 4.16, in addition to airspeed it is not unusual also to find a numerical readout of Mach number. A warning bar indicates maximum safe speed. Bugs may be set around the circumference of the instrument to indicate placard speeds.

The Vertical Speed Indicator (VSI)

Static pressure is fed to both sides of the capsule, but one side is fed through a fine orifice. When flight is straight and level, the pressure on both sides of the capsule is the same and the instrument reads zero. If the aircraft ascends or descends, the static pressure will change instantaneously on the side of the capsule directly fed, and the orifice bleeds air at a predetermined rate to the other side, resulting in a

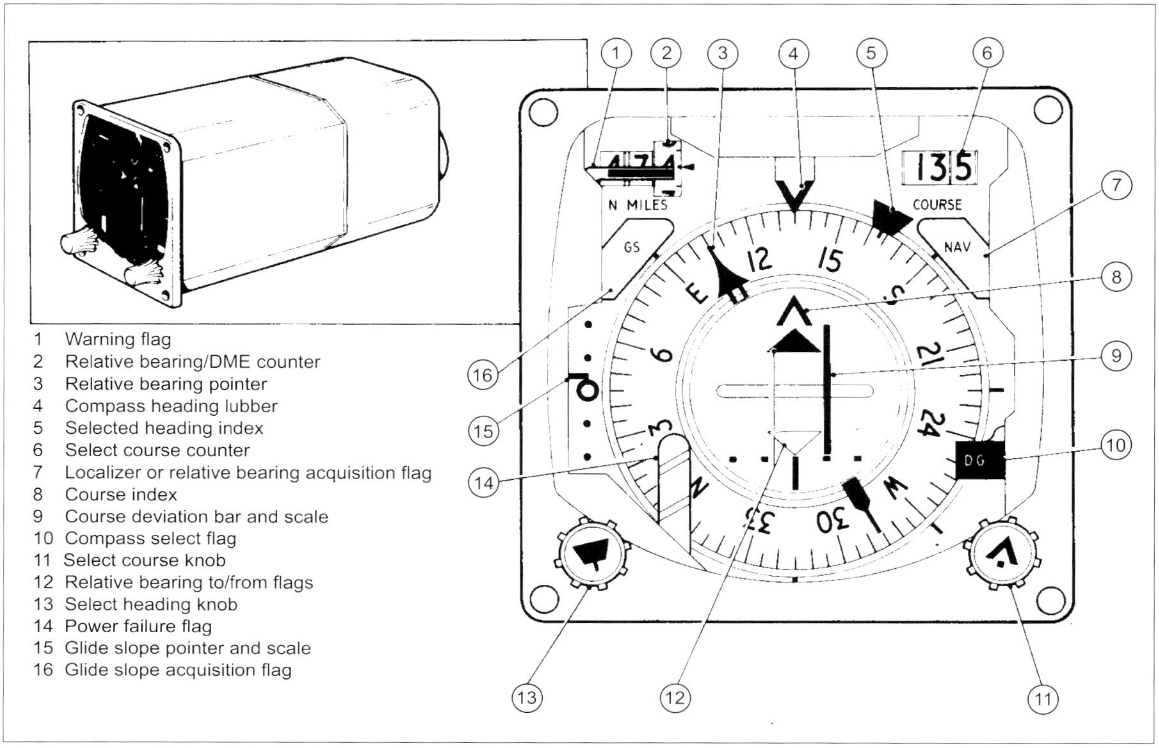

Fig. 4.17 Horizontal Situation Indicator (HSI) (Smiths Aerospace)

pressure difference. The resultant capsule deflection moves a pointer against the climb/dive scale of the instrument.

The Horizontal Situation Indicator (HSI)

The HSI shown in Fig. 4.17 functions as a magnetic compass or as a directional gyro indicator, depending on the mode selected, and indicates:

- *Aircraft heading* which is indicated by rotation of the compass card.
- *Aircraft course* which registers against and rotates with the compass card.
- *Selected course* which is set relative to the compass card. A drum counter display of selected course is also incorporated.
- *Selected heading* which registers against and rotates with the compass card, set relative to the compass card by means of the heading selection knob.
- *Relative bearing* which is displayed by the relative bearing pointer driven independently of the compass card from the radio navigation equipment.
- *Deviation in azimuth* from a selected relative bearing radial or ILS localizer beam.
- *Deviation in pitch* from the ILS glidepath.
- The *DME counter* displays the distance of the aircraft from a beacon.
- A *to/from flag* indicates the aircraft direction relative to the beacon.

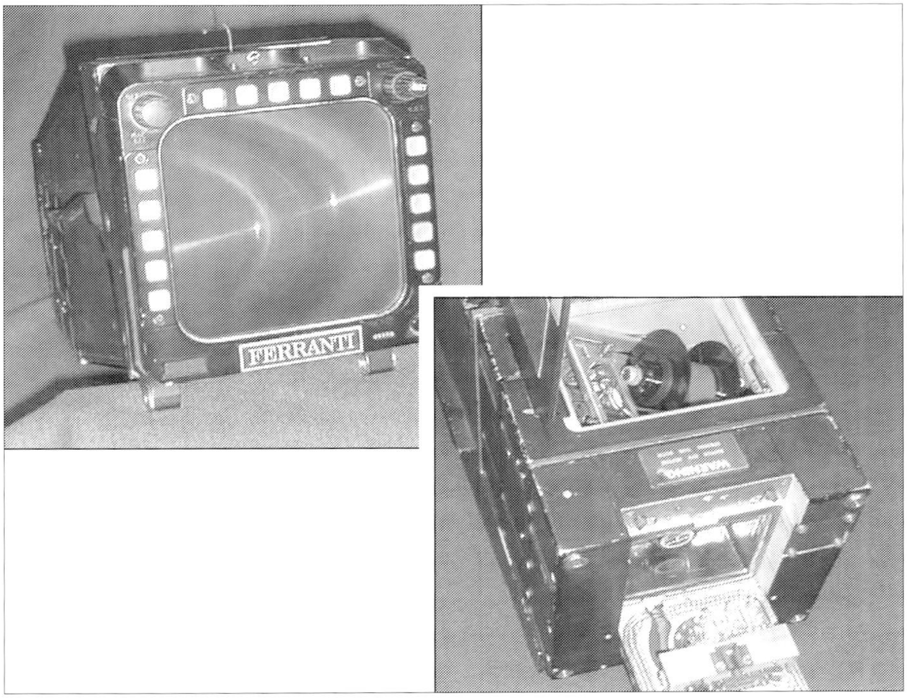

Fig. 4.18 Filmstrip moving map display (BAE SYSTEMS)

The Film Strip Moving Map Display

The moving map display in 1950s fast jet fighter aircraft made the transition from electromechanical instruments to electro-optical displays. The Ferranti Combined Map and Electronic Display (COMED) shown in Fig. 4.18 typified the equipment.

Conventional aviation paper maps/charts were photographed in overlapping lateral strips onto a 35 mm colour filmstrip which was then rear projected into a translucent screen to produce a 6×6 in image. The film traction mechanism owed much to the principles of the electromechanical ADI attitude ball. Film of 57 ft length was wound from spool to spool with a fast and accurate along-film transport mechanism representing east–west motion of the aircraft. North–south motion tracks across the filmstrip. When the edge of the filmstrip is reached, the transport mechanism fast-winds to seek the next adjacent lateral map strip. Rotation of the mechanism oriented the map heading-up, track-up, or north-up. The map moves to keep ownship in the centre of the screen, or, more usually, near the bottom of the screen with the map rotated or oriented to show current track vertically up the display in order both to be intuitive and to maximize the view ahead.

A monochrome green CRT image could be optically combined with the map image to overlay onto the map a planned route or an image from ground mapping radar, the latter of which should correlate with the map if the navigation computing is accurate.

The map was illuminated with a 50 W halogen lamp, almost identical to a car headlamp but accurately mounted on a base for correct alignment of the filament.

The lamp life was only around 200 h, and so three lamps were provided on an autochange mechanism. Dimming for night-time use was electrical, supplemented by a day/night mechanically moving filter with transmission of around 1 per cent to prevent the light becoming too red at low drive levels.

The map was projected onto a 2 in diffusing screen with a lens of ×6.8 magnification. The screen vibrated in its own plane, to eliminate graininess. The viewing optics had a ×2.2 magnification lens creating a real image of the 2 in vibrating screen. This real image is in the plane of the display front face where there is a transparent field lens. The field lens does not magnify or distort the image, but determines the exit pupil from which the image can be seen. The field lens is a 6 in diagonal Fresnel lens with 200 grooves per inch. It images the ×2.2 magnification lens into the plane of the pilot's eye, forming an exit pupil in space around the pilot's head. This exit pupil is about 8 in across, and the whole of the map image can be seen from within the exit pupil. Outside the pupil there is no image, and no stray light is emitted into the cockpit. Furthermore, sunlight can only enter the system to illuminate the diffusing screen, via this same exit pupil. Since this is usually blocked by the pilot's head, image contrast in maintained. Cardinal point performance is as follows:

Screen area	6 in
Luminance	2000 fL (6800 cd/m^2) with no film
Film transmission	Typically 30%
Contrast	10:1 in full sunlight of 10 000 fc (1 080 000 lux)
Resolution	150 line-pairs/in

References

(1) **Coombs, L.F.E.** (1990) *The Aircraft Cockpit*, Patrick Stephens Limited.

(2) **Chorley, R.A.** (1976) Seventy years of Instruments and Displays, 3rd Folland Lecture, Royal Aeronautical Society, February.

(3) TSO-C113, Technical Standard Order: Airborne Multipurpose Electronic Displays, Federal Aviation Administration.

(4) TSO-C3d, Technical Standard Order: Turn and Slip Instruments, Federal Aviation Administration.

(5) TSO-C4c, Technical Standard Order: Bank and Pitch Instruments, Federal Aviation Administration.

(6) TSO-C52b, Technical Standard Order: Flight Director Equipment, Federal Aviation Administration.

Chapter 5

The Head-Down Display

Introduction

The Multifunction Head-Down Display, variously called the MFD, the MHDD, the MPD (Multipurpose Display), and the MPCD (Multipurpose Colour Display), provides the flight deck/crewstation/cockpit designer with a flexible display medium on which to present data in a variety of ways according to the information needed by the crew for the current phase of the flight or mission.

This chapter describes Head-Down Display MFD implementations in the major technologies used to date and outlines current display format philosophies. Chapters 9 and 10 discuss potential future display technologies and future format developments respectively.

At the outset of the electro-optical era, display capability was limited primarily to that able to be provided by the monochrome (usually green) CRT. Therefore, although early military cockpits used the CRT for textural/symbolic systems data and for sensor video, an ingenious combination of mechanical and electro-optical technology was required to make a full colour topographical moving map. The map itself was photographically captured on a 35 mm filmstrip and then projected onto a rear view screen. Symbology written on a monochrome CRT was optically combined with the map image to provide an overlay of additional flight and tactical information – refer to Chapter 4. This moving map device in its various guises served the military community well for many years.

The next significant breakthrough came with the application of the colour shadow mask CRT to airborne environments. The shadow mask CRT is arguably the best example of Commercial Off-The-Shelf (COTS) technology applied to airborne applications. However, it would take significant developments (albeit evolutionary rather than revolutionary) to make the shadow mask CRT both bright enough and sufficiently robust to be able to be used in airborne applications, civil and military. Once this was achieved, the 'glass' flight deck as we know it today was truly realizable.

However, CRT technology is bulky, heavy, and power hungry and requires extremely high voltages to operate at high brightness, posing serious limitations to the crewstation designer. The advent of AMLCD technology seemed to offer the promise of eliminating all those problems. In the event, AMLCD technology has proved as difficult if not more difficult to apply to airborne applications. Although small (i.e. flat), light and low voltage, the visual performance of an AMLCD varies with viewing angle (particularly an issue for cross-cockpit monitoring), and it is temperature sensitive and fragile. Furthermore, as was learnt very painfully, extreme attention to manufacturing process control is required to achieve consistent quality product. This latter only comes with years of experience producing high-volume product. Custom-size low-volume aerospace product is to be avoided if at all possible.

These problems are now largely solved, and the AMLCD is the leading technology, often replacing CRT when obsolescence issues arise. Display sizes for new crewstation designs are transitioning from the custom square format sizes historically used in aircraft to rectangular laptop PC and professional/industrial glass sizes. The focus now is to ruggedize COTS AMLCD devices, not design custom aerospace glass.

The CRT multifunction Head-Down Display

The cathode ray tube was the first fully flexible display device to be used for primary flight displays in the cockpit. Fixed-format Light-Emitting Diode (LED) displays were used before the CRT, but for engine instruments only. Although early research used monochrome CRTs, the added dimension of colour was soon introduced once it was established that the shadow mask colour CRT extensively used in domestic television could be applied to the airborne cockpit environment.

CRT MFD: principles of operation

The block diagram of a typical colour CRT technology based MFD is shown in Fig. 5.1. In this case the diagram applies to shadow mask CRT based equipment operating in raster and stroke modes. A monochrome CRT, raster-only design can omit certain of the complexities of the shadow mask design with a much simplified deflection amplifier.

The functional elements of the CRT MFD comprise:

- Shadow mask CRT.
- High- and low-voltage power supply.
- X and Y deflection amplifier.
- Video amplifier.
- Ramp generator.
- Encoder.
- Keypanel.
- Microcontroller.

The shadow mask CRT

The shadow mask CRT shown in Fig. 5.2 comprises an evacuated glass bulb in which an electron gun emits electrons at high velocity to impact on a phosphor

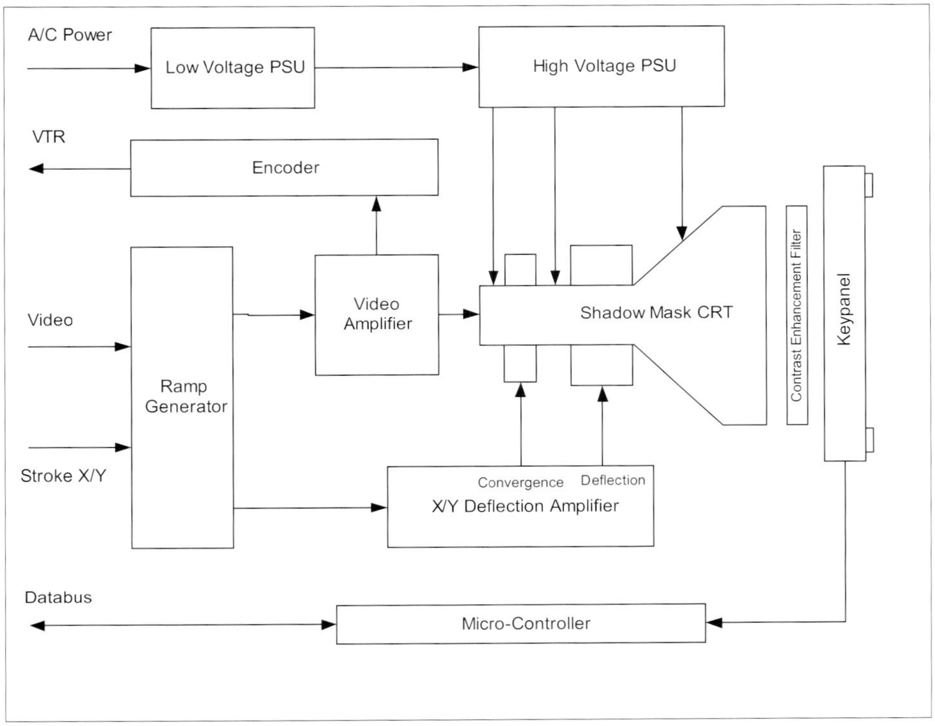

Fig. 5.1 Colour shadow mask CRT MFD block diagram

screen. The electron gun comprises three cathodes, today usually the in-line configuration, though early applications used the delta configuration.

Electron emission is modulated by three grid electrodes, focused by a multistage electron lens and finally accelerated towards the faceplate through a shadow mask. A series of red, green, and blue phosphor dots are deposited on the CRT faceplate in a process that uses the shadow mask itself to maintain precise registration. The electron gun, shadow mask, and phosphor dot geometry are arranged so that electrons emitted by each cathode only illuminate phosphor dots of its designated colour. The composite beam bundle is deflected in the X and Y axis by a magnetic yoke to scan the CRT display surface. Modulating the individual beams produces a full colour image comprising three registered primary colour images. The shadow-mask CRT requires specialized high-voltage power supplies.

A typical CRT for civil transport application primarily operates in stroke mode. It uses a final anode potential of around 18 000 V (18 kV) and 6–8 kV focus potential. Beam currents are of the order of 600 μA per gun. With a shadow mask transmission of less than 30 per cent, 20 W of power is lost in the mask itself, which must be radiated to the bulb. Spring attachments maintain the mask in tension and in registration with the phosphor dots.

The display of a daylight-viewable, full colour topographical map in raster mode in the high-ambient illumination conditions to be found in a fast jet fighter requires a considerably brighter shadow mask CRT. A specialized-custom CRT was developed by Tektronix (later Planar) with funding from industry and the US DoD. The CRT used in the AV-8B, F/A-18, the Tornado, and Eurofighter Typhoon uses a novel electron gun, called the periodic focus gun which operates at

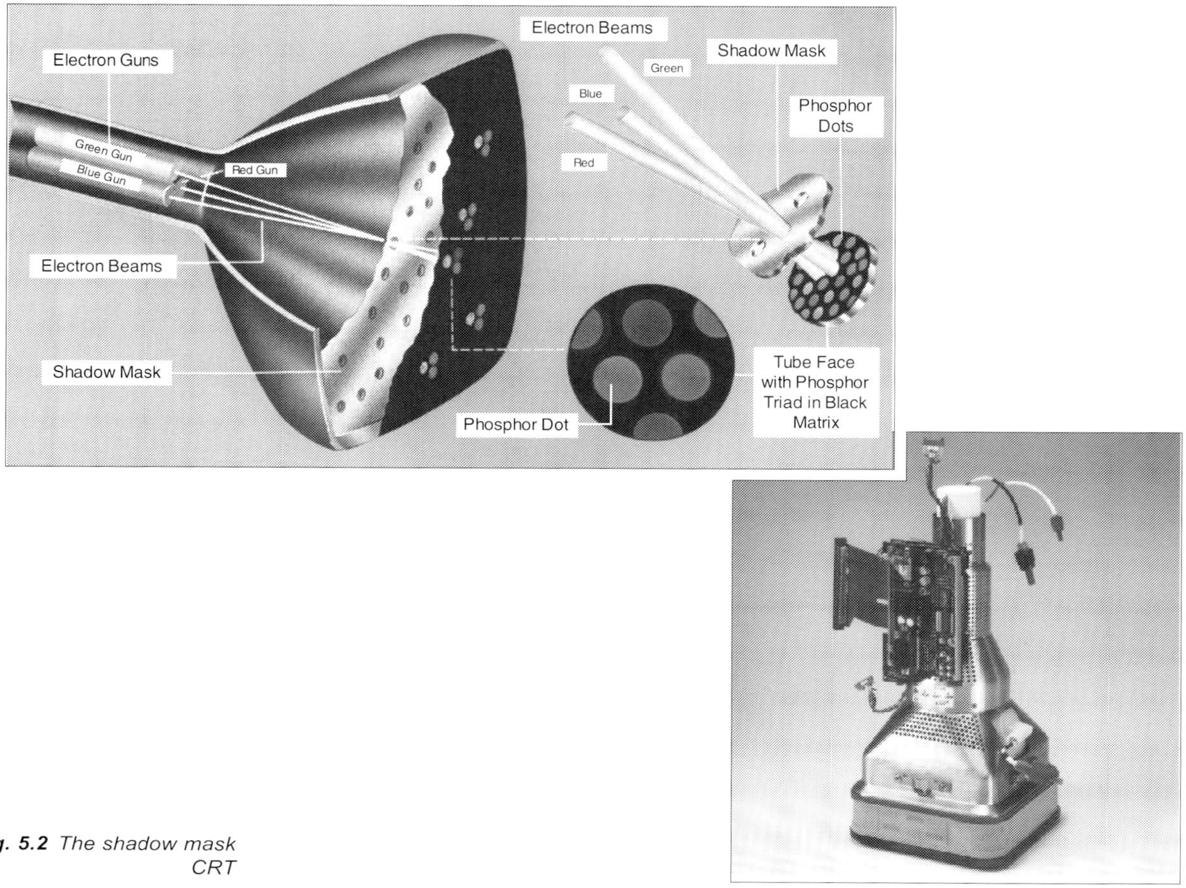

Fig. 5.2 *The shadow mask CRT*

28 kV and sources beam currents of 2 mA per gun while still maintaining a high resolution. Thus, this CRT is some 5 times brighter than its civil transport counterpart.

At full brightness white field, some 70–80 W of power is dissipated in the shadow mask of this high-performance CRT, so an entirely new shadow mask mounting method had to be devised, called the taut mask. In this arrangement the shadow mask is welded when under high tension to a steel mounting ring. The ring is maintained in precise registration with the flat phosphor faceplate by means of an accurately machined ceramic 'Q' spacer (see Fig. 5.3). Now, as the mask heats up due to beam current, tension in the mask decreases, but the shadow mask retains precise registration with the phosphor dot screen. The mask material is carefully selected to have the best compromise in characteristics in order to be able at one and the same time to be precisely etched to produce the perforated mask structure, to have sufficient tensile strength, to be a good radiator of heat, but yet to have a low coefficient of expansion. During development, the material being sought was often referred to as 'unobtainium'. It is in fact a steel/invar alloy.

At very high beam currents, the mask tension will become zero, the so-called wrinkle point, and registration is lost. The mask will now oscillate if vibrated, and

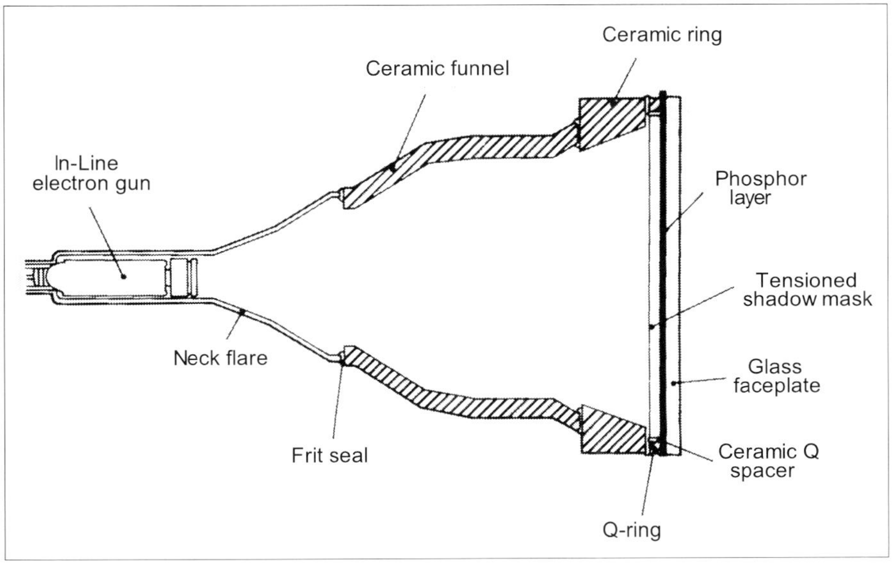

Fig. 5.3 *Taut mask CRT cross-section (Planar Advance)*

the screen will be covered in psychedelic colours as the electron beams land on the wrong colour phosphor dots. The effect, though entertaining, is not damaging, and correct performance returns if the beam current is reduced. The 'wrinkle' point is typically at 80 W. Special protection mechanisms have to be built into the deflection amplifier to detect scan failure. At these high beam powers, failure of a single deflection axis will rip the mask in a few seconds, and total scan failure will burn a hole in the mask in milliseconds and bore a hole through the glass faceplate in seconds.

In order to prevent disturbance of the beam geometry by external magnetic fields, the CRT bulb is encased in a mu-metal shield. The shield also incorporates damping mechanisms to ruggedize the CRT to the airborne environment. In military applications it has usually been necessary to add additional antivibration mounting. This is best achieved by attaching resilient mounts between the CRT CofG and the MFD chassis structure.

The shadow mask CRT is a large heavy device, and not easy to package. Typically, a 6.25×6.25 in CRT is 35 cm (14 in) long and weighs 5 kg (11 lb), including magnetic components and shield.

The phosphors deposited on the CRT faceplate emit light in the red, green, and blue primary colours. The red and green are narrow-band, short-persistence ($<10\,\mu s$) phosphors; the blue is broadband, longer-persistence (1 ms) phosphor.

The gamut in which colours can be produced by a shadow mask CRT is best described in CIE u' v' colour space. By suitable combination of cathode drive voltages, any colour can be produced within the triangle described by the three primary colours:

	u'	v'	Dominant wavelength (nm)
Red	0.42	0.52	610
Green	0.14	0.55	550
Blue	0.19	0.134	460

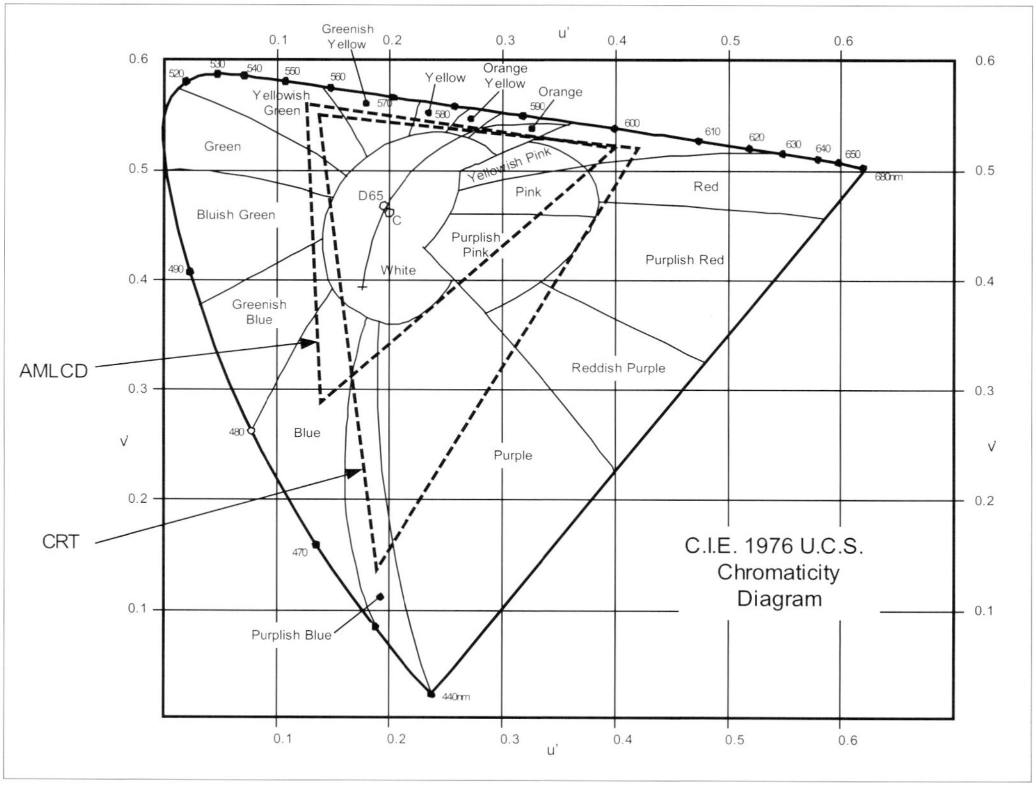

Fig. 5.4 CRT and AMLCD colour gamut

Figure 5.4 shows the colour gamut of a shadow mask CRT. Also shown is the gamut available from an AMLCD for comparison. The CRT has a purer blue, though it can be argued that this is of little operational advantage.

High- and Low-Voltage Power Supply (HVPSU and LVPSU)

The LVPSU conditions the aircraft main voltage supplies (usually 115 V AC, 400 Hz, three-phase) to several regulated DC voltages for use within the equipment. The LVPSU incorporates transient and spike suppression and provides short-circuit and overvoltage protection.

The HVPSU provides the specialized voltages required for the CRT electrodes. Special attention must be paid to avoid high-voltage corona discharge which may occur under adverse low-pressure and high-humidity operating conditions. No matter how well the CRT is built, there is always the risk of internal flashover between electrodes within the evacuated glass envelope. The HVPSU must incorporate protection devices to prevent damage to itself and to the other electronics should a flashover occur.

X/Y deflection amplifier

To produce a high-brightness, daylight-viewable, high-quality text and symbolic display, the shadow mask CRT is usually operated in stroke (cursive) mode, that

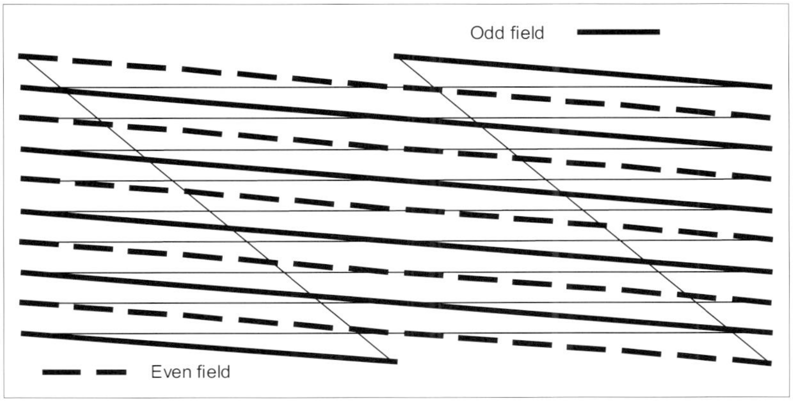

Fig. 5.5 *Video sync and interlace*

is, the CRT beam is caused to trace out the desired symbology as a pen plot. The beam current is modulated 'on–off' as raising/lowering the pen, 'off' when moving (slewing) to the next symbol. In this way, the beam can be moved more slowly (typically 10 times more slowly) than when producing a raster scan, and the image is correspondingly brighter. Of course, as with a raster, the image is rewritten frequently and at a sufficient rate to avoid perceptible flicker (usually reckoned to be greater then 60 Hz; 75 Hz preferred). Format content is therefore limited by the deflection writing rate.

Sometimes the writing rate is varied to adjust relative brightness of symbols, a slower rate emphasizing critical symbols such as priority targets or threats. Sometimes symbols are 'in-filled' with local raster-like strokes to shade an area. The PFD attitude ball sky/ground shading is an extreme example of this. Often the image is defocused for the period of time the shading is being drawn so that the local raster structure is not visible.

When displaying a topographical map or sensor video, the CRT is operated in conventional raster mode, that is, the CRT beam traces the whole display surface area from top to bottom as a series of left to right horizontal lines, then retraces back to the top to start again. To halve the bandwidth, early broadcast TV video systems used a technique called interlace, that is, the raster is drawn as two fields, the odd field and the even field, which interleave one another as shown in Fig. 5.5. Each video field is drawn at 60 Hz (50 Hz in the United Kingdom/Europe) to avoid flicker, but the whole image is drawn at 30 Hz (25 Hz in the United Kingdom/Europe). In a raster video system it is not necessary to transmit the raster as well as the video, only synchronization pulses to tell the receiver when to start a line and when to start a field. The sync pulses can be transmitted encoded into the video as composite sync.

The original broadcast TV standards set generous line and field retrace times to allow the then available electronics time to recover the beam to the start of the next line or field. Today, those constraints are removed, but the standards remain, so the video field retrace time can be used to write stroke symbology over the raster. The slower stroke writing rate usefully emphasizes the stroke symbology. This mode of operation is called 'hybrid'. Typical hybrid X, Y and video waveforms are shown in Fig. 5.6.

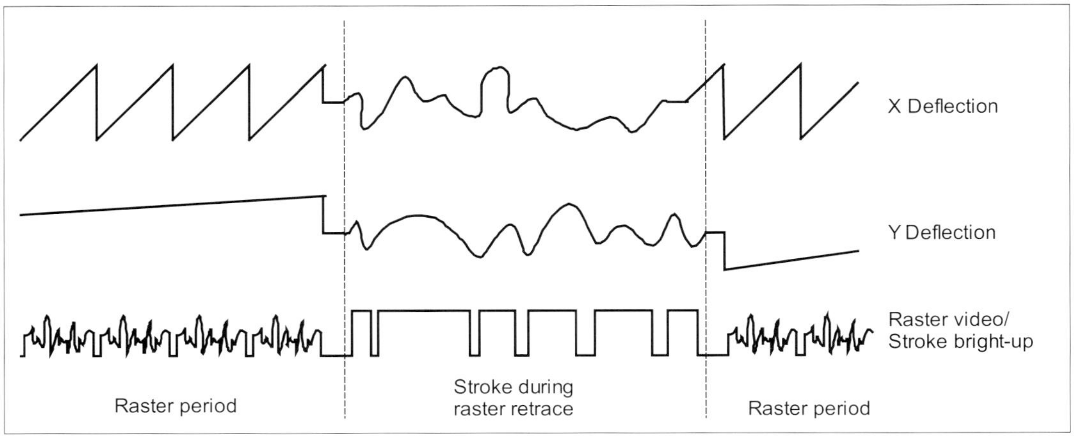

Fig. 5.6 *'Hybrid' (stroke+raster) operation*

The requirements for the 'hybrid' deflection amplifier are therefore considerably different to those for a domestic broadcast receiver operating in raster only. The 'hybrid' deflection amplifier is a high-bandwidth (typically >1 MHz), high-precision power amplifier capable of delivering the required current (usually several amps) into the inductive X and Y magnetic deflection yokes placed around the neck of the CRT.

In addition, and in order to maintain exact registration of the images produced by the three electron beams, the deflection amplifier must generate and apply convergence correction currents to the magnetic convergence coils placed at the electron gun end of the CRT. The convergence terms are often complex second- and third-order combinations of the X and Y deflections. The coefficient of each term varies from CRT to CRT and must be adjusted and set during final test of the MPD. High-precision 'in-line' electron guns and so-called 'self-convergent' deflection yokes can yield significant simplification, but the degree of convergence precision required in the airborne applications is greater than for domestic TV, and so these electronic corrections are seldom eliminated completely.

The video amplifier

The video amplifier provides the voltage control signal to the CRT grid (or, more usually, the cathode) in order to modulate the beam current.

In stroke mode, the beam current is set proportionally to the ratio of contribution from each gun to achieve the desired colour. Usually, a set of 16 colours is more than sufficient for symbolic images. In raster mode the beam current carries the video image.

The video amplifier also applies the brightness and contrast adjustments to the CRT drive signal. The airborne MFD operates under an extremely wide range of ambient illumination, from full sunlight to night-time. At night the user may well wish to use the display at very low brightness levels in order that his eyesight remains night adapted. The airborne CRT has to cover a brightness dynamic range of more than four orders of magnitude and maintain colour balance between the three guns (colour tracking) over this whole range. In order to do so it is necessary to monitor the performance of each electron gun and compensate for its change in emission (called cut-off) as the CRT ages.

Ramp generator

A typical MFD will receive external drive signals in a variety of types. The ramp generator decodes these signals and generates the appropriate waveforms to deflect the CRT beam and modulate the cathode/grid.

For stroke operation the MFD receives analogue X and Y deflection signals as low-level voltages which directly describe the symbol shapes. A binary code describes the symbol colour.

Raster video is received as analogue R:G:B signals with separate sync (line and field) from which the ramp generator detects the video line standard and internally generates the X and Y ramp scan waveforms. Typical video standards include:

- 525 line 30:60 Hz (based on US domestic TV standard).
- 625 line 25:50 Hz (based on UK/EU domestic TV standard).
- 875 line 30:60 Hz (high-resolution sensor video – monochrome).

Encoder

The encoder function reconstructs the screen image and outputs it in suitable video format to a video tape recorder for ground replay post mission.

Keypanel and microcontroller

The microcontroller performs various housekeeping and Built-in-Test (BIT) monitoring which it reports to the aircraft via the data bus. The microcontroller also scans the keypanel for crew key presses, which after debounce it transmits on the data bus. Any commands from the data bus are interpreted and actioned, including keypanel legend messages.

Shadow mask CRT characteristics

The shadow-mask CRT has a number of characteristic features that require careful attention if acceptable performance is to be obtained in aerospace applications:

- *Brightness*. In video applications the CRT image is developed as a raster scan, that is, the CRT beam traces out the whole screen from top left to bottom right. The information content is carried entirely in the modulation of the beam. The European TV standard has 575 active lines and is painted in two interlaced fields of 288 lines each at 50 Hz per field, 25 Hz per full frame. PC applications use higher-resolution line standards (XVGA is 768 rows × 1024 columns) at higher non-interlaced frame rates, but still raster scanned. As previously described, for airborne applications, raster scan will simply not produce an image bright enough to be seen under the adverse lighting conditions to be found in the cockpit. The image is written in stroke format. The information content is thus now primarily in the beam deflection, and the beam modulation merely defines the information colour. This mode of operation demands a complex, high-precision, and high-bandwidth deflection amplifier to drive the X and Y CRT deflection coils.
- *Contrast*. Unilluminated by the electron beam, the phosphor dots are a whitish-grey colour. A contrast enhancement filter has to be fitted to the CRT

to reduce the sunlight reflections that otherwise would occur in the adverse lighting conditions of the cockpit. While the filter attenuates light emitted by the CRT, sunlight reflected by the phosphors has to pass through the filter twice and is therefore attenuated twice. The filter has a special spectral transmission characteristic, called a triple-notch filter, which favours the spectral peaks of the CRT phosphors while attenuating other frequencies. The effective two-pass transmission to reflected sunlight is less than 4 per cent; the single-pass transmission to the CRT phosphors is better than 25 per cent.

- *Colour purity*. Any errors in registration between the electron beam and the phosphor dots will produce a landing error partially exciting an adjacent phosphor dot of the wrong colour. The effect is to produce impure colours and can be very psychedelic, but not helpful. Impure colours can be produced by small residual magnetic fields – changes in the earth's magnetic field are sufficient; hence the protection afforded by the mu-metal shield. Adjustable permanent magnets around the CRT neck neutralize assembly tolerances. Additionally, a degaussing field is often applied at switch on. A degaussing field is an alternating magnetic field that rapidly decays to zero, removing any residual magnetism in the process.
- *Convergence*. It is essential to maintain the constituent parts of the image in perfect registration; that is, the three electron beams must converge at all points on the CRT faceplate. This parameter is even more critical for a symbolic image than for video. Any misconvergence of, for instance, a white line would cause it to diverge into its red, green, and blue components with irritating and potentially confusing results. In a shadow mask CRT, convergence is corrected by static and dynamic electromagnets placed close to the electron gun. The dynamic terms are complex second- and sometimes third-order equations in both X and Y deflection components. The convergence terms must be adjusted for each and every CRT. The bandwidth of the convergence amplifiers must be commensurate with the bandwidth of the main beam deflection amplifiers.
- *Line width*. Symbol line width is significant in the context of display human factors and must be controlled. The line must be neither too broad nor too narrow. The symbology line width in a stroke-written CRT is a function of the electron beam spot width. At high brightness, with corresponding high electron beam current, the CRT spot width will grow with a tendency to blur or flare. Careful design of the electron lens, control of manufacturing tolerances, and electron gun alignment are required to mitigate against this effect. Conversely, at low brightness and hence low drive current the spot size may be too small and the electron beam may have to be defocused to obtain the desired line width. Any change in electron lens drive voltages to defocus the beam will have an impact on the scaling of convergence and purity correction terms, which have to be compensated accordingly. Background shaded areas (such as attitude sky–ground) are generally produced by local raster in-fill and defocused to blur the raster line structure.
- *Greyscale/colour tracking*. Unlike a domestic TV or PC monitor, the cockpit display is used over a wide range of ambient lighting conditions and the display brightness range required is more than 10 000:1. The ratio of electron beam drive currents must be maintained across this wide dynamic range to

assure colours track correctly. Careful design is required of cathode drive amplifiers, which must continually monitor and adjust their parameters to track cathode performance as the CRT ages.

- *Flicker/refresh rate*. As previously discussed, CRT phosphors have short persistence. In order to avoid flicker in peripheral vision it is necessary to refresh the display at rates in excess of 60 Hz. In a stroke-written display it is also essential that spatially related elements (e.g. scales with pointers) are also temporally related (i.e. written close together in time). If this precaution is not taken, then, under vibration conditions, image elements may appear to separate with the potential to misread information. The effect is created in the human brain, not in the display itself, and is due to vibration of the eyeball.
- *Compatibility with night vision goggles*. The phosphors employed in a colour CRT are rare-earth type with a combination of broadband and narrow-band spectral emissions. Although the majority of the spectral emission is at the primary colour wavelengths of red, green, and blue, there is some significant radiation outside the visual spectrum in the infrared band. To be compatible with night vision goggles, this radiation must be significantly reduced. To compound the problem further, night vision goggles themselves are sensitive to some light in the red part of the visual spectrum. To achieve full compatibility it is necessary to adopt a combination of strategies that include applying a filter to the goggle to reduce its sensitivity to visual red, selecting a red phosphor whose primary emission is outside the pass band of the goggle, and fit a filter to the display to eliminate emissions in the IR pass band of the goggle. This subject is discussed in more detail in Chapter 12.

The F/A-18 and AV-8B Multipurpose Colour Display (MPCD)

The MPCD shown in Fig. 5.7 uses a 5×5 in taut shadow mask CRT to achieve a full colour, high-brightness, high-resolution display in stroke, raster, and hybrid modes. Cardinal point performance is as follows:

Usable screen area	5×5 in square format	
Luminance:	Stroke	Raster
white	570 fL	260 fL
red	120 fL	45 fL
green	400 fL	200 fL
blue	50 fL	15 fL
Contrast ratio:	Stroke	Raster
white	6.7	3.6
red	2.2	1.4
green	5.0	3.0
blue	1.5	1.1
Stroke writing speed	0.8 mm/μs	
Line width	0.018 in (0.46 mm) to 50% point	
Resolution	65 line pairs/in	
Video	R:G:B generally to EIA-RS-170, 525 line 30:60 Hz interlaced, 625 line 25:50 Hz interlaced, 875 line 30:60 Hz interlaced	
Dimensions	6.7 in wide×7.01 in wide×17.3 in deep	
Weight	24 lb	

Fig. 5.7 *F/A-18 and AV-8B multipurpose colour display (MPCD) (Smiths Aerospace)*

Power	115 V, three-phase, 400 Hz, 180 VA
Dissipation	165 W at maximum brightness
MTBF	3500 h

AMLCD multifunction Head-Down Display

Active Matrix Liquid Crystal Display (AMLCD) is now the accepted technology for all new applications. While it is true that the AMLCD overcomes many of the 'features' that make CRT a difficult technology for airborne applications, the AMLCD has 'features' of its own that have presented significant challenges.

The principle advantages of the AMLCD (with backlight) are:

- Significantly less depth than a CRT and easier to package (typically <60 mm).
- Significantly less weight than the CRT (typically <1 kg).
- Significantly less power than a CRT (typically <20 W).
- No high voltages.
- No magnetic components and no influence from external magnetic fields.
- Perfect registration (no misconvergence).
- Fixed pixel (spot) size at all display brightness levels.
- Perfect colour tracking (except perhaps some minor influences from the backlight at low temperatures).

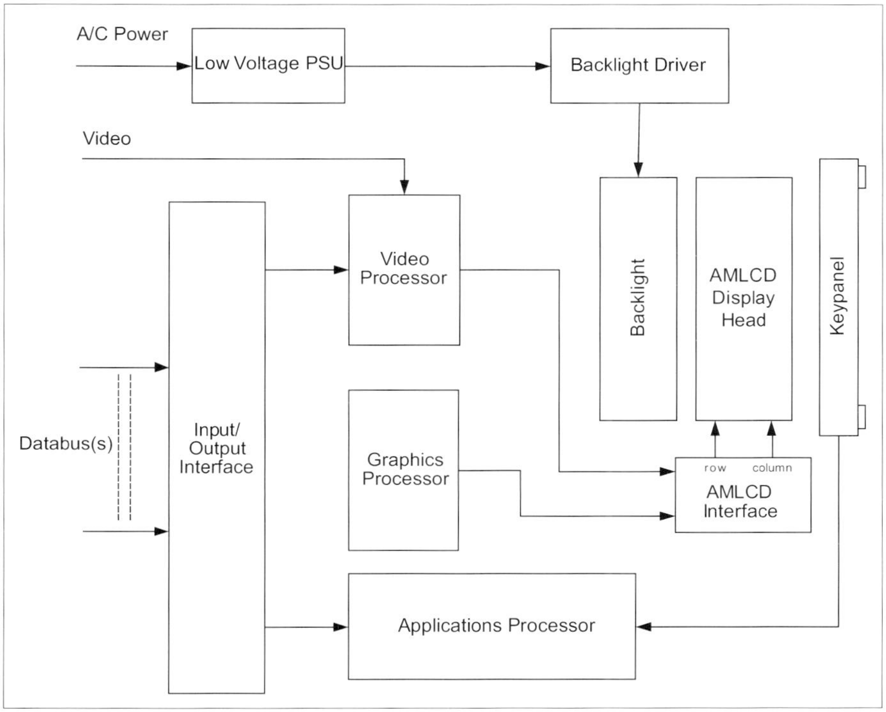

Fig. 5.8 Active matrix liquid crystal display MFD block diagram

AMLCD HDD principles of operation

The block diagram of a typical colour AMLCD technology based MFD is shown in Fig. 5.8. In this case the diagram refers to fully integrated AMLCD based equipment with integral data acquisition, data processing, and graphics processing in addition to the AMLCD and its basic support functions. This architecture is described more fully in Chapter 11, where it is described as an Integrated Display Unit (IDU).

The functional elements of an IDU comprise:

- The Active Matrix Liquid Crystal (AMLCD) Display Head Assembly (DHA).
- Backlight and associated backlight driver.
- AMLCD interface.
- Video processor.
- Graphics Processor.*
- Input/Output (I/O) interface.*
- Applications processor.*
- Keypanel.
- Chassis, power supply and interconnect.

Note: a dumb AMLCD MPD without integral graphics processing will omit the asterisked functional elements.

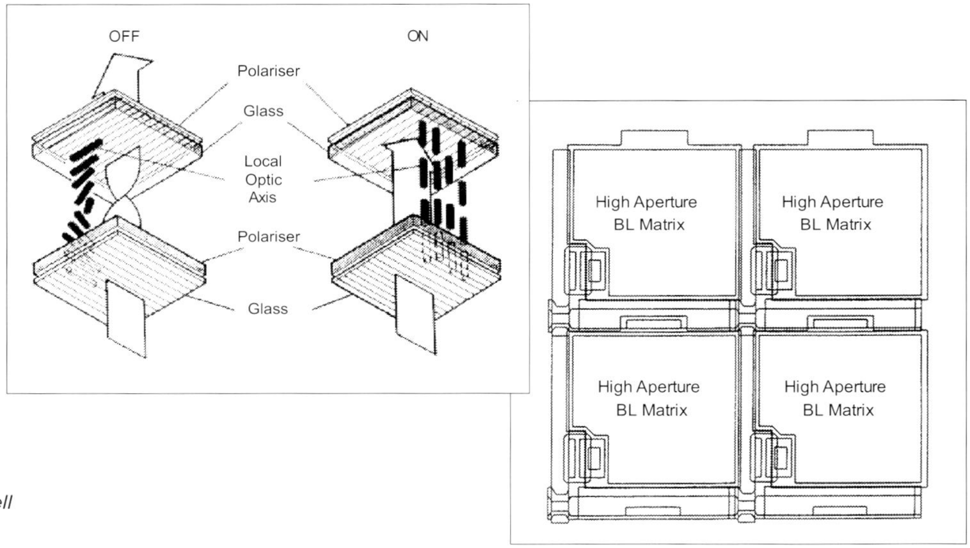

Fig. 5.9 *The AMLCD cell*

The AMLCD display head assembly

In an AMLCD, liquid crystal (LC) material is introduced between two glass plates (known as the active plate and the passive plate for reasons we shall explain later). The spacing between the glass plates is critical and is around 4 μm (LC material specific).

The following is a much-simplified explanation of the operation of the cell. Reference should be made to Fig. 5.9.

Liquid crystal material is a viscous organic fluid containing long polymer chains, which have the property of rotating the plane of polarization of light according to their alignment axis. The alignment axis can be changed with electrical bias.

The plane of polarization of light alters as it passes through the cell. Placing two crossed polarizers either side of the cell (bonded to the glass plates) makes the cell into a light valve.

With the cell unbiased, light is plane polarized by the first polarizer, its plane of polarization is rotated by 90° as it passes through the LC material, and it then passes through the second polarizer. In the biased state, light is not rotated by the LC material and does not pass through the second polarizer. Greyscale is obtained by applying an intermediate voltage which partially aligns the molecules so the cell is partially transmissive.

Accurately maintaining the cell gap is vital for correct operation. To achieve this, the LC material is filled with very accurately machined glass beads in suspension which become randomly distributed throughout the cell and are sandwiched between the glass plates under carefully controlled pressure during manufacture. It is easy to illustrate the sensitivity in performance to cell gap width by simply pressing lightly on any AMLCD and witnessing the resultant colour fringing spreading from the pressure point.

Row and column address lines are fabricated onto the active plate (prior to assembly) to form a matrix and an amorphous silicon Thin-Film Transistor (TFT) is fabricated at each intersection. The processing steps are similar to any silicon foundry process except in this case onto large glass sheets. The row and column drive signals to the TFT control charge applied across the local cell defined by the TFT itself plus an associated pair of transparent electrodes placed opposite one another on the active and passive plates. This charge is the bias for the local light valve. Note that there is only one LC cell which covers the whole of the glass plate; the TFT influences the operation of LC molecules in its local environment to form a subpixel.

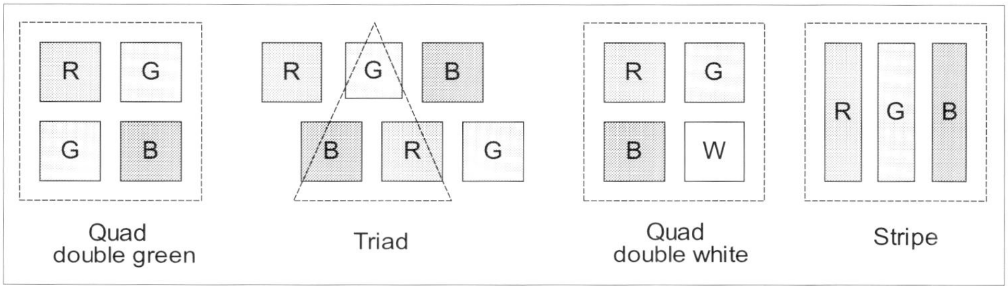

Fig. 5.10 *Colour group pixel structures*

Red, green, and blue colour filters are deposited on the passive plate in exact register with the subpixels to form a colour group or pixel. A number of colour group arrangements have been popular, as shown in Fig. 5.10, but now the accepted norm is the R:G:B stripe.

A typical XVGA compatible panel is shown in Fig. 5.11. This panel has been ruggedized for the airborne environment. It is 6.25×6.25 in square and has 768×768 R:G:B pixels; a total of 1.8 million pixels.

The TFT row and column address lines are brought out to the edge of the glass plate, that is, a total of around 1700 connections for the XVGA panel described. The lines are driven by row and column driver Integrated Circuits (ICs). Drive voltages are typically in the range -5 to $+15$ V.

Fig. 5.11 *AMLCD panel with drivers (Korry Electronics)*

The driver ICs are mounted on printed circuit boards which are folded back at 90° to the AMLCD glass panel. The ultrahigh-density interconnect between the drivers and the AMLCD glass panel is made with Tape Automated Bonding (TAB). The TAB process uses copper traces in a flexible polyamide base layer. The driver ICs are mounted on the TAB. Connections are made between the TAB and the glass panel using a pressure-sensitive anisotropic adhesive. The adhesive contains small silver balls which touch each other under pressure and make contact between circuit traces in the TAB and circuit traces on the glass. The balls do not touch in the axis perpendicular to the pressure axis, and isolation is maintained between adjacent traces. Once the connections are correctly made, the adhesive is heated and permanently sets.

The backlight and backlight driver

The AMLCD is a light valve but with a typical transmission of less than 7 per cent for a colour panel. To obtain a usable display requires the addition of a bright backlight (see Fig. 5.12). The fluorescent lamp is the technology of choice, but for airborne applications it has to be much brighter than the lamp in a laptop PC. Also, the light source has to be dimmed for night-time use.

There is an important distinction between the CRT and the AMLCD. In the CRT the image and the light source are one and the same, namely the phosphor light emitter. In the AMLCD the image is separated from the light source, the image is made in the LCD; the backlight is the light source.

A number of fluorescent lamp backlight configurations have been used: hot- or cold-cathode, single, or multiple lamps.

At first there was considerable concern over the reliability and longevity of fluorescent lamp cathodes. Although fluorescent lamps have a long history, they had never been used in airborne applications and had never been dimmed. Therefore, early designs had multiple lamps since failure of the backlight would be a single point failure that would result in complete loss of the display (as it had been for the much more complex CRT, of course).

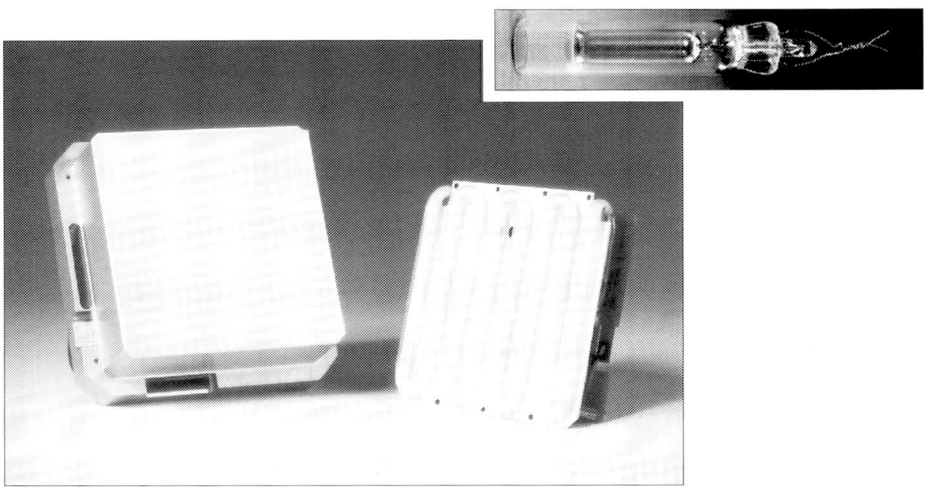

Fig. 5.12 *Single, cold-cathode, serpentine backlight (Korry Electronics)*

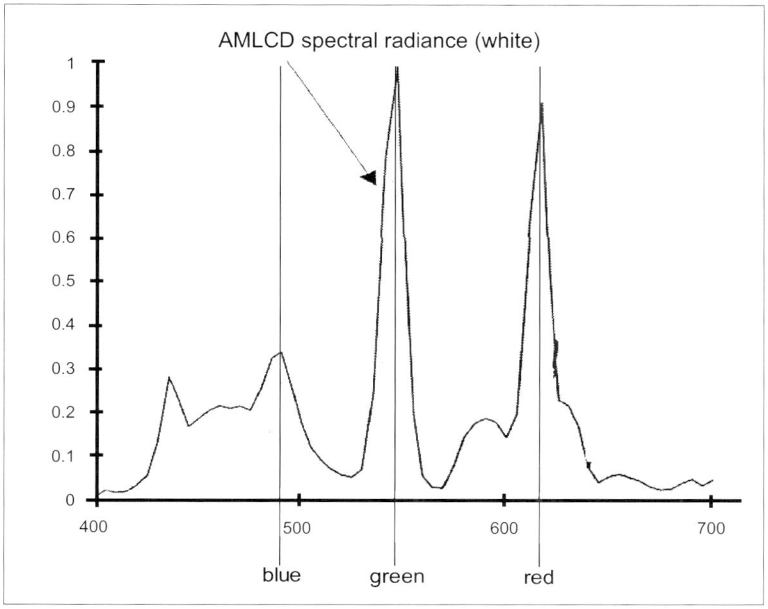

Fig. 5.13 AMLCD spectral emissions

Extensive life testing coupled with materials research and development of sophisticated drive circuits has now established the fluorescent lamp as a reliable and long-lived product. Lifetimes in excess of 16 000 h have been demonstrated which exceeds by a significant margin the lifetime of the CRT. The single, serpentine ('S'-shaped), cold-cathode lamp has become the configuration preferred by most designers (**1**). An optical stack is placed between the fluorescent tube and the backlight to provide a uniform diffuse light source to back-illuminate the AMLCD.

The fluorescent lamp works by striking an arc in a mercury vapour. The arc emits ultraviolet light which in turn is converted to visible light by phosphors deposited in the inside of the fluorescent tube. The lamp phosphors are a mix of the same red, green, and blue phosphors used in the CRT, but in this case the lamp produces a white light whose colour temperature is defined by the mix ratios of the three phosphors. The mix ratios are selected to match the AMLCD colour filter transmissions so that, with all pixels 'on', the display is white.

The colour filters are broadband. The colour gamut of the AMLCD is similar, therefore, to the CRT except that there is always some leakage from the off-pixels in the subgroup which slightly desaturates the primary colours. This effect can be seen in the CIE u′ v′ diagram (Fig. 5.4). The desaturation of blue is most apparent from the diagram, but it is not of any real significance in cockpit instrument applications. The display spectral emissions shown in Fig. 5.13 are therefore a combination of the backlight spectral emission and the transmission of the filters.

The luminance of the backlight is controlled by the backlight driver function. For normal daytime operation, the backlight is operated as a conventional fluorescent tube; that is, the arc is struck and the processes described above produce light. Figure 5.14 shows a typical lamp current waveform. The frequency of the current waveform affects the efficiency of the lamp – for optimum efficiency the frequency should be in the tens of kHz range. To dim the lamp, the HF pulses

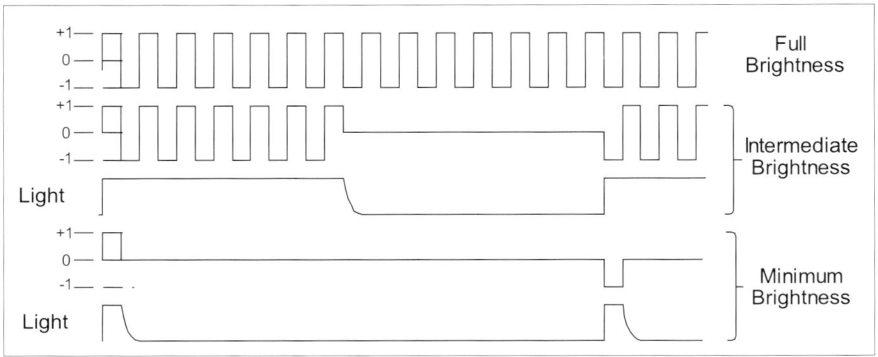

Fig. 5.14 *Typical fluorescent lamp current drive waveforms*

are gated to provide pulse width modulation of the lamp. To prevent flicker, a low-frequency repetition rate of greater than 60 Hz is chosen. With good design, this method can achieve good control of the lamp over a dimming range in excess of about 4000:1, which is sufficient for daytime operation and for civil transport applications.

For military night-time use, a wider dimming range (typically 20 000:1) is required and a number of techniques have been used, including:

- *A second (dimmer) lamp*. Light from the second lamp is usually coupled into the backlight optics from the sides. The loss in efficiency in night-time use is not important. The lamp is usually an alternate technology (e.g. tungsten filament or LED), and is often filtered to achieve compatibility with night vision goggles – see Chapter 12.
- *RF coupling*. An RF field coupled to the lamp will also raise the energy state of the gas, causing it to emit UV light and hence the lamp phosphors to emit visible light as before. For the low luminance required for night-time use, the field coupling can be inefficient, though it does have to be uniform to achieve a uniform light output over the whole tube surface. The RF field strength is low and can be contained within the required EMC emission levels. Compatibility with night vision goggles is achievable with the insertion of an IR blocking filter between the backlight and the AMLCD.

AMLCD interface

The AMLCD interface formats pixel data for the architecture of the specific AMLCD to be driven and performs a number of ancillary functions required to support the AMLCD. The extent and nature of these ancillary functions depend on the nature and intelligence of the AMLCD device selected for the application.

Early AMLCD devices had little intelligence and there was no interface standardization in the industry. Typically, the AMLCD Interface:

- Reformats and reroutes image pixel data into the data streams required by the AMLCD row and column drivers.
- Generates the specialized voltages and currents required by the AMLCD and provides safe on–off power sequencing.
- Generates the voltages required to control the AMLCD grey-level transmission curve. This characteristic varies from AMLCD to AMLCD and varies

with temperature. The AMLCD interface stores personalized characterization data obtained during optical testing, including temperature compensation.

Modern AMLCDs require little external interfacing, offering to the display integrator one of a few now standardized interface types. These include R:G:B parallel and high-speed LVDS serial interfaces. The 'on-board' AMLCD drive circuits provide for specialized voltage generation and temperature compensation. Improved production methods now mean that there is little variation in performance AMLCD-to-AMLCD and individual AMLCD characterization is seldom necessary for all but the most demanding applications.

These improvements simplify the integration for the display integrator but leave little opportunity to optimize the AMLCD performance if it is going to be operated outside the manufacturer's tested performance parameters. This is a likely situation if a COTS AMLCD is being uprated for use in all but the most benign aerospace applications.

Video processor

The video processor selects the video source (if more than one), captures it, and reformats it to map to the AMLCD pixel structure. Although for best image reconstruction it is desirable for there to be an exact 1:1 mapping between the video image source and the AMLCD pixel structure, in practice this is rarely the case. A COTS AMLCD pixel structure is likely to be one of the PC video graphic

Table 5.1 Video graphics adaptor standards

Video graphics adapter standard	Rows	Columns
VGA	480	480
SVGA	600	800
XVGA	768	1024
SXVGA	1024	1280

Table 5.2 Common video standards

Video standard	Number of active lines
525 line; 30:60 Hz interlaced	485
625 line; 20:50 Hz interlaced	575

adapter formats, whereas on-aircraft video is likely to be in one or more of the traditional TV-like video formats. The disparity is shown in Tables 5.1 and 5.2 respectively.

The image processing steps include:

- *Image capture*. The capture process digitizes the video image and stores the scene in a video memory. The resolution must be sufficient to capture the full range of image greyscales — usually 8 bits is sufficient (256 grey levels per colour or 16 million colours for a R:G:B video image). The sampling rate must be sufficient to support the highest screen (AMLCD pixel) resolution of

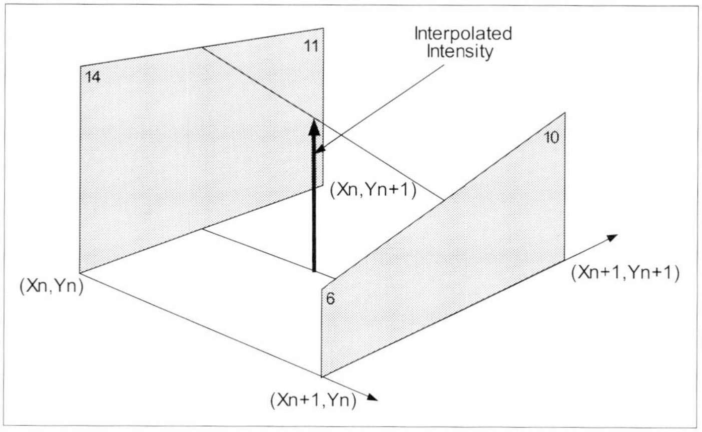

Fig. 5.15 Bilinear spatial interpolation

the image. For instance, to display the video at full size on an XVGA screen requires at least 1024 samples to be taken during the video horizontal line time. To avoid sampling artefacts, the video signal should be prefiltered in the analogue domain prior to digitization to remove any image frequencies above the Nyquist frequency of the digitizer.

- *Image transformation/scaling/resizing*. The image translation, scaling, and resizing process sequentially accesses the video memory at the desired final location of each pixel on the screen. This may not be a precise correspondence, and so to obtain a high-quality image it is usual to access the video memory at locations surrounding the true pixel location and interpolate its value from the value (grey level) of the video image at those locations. This process is called spatial interpolation and one form of it, bilinear interpolation, is shown pictorially in Fig. 5.15.
- *Image deinterlacing*. A final complication arises from the interlaced nature of the video signal from most sensor sources. AMLCD devices are in general addressed in a non-interlaced manner. Every pixel is refreshed at the manufacturer's recommended rate, usually 60–70 Hz. Interlaced video refreshes half the image at field rate (60 Hz for 525 line video) and the whole image at frame rate (30 Hz for 525 line video). One complete frame of video is not the sum of the two fields (odd and even) since data for the second field will be sampled from the scene one field time later than data in the first field. Moving objects in the image will be in a slightly different location. Simply summing the odd and even field together and then playing them twice to the AMLCD results in tearing artefacts on vertical edges of moving objects. Deinterlacing is achieved by temporal interpolation between captured images in more than one field. There are a number of algorithms for this, all variations of the same principle, the better and more complex requiring the captured image to be sampled for more fields. This increases cost, memory size, processing power, and data latency (the delay between scene capture and scene replay). In all cases the appropriate interpolation algorithm needs to be selected for the application depending on image type, anticipated rate of image change (motion), purpose, and tolerance to latency. Fortunately, there

are a few COTS devices available that, together with copious memory, can perform this function.

Graphics processor

The graphics processor constructs the graphics image in real time, combines it with the video image, and transfers the result to the AMLCD interface for reformatting/transmission to the AMLCD. The AMLCD image can in general be arranged to be:

- Graphics only (on a black background).
- Video only.
- Graphics overlaying the video background. In this case, it is not uncommon to blank or reduce the intensity of the video image behind the graphics image to improve the legibility (or contrast) of the graphics image.

Table 5.3 Typical graphic primitives

Graphic element	Attributes
Lines	Start, end (or length), angle, width
Circles/ellipses/arcs	Centre(s), angle, radius, arc length, width
Alphanumerics	Font, style, size,
Icons	Size, in-fill

The graphics image is constructed as a series of graphic primitives, examples of which are given in Table 5.3. Graphic primitives can be written to any point on the AMLCD surface, and (in general) rotated to be at any angle.

The graphic image is generally constructed in a series of planes or levels that overlay one another with assigned priority. For example, the attitude portion of the Primary Flight Display (PFD) has the following prioritizations:

- Priority 1 – Aircraft reference symbol.
- Priority 2 – Flight director.
- Priority 3 – Attitude bars (pitch and roll).
- Priority 4 – Sky/ground shading (attitude ball).

Icon and alphanumeric character sizes, line widths, and colour are carefully chosen to aid rapid acquisition by the crew in accordance with the general guidelines described in Chapter 12. It is important to avoid any motion artefacts (jerkiness or ratcheting) of graphic elements as they move and/or rotate. Any such artefacts can be distracting to the crew, especially in peripheral vision. Therefore, high-quality displays generally employ antialiasing techniques to ensure that graphic elements have smooth edges in spite of the discrete pixel structure of the display.

The most effective form of antialiasing is to emulate the Gaussian spot profile of the CRT electron beam and plot that profile to the screen at subpixel accuracy. Figure 5.16 shows a typical Gaussian spot profile in relation to screen AMLCD

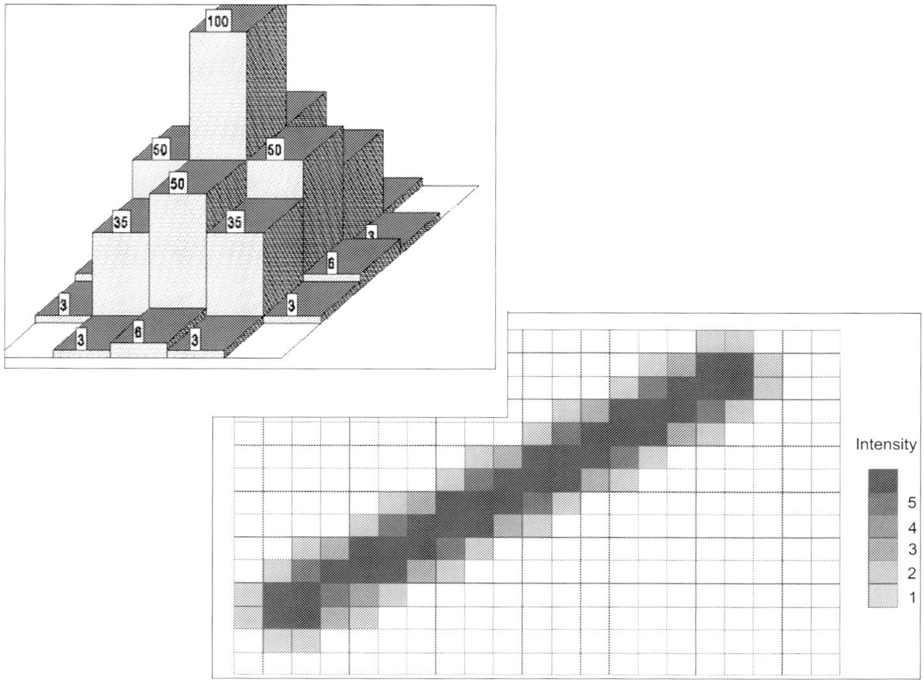

Fig. 5.16 *Gaussian spot profile overlaid to form an antialiased line*

pixels and the outcome of plotting that profile (as a series of dots constituting a straight line) to the screen, at subpixel accuracy, to produce an antialiased line.

Once again, today there are COTS graphics chipsets available that are able to draw antialiased graphics in real time. These chip sets can be programmed in high-order graphics languages, such as Open-GL, using appropriate display development tools such as VAPS.

Applications processor

The applications processor is a general-purpose processor with sufficient memory and throughput to be able to compute and manage the display format generation in real time. Current-generation designs use COTS processors and their associated management devices plus copious memory. The application software is likely to be written in a high-order language such as C++ and run under the control of a COTS real-time operating system (RTOS).

Input/output interface

The I/O interface will be aircraft specific, but typically will comprise:

- Discretes: 28 V/open circuit to interface to crew controls, knobs and switches, and mode annunciator panels.
- Data bus: ARINC 429 (or possibly AFDX) in a civil transport application. A combination of ARINC 429 and MIL-STD-1553B in a military application.
- RS422/232 for debug and software downloads.

Keypanel

Surrounding the AMLCD screen is a set of multifunction keys. Key legends are usually displayed on the AMLCD surface adjacent to each key. The keys communicate crew selections and commands to the display system, and hence to the avionics system via the databus. The keys and/or the bezel housing the keys are usually lit for night-time use.

AMLCD characteristics

There are a number of characteristic features of the AMLCD, different to those of the CRT, that require careful attention if acceptable performance is to be obtained in aerospace applications:

- *Brightness*. The AMLCD is a light valve. The display brightness is controlled by the backlight. Now a mature technology, a single, serpentine fluorescent lamp is able reliably to provide an efficient, bright, and dimmable light source.
- *Viewing angle*. The optical performance of an AMLCD is critically dependent on the cell gap. At increasing viewing angles from the normal to the display, the optical cell gap effectively increases and the optical performance of the cell degrades. It no longer acts as a perfect light valve, and some light 'leaks' through the cell. This light leakage reduces the display contrast. This can be seen in a typical isocontrast plot shown in Fig. 5.17. The viewing angle requirements of a typical crewstation installation are shown in Fig. 5.18. This diagram shows a composite of the requirements for a civil transport aircraft and for military aircraft. In the civil transport aircraft there are two requirements, one for the displays directly used by the crew member, the second for cross-cockpit viewing of the displays used by the other crew member. In virtually all crewstation designs the displays are installed so that they are viewed from above the normal (perpendicular) to the display surface. A primary reason for this is simply the installation constraints, but it also

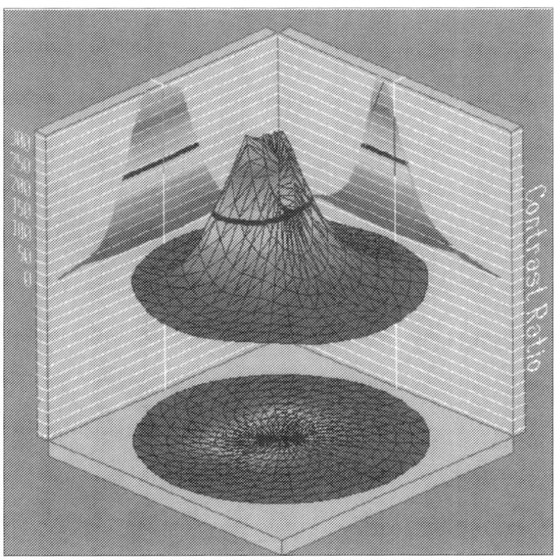

Fig. 5.17 Typical AMLCD isocontrast plot

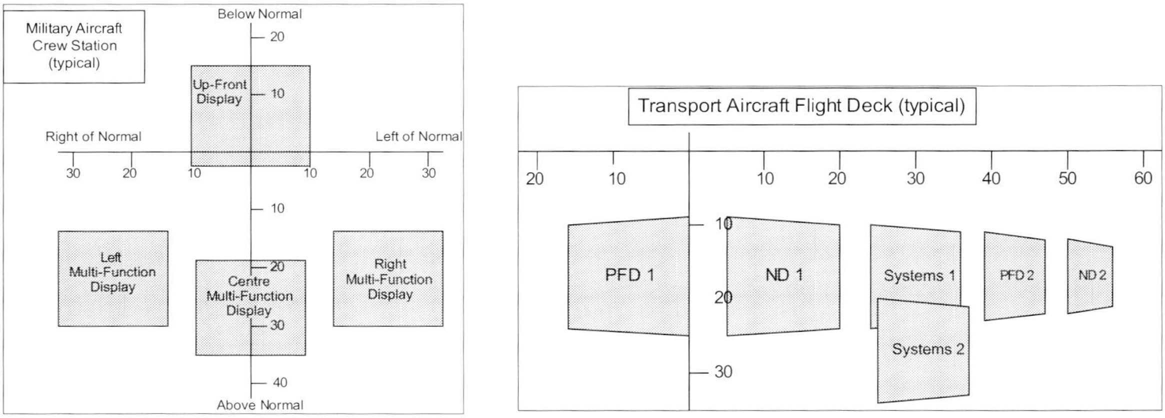

Fig. 5.18 *Typical viewing angle requirements*

ensures that the specular reflection of the sun's orb is not visible. This is the opposite configuration to that usually adopted for laptop PC screens where the display is usually viewed from just below the normal (perpendicular) to avoid reflections of the user's clothing.

- *Contrast*. The specular and diffuse reflectance of an AMLCD is low, principally a result of its low transmission. A contrast enhancement filter is not required. However, it is usual to bond a cover glass to the AMLCD to protect and ruggedize it for the airborne environment. This glass has an antireflective coating. For adequate visual performance a contrast ratio of 50:1 is desirable, but 30:1 is acceptable over the normal viewing cone. The need for the crew to monitor primary flight information across the cockpit demands a horizontal viewing angle in excess of $\pm 50^{\circ}$. This represents a considerable challenge for AMLCD technology. Much work has been done with special optical compensation films which are often laminated to the AMLCD front glass to improve the viewing angle, especially to facilitate cross-monitoring of the displays on a civil flight deck by both crew members. A contrast ratio of 10:1 is just adequate. Technology and process control continue to make significant advances in this area.
- *Greyscale*. Greyscale is obtained by partially switching the cell between its 'on' and 'off' states. The greyscale transmission curve is not linear. It is viewing angle dependent and requires temperature compensation since it is sensitive to temperature. It varies from AMLCD to AMLCD due to manufacturing tolerances and requires each AMLCD to be characterized and its 'personality' locked into the drive electronics. For the symbolic images used in primary flight instrumentation displays it has been demonstrated that 16 grey levels (4 bits) per colour are sufficient, a full colour video image requires 256 levels per colour (8 bits).
- *Black level uniformity*. This characteristic has been perhaps the most problematic feature of the AMLCD (second only to viewing angle). It is particularly a concern for symbolic images painted on a black background; it is not so significant for video images which typically have little black content. The display transmission in the black (driven) state is sensitive to a wide range of

factors. Any impurities in the materials, any leaching of compounds to contaminate the LC material, any variation in cell gap introduced during the sealing process, and any non-uniform pressure applied by the bonding of cover glasses or attachment to mounting features can all cause a local degradation in the black level. The result is a blotchy black background which can be most distracting. This feature, more than any other, has received adverse criticism for the AMLCD, especially when compared with a CRT, which, since devoid of any excitation in the black background, is perfect in this regard. Fortunately, improvements in materials and in process control, driven by the volume laptop PC market, are overcoming this deficiency.

- *Response time*. Unlike the CRT, the response time of an AMLCD is long, typically 20–30 ms. The concern therefore is not flicker, but smearing of rapidly moving characters, and careful choice has to be made of LC material to minimize smearing without degrading top-end temperature performance.
- *Thermal management*. Many of the parameters associated with the performance of the AMLCD are temperature sensitive, and it is necessary actively to manage the thermal environment of the AMLCD. Indeed, the LC material will not function at low temperatures (the molecules will not twist), and at high temperatures control is lost and the display clears. Fortunately, neither effect is irreversible. The ideal operating point is in the region of 30–40 °C. Therefore, a rear cover glass is bonded to the AMLCD that has an Indium–Tin Oxide (ITO) heater deposited on it. ITO has a transmission to visible light of about 95 per cent. The ITO heater forms part of a control loop (often also including blown air) that rapidly raises the temperature of the AMLCD at cold start and then maintains the operating temperature at the desired operating point. In high-performance designs the heater comprises several separately controlled elements to facilitate rapid safe heating of the AMLCD from low temperatures while maintaining a uniform temperature across the AMLCD.
- Typical AMLCD performance with temperature is shown in Table 5.4. Bilevel (that is 2-state on/off) performance means the AMLCD will support a textural image with the declared update rate. Greyscale performance means the AMLCD will support a video image with the declared update rate. Refresh rates faster than the stated update rate will produce smearing of the image. The performance of the fluorescent lamp is temperature dependent also. Mercury will not vaporize at low temperatures and the arc cannot be struck. As can be seen in Fig. 5.19, the fluorescent lamp operates most efficiently at temperatures in the range 40–50 °C. Actively managed heaters raise and maintain the operating temperature of the lamp also.

Table 5.4 AMLCD performance with temperature

	Temperature (°*C*)	
Update rate (Hz)	**Bi level**	**Greyscale**
5	−15	+5
10	0	+20
30	+15	+30

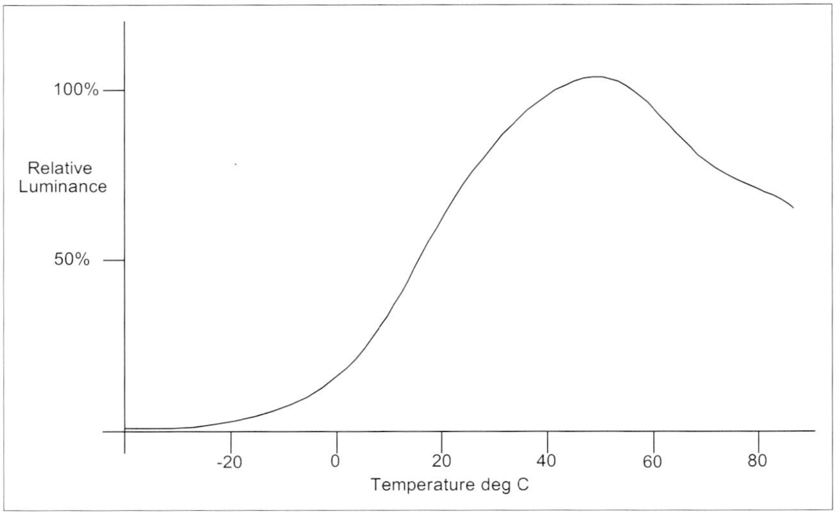

Fig. 5.19 *Fluorescent lamp performance with temperature (Smiths Aerospace)*

Table 5.5 Typical AMLCD cold-start strategy

Phase	Time	Strategy
1	0–1.5 min approx.	Unit non-functional (no display). All available power directed to the lamp and AMLCD heaters
2	1.5–3.5 min approx.	At 60 s the lamp has reached+10 °C and can be safely lit. AMLCD> −15 °C (bilevel response). Unit turned on (lamp set to maximum brightness and power). Lamp heater power reduced to maintenance level, temperature continues to rise on account of self-heating
3	3.5–4.5 min approx.	Lamp reaches maximum efficiency. AMLCD reaches+30 °C (greyscale response). AMLCD heater switched off
4	> 4.5 min	Lamp heater turned off

For a military display with a requirement for a rapid cold-start performance, it is necessary to adopt a sophisticated closed-loop cold-start strategy to manage the process safely and in the minimum time. A typical strategy is shown in Fig. 5.20 which shows cold start from −40 °C. Usable bilevel (text) performance is achieved within 90 s, and full-spec video performance is achieved within 5 min. The strategy is subdivided into four phases as described in Table 5.5.

Extremely high non-operating temperatures can be reached in a military fast jet bubble canopy cockpit when the aircraft is parked in the hot desert sun. Tests on an AV-8B indicated that in the mid-afternoon, when the outside air temperature reaches +40 °C, the air inside the cockpit could reach +90 °C. If the aircraft is parked facing east, then the sun will fall on the display, which will absorb radiance from the sun, raising the temperature of the AMLCD itself to above +110 °C until shaded by other features in the cockpit. This situation is shown in Fig. 5.21.

Fortunately, this is a worst-case condition, but may be expected to be encountered several times during the service life of a display. To survive this scenario, careful attention needs to be paid to materials and process to assure no

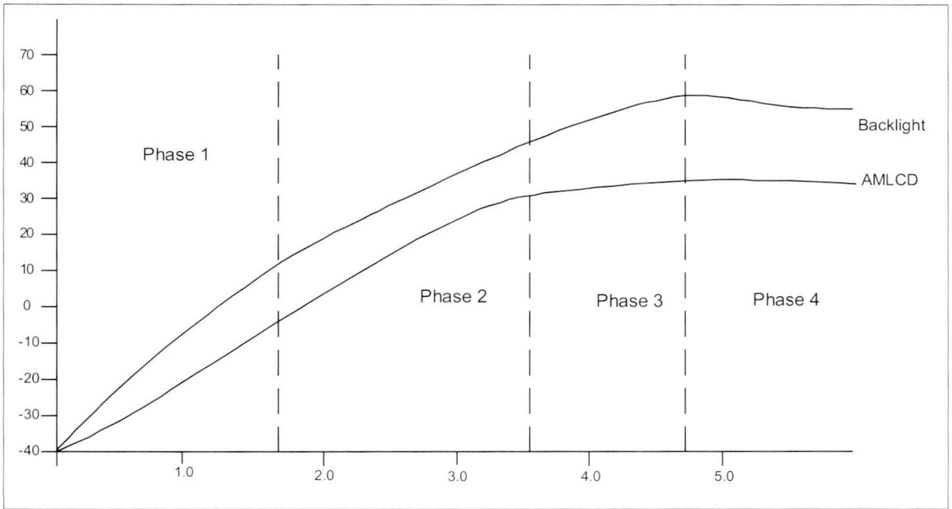

Fig. 5.20 *Typical AMLCD cold-start performance from −40°C (Smiths Aerospace)*

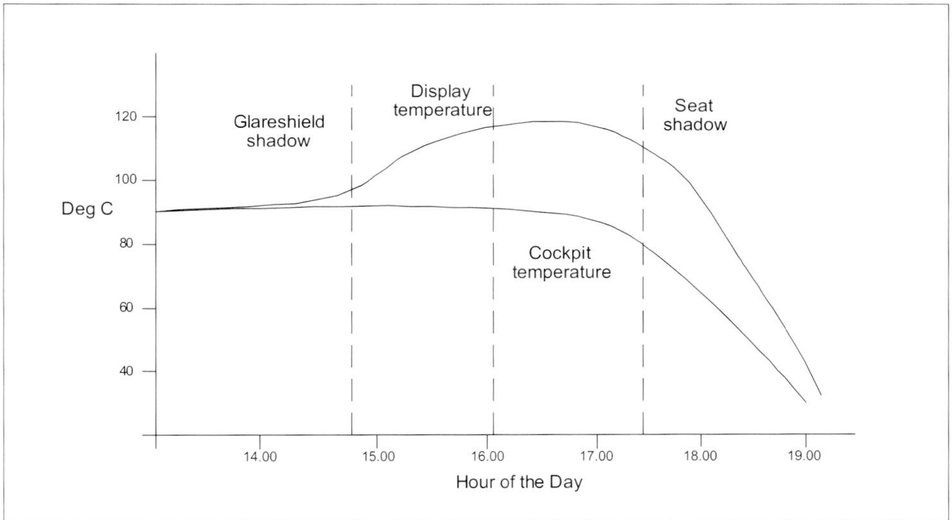

Fig. 5.21 *AMLCD non-operating cockpit temperatures (Smiths Aerospace)*

sudden and irreversible damage is done to the display. However, slow degradation after prolonged exposure to these temperatures is inevitable. High-risk components and materials are the polarizers, the glass bonding adhesive, the AMLCD sealing, and the TAB interconnect anisotropic adhesive. All these materials can be carefully selected and the processes closely controlled for a full custom aerospace designed AMLCD, but that is not the case for a COTS AMLCD. For this reason it is wise extensively to test COTS AMLCDs in a high-temperature environment in order to assure their long-term reliability for the application environment.

AMLCD sourcing

While, as can be seen, there have been significant technical difficulties in applying AMLCD technology to the flight deck, these are now largely overcome and

AMLCD displays are the de facto standard. However, no less a problem has been the sourcing of that technology.

Airborne AMLCD displays have fairly obviously sought to retain the pre-existing electromechanical and CRT instrument sizes of 3ATI, 5ATI, ARINC C (6.25×6.25 in square), and ARINC D (6.7×6.7 in square). These are all unique to the aerospace industry. Therefore, it was appropriate to look to custom houses (as the industry had for CRT) to fabricate AMLCDs. This route was mandated by the Military, who wanted an 'on-shore' (US or European) AMLCD source, not a Far East source, and was reinforced by the Far East AMLCD glass foundries who showed a complete lack of interest in the custom specifications and sizes of aerospace glass formats; their interest was focused on the then fledgling but soon to be high-volume laptop PC market.

A number of custom 'on-shore' AMLCD glass foundries were established – some with significant government help, some under the umbrella of aerospace equipment suppliers, and some stand-alone. One by one they have failed. Mostly the reasons have been technical, but driven by lack of volume to make good business sense. As described above, the critical performance parameters of a high-quality AMLCD are heavily dependent on good process control. Good process control only comes with volume manufacturing experience. Volume manufacture is not compatible with aerospace glass requirements.

Today there remains only one custom aerospace glass foundry, and that under the umbrella of a European aerospace equipment supplier. All other aerospace glass is obtained in the Far East. It is customized for aerospace applications by specialist organizations that have acquired the skills to post-process the bare glass cell by the addition of bonded cover and heater glass, the TAB attachment of row and column drivers, and the mounting of the end product in a rugged frame.

There still remains the fundamental issue of the glass size. By bundling the whole aerospace industry requirements together, it has been possible to convince a few Far East AMLCD glass foundries to process high-volume batches of common aerospace sizes. The ruggedization specialist then finishes the glass as and when required.

Some organizations claim to have cut COTS laptop PC glass down to the desired aerospace size. To be viable this requires accurate knowledge of the interconnect architecture which the original manufacturer is unwilling to provide or to guarantee not changing without warning. It also requires a process that can cut and reseal the LC cell without stress to the critical glass elements or without introducing contaminants. Finally, it has to be proven to be environmentally sound and reliable in the long term.

In recognition of all these factors, Boeing have mandated that the flight decks of all new aircraft, civil and military, will use common commercial standard glass sizes, not custom aerospace sizes.

COTS AMLCDs ruggedized for airborne use

High-volume COTS AMLCD displays now have optical performance characteristics that are appropriate for use in airborne applications. The process to ruggedize them to the environment generally follows the steps identified below.

1. ***Select an appropriate COTS AMLCD product.*** This is the difficult part. The COTS AMLCD cell must have the appropriate intrinsic optical qualities for the application. These include:
   - usable screen area,
   - resolution (pixel structure),
   - grey shades (number of colours),
   - viewing angle/contrast ratio from the design eye for the desired installation (note earlier comments about the different configuration in vertical viewing angle between airborne crewstation and laptop PC displays: laptop PC glass is designed to be viewed from below the display normal, crewstation displays are usually viewed from above the display normal; in some instances it may be advantageous to rotate the COTS glass).

   In addition, the AMLCD cell must have the appropriate intrinsic performance over the temperature range of interest in the application. Mostly this is determined by the choice of the liquid crystal material which is often a compromise between the high-temperature clearing point (the temperature at which the cell fails to operate and becomes clear) and the response time to be able to support the display image update rate at low temperature without smearing. The other materials and processes also need to be appropriate to support reliable operation through life in the likely operational environment, particularly low temperature and high temperature exposure with solar loading. Confidence in these latter aspects can only be gained through extended life testing of the chosen AMLCD. The selected product must be constructed in a manner and have a profile that facilitate the ruggedization processes that will subsequently be applied. Finally and most importantly, the source of the product must be secure. Many of the ruggedization process steps rely on aspects of the product design and not on the supplier data sheet which may be changed without notice. These aspects need to be carefully considered when determining the parts procurement policy.
2. ***Foldback the driver circuit boards.*** In some circumstances it may be necessary and appropriate to remove the driver Printed Wiring Boards (PWBs) from the AMLCD cell and replace them with alternative driver PWBs with improved performance or appropriate physical profile. In any event, a TAB repair capability is advisable. The driver PWBs are connected to the glass by a process called TAB bonding of the flexible interconnect to the glass using a conductive anisotropic adhesive. This process allows a limited 'repair/rework' operation under carefully controlled conditions. The original TAB bond is simply removed by peeling away the flexible interconnect. After cleaning the glass surface, the new interconnect is attached by the conventional TAB bonding process using appropriate pressures and temperatures to cure the adhesive.
3. ***Bond a rear heater glass.*** To achieve satisfactory performance at low temperatures it is necessary to heat the AMLCD. To achieve this, an Indium–Tin Oxide (ITO) coated heater glass must be bonded to the rear surface of the AMLCD glass cell. Connections are made to the ITO coating and a current is passed through it to heat the glass.
4. ***Bond a front AntiReflective*** (AR) cover glass. It is usually necessary also to bond an AR cover glass to the front of the AMLCD cell to reduce the intrinsic

specular and diffuse reflectance to acceptable levels for airborne use in direct sunlight. The bonding of the heater and AR cover glasses must be done with extreme care and has historically proven to be extremely troublesome. It is essential that the bonding process does not impart any pressure on the AMLCD cell which may alter the cell gap. Any variation in cell gap will manifest itself as a reduction in local 'on-state' transmission and degrade the black-level uniformity, producing a 'blotchy' appearance to the display. This is particularly distracting for 'instrument'-type display formats which are generally 'white-on-black' images.

5. ***Replace the backlight.*** The backlight of COTS displays intended for use in laptop PCs or office workstations has insufficient brightness for airborne crewstation displays viewed in full sunlight. It is necessary to relamp the display with a backlight that has sufficient intensity and dimming range, and possibly also compatible with night vision goggles. This backlight is inevitably physically larger then the original backlight, and hence the need to be able to fold back the driver PWBs to accommodate it.
6. ***Mount in a rigid protective support frame.*** The AMLCD, now with bonded heater and AR cover glass, the driver PWBs, and the backlight require to be mounted on a rigid support frame. The support frame must:
   - clamp the AMLCD cell on one of the glass substrates (usually the AMLCD active plate), taking great care to ensure that no stress is applied to the AMLCD cell which may distort the cell gap for the reasons identified above;
   - support the driver PWBs, taking great care not to stress the TAB interconnect but support it in a manner that ensures its long-term reliability;
   - support the backlight, often itself an integral part of the support frame;
   - provide features that, when mounted in the final display unit, prevent the ingress of moisture and contaminants;
   - in severe environmental applications, provide vibration and shock isolation between the whole assembly and the final display unit.

Examples of AMLCD HDD displays

Integrated display unit multifunction display

An example of a fully integrated multifunction display is shown in Fig. 5.22. This unit is intended for dual-use operation in rotary wing civil and military applications. Although shown with a 6.25×6.25 in square format AMLCD, the unit is a modular construction and has been designed to accommodate other AMLCD glass sizes by replacing the front display head assembly module with, for example, 10.4 and 15 in rectangular format COTS AMLCDs. Cardinal point performance is as follows:

Usable screen area	6.25×6.25 in, 10.4 and 15 in options
Resolution	768×768 (XVGA) stripe pixels
Grey levels	256 per colour, 16 million colours
Luminance (maximum)	350 fL white (maximum)
Luminance (minimum)	0.05 fL (dimming range > 10 000:1)
NVG compatibility	NVIS Class B

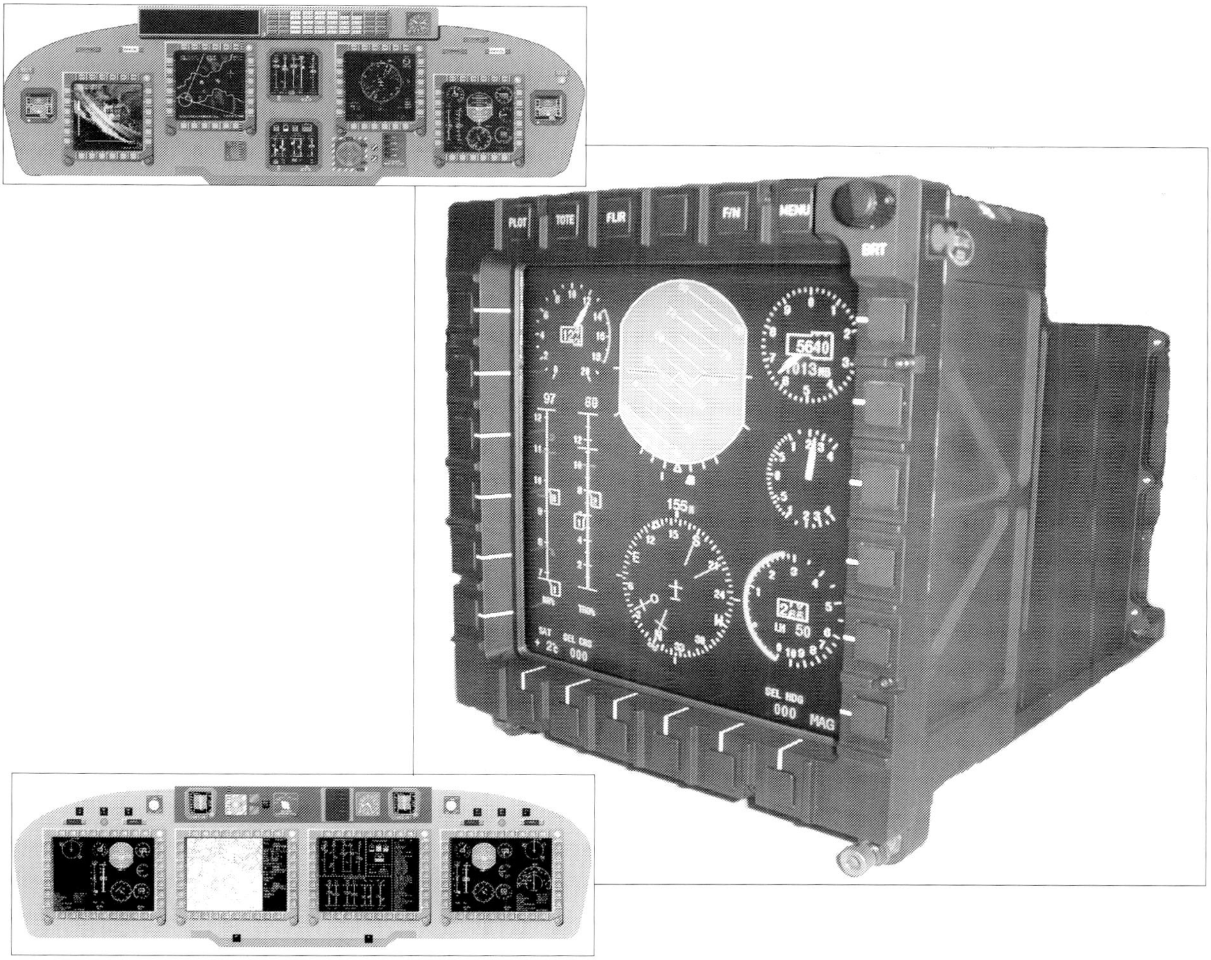

Fig. 5.22 Integrated display unit (Smiths Aerospace)

Viewing angle	±55° horizontal
Interfaces (video)	525 line 30:60 Hz, 625 line 25:50 Hz
Interfaces (digital)	ARINC 429 and MIL-STD-1553B
Processing	Integral antialiased COTS graphics processor, integral application processor

5ATI display unit

The 5ATI display unit installed in the KLM classic 747 aircraft is typical of current state-of-the-art fully integrated smart AMLCD instruments. The display unit, shown in Fig. 5.23, is designed to replace electromechanical ADI, HSI, and engine instruments in 'legacy' or 'classic' aircraft. It is fully compatible with present and future aircraft systems and provides significantly improved reliability and operational flexibility when compared with the electromechanical instrument it replaces. A single part number fits all applications; the display unit detects its location in the instrument panel and configures itself accordingly.

Fig. 5.23 5ATI display unit (Smiths Aerospace)

The display unit includes interfacing and data processing to match a wide range of signal types including:

- Synchro and resolver.
- Analogue.
- Digital encoder.
- Discretes.
- Serial digital data bus (ARINC 429).
- Map (ARINC 568).
- Weather radar (ARINC 453).

These interfaces can be configured to match the interfaces of the electromechanical instruments being replaced, providing form and fit interchangeability with increased functionality.

The integral graphics processing constructs and writes symbology onto the full colour active matrix liquid crystal display. The display formats include:

- EADI.
- EHSI – compass rose, map, plan, arc, TCAS.
- Engines – vertical tapes, round dials.
- EADI/EHSI combined format.

Display formats mimic the presentations of the electromechanical instruments they have been designed to replace to aid pilot transition. New functionality replicates established practice for 'glass cockpit' instruments. The map display provides enhanced horizontal situation awareness and is fully compatible with present and future Global Positioning Systems (GPSs) and Flight Management Systems (FMSs). The map is a key element in the modern cockpit for the newly emerging concept of 'free flight' as part of the Future Air Navigation System (FANS-CNS/ATM). Weather radar can optionally be presented on the map format.

A functional block diagram of the unit is shown in Fig. 5.24.

The 5ATI subassemblies, shown in Fig. 5.25, are housed in a chassis that is mechanically interchangeable with its electromechanical counterpart. The chassis comprises a rear card cage complete with a printed wiring interconnect harness

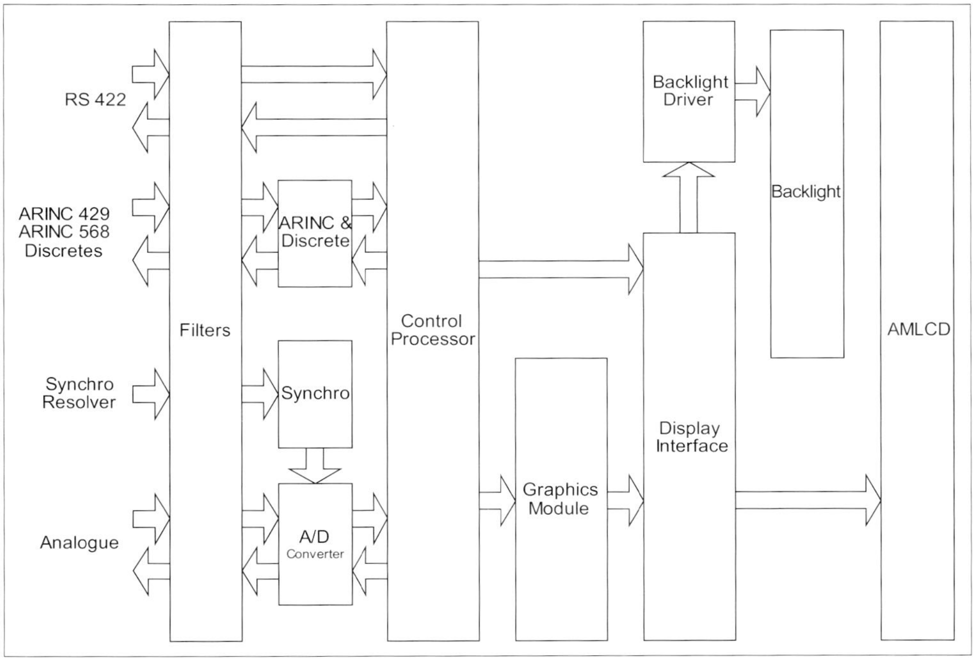

Fig. 5.24 *5ATI block diagram (Smiths Aerospace)*

Fig. 5.25 *5ATI subassemblies (Smiths Aerospace)*

into which the electronic modules are assembled. The AMLCD, complete with backlight, is securely attached to the front of the chassis. An optional slip indicator (passive bubble type) can be simply attached to the instrument front face if it is to be installed in the ADI location. Cardinal point specifications are as follows:

Dimensions	127 mm × 127 mm × 229 mm deep (5 in × 5 in × 9 in)
Weight	3.0 kg (8 lb)
Power	25 W typical, 35 W full brightness
Interfaces	RS422, ARINC 429/568, ARINC 453, synchro, resolver, analogue, digital encoder, discretes
Brightness	0.1–200 fL continuously dimmable (including autobrightness)
Screen area	102 mm × 102 mm (4 in × 4 in)
Reliability	15 000 h MTBF
Maintenance	On-condition (no routine maintenance)
Certification	TSO-C3d, TSO-C4c, TSO-C6d, TSO-C52b and TSO-C113 [see references (**2**) to (**6**)]

Head-Down Display format examples

The following indicate typical examples of display formats.

Primary flight display

Typical Primary Flight Display (PFD) formats are shown in Fig. 5.26. The PFD provides (**7**):

- Standard 'ADI type' attitude information.
- Speed information.
- Electronically defined flight path deviations including Localizer/Glideslope and VOR radial deviation.
- Flight director command bars.
- Lateral acceleration (equivalent to conventional ball, located on bottom of ADI).

The PFD differs from earlier instruments in the following major innovations:

- Improved speed information through the inclusion in the speed scale of critical speeds together with ranges and margins. This information is displayed on a 'need to know basis' only when necessary.
- Speed trend: an upward or downward pointing arrow originating from the 'present speed' marker indicates the speed that the aircraft will attain in the next 10 s, taking into account its present attitude and power setting.
- Improved flight path information displays through heading reference, the inclusion of Flight Path Vector (FPV) and of a Flight Path Target (FPT) especially intended to improve non-precision approach accuracy (when the flight director mode is not engaged).
- Improved navigation through the inclusion of radio altitude and Decision Height (DH), selected altitude or flight level, and deviation.
- Inclusion at the top of the screen of the Automatic Flight System (AFS) Flight Mode Annunciator (FMA) and the resulting autoland capability.

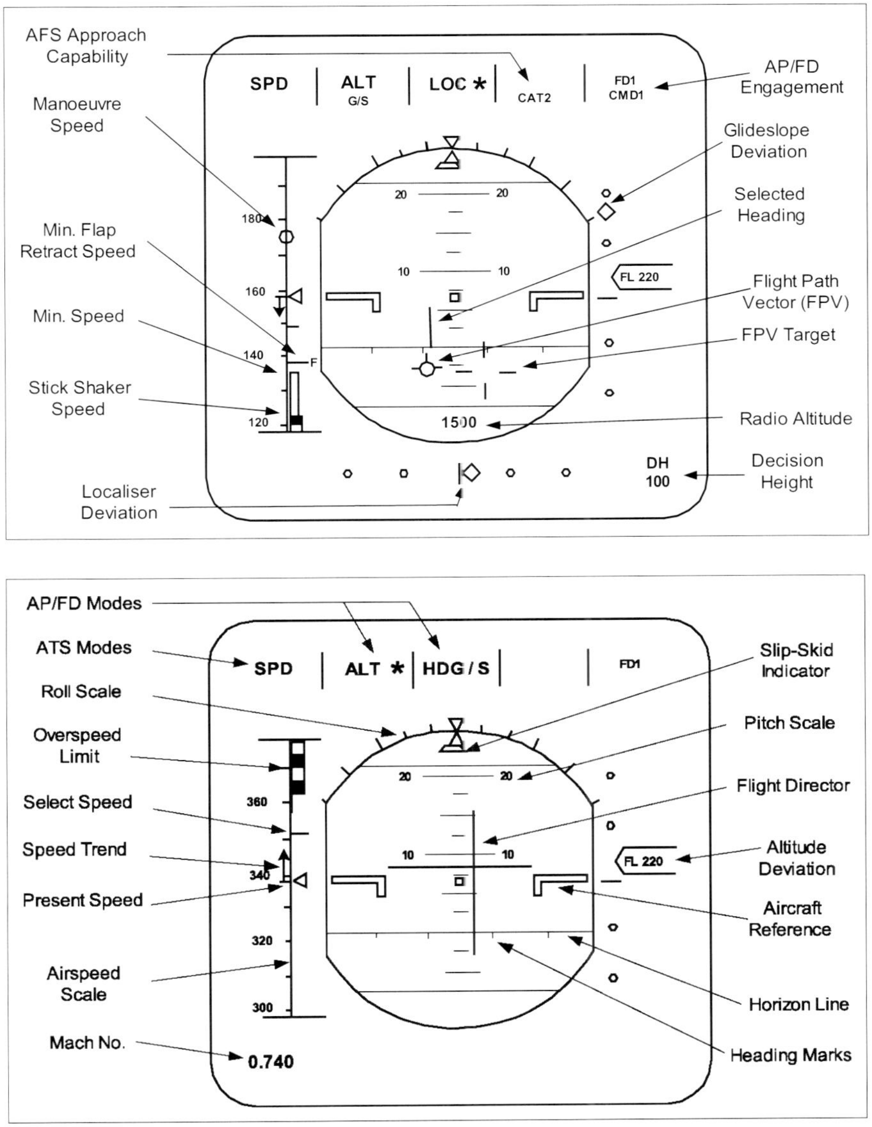

Fig. 5.26 *Typical primary flight display formats*

Navigation display

Typical Navigation Display (ND) formats are shown in Figs 5.27 and 5.28. Four modes are available, of which PLAN mode pictorially presents the flight path centred on a chosen waypoint, with North uppermost. MAP mode presents the compass rose in the upper arc and displays a plan view of the flight path originating from the present position of the aircraft. ARC mode is similar to MAP, but without flight plan data, while ROSE mode is an electronic copy of the electromechanical HSI.

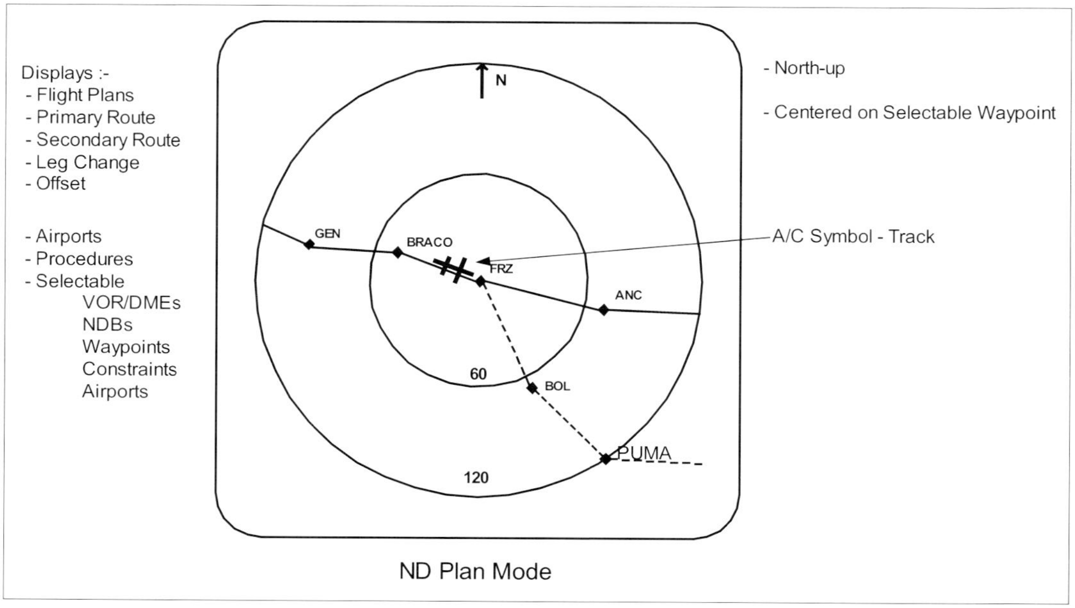

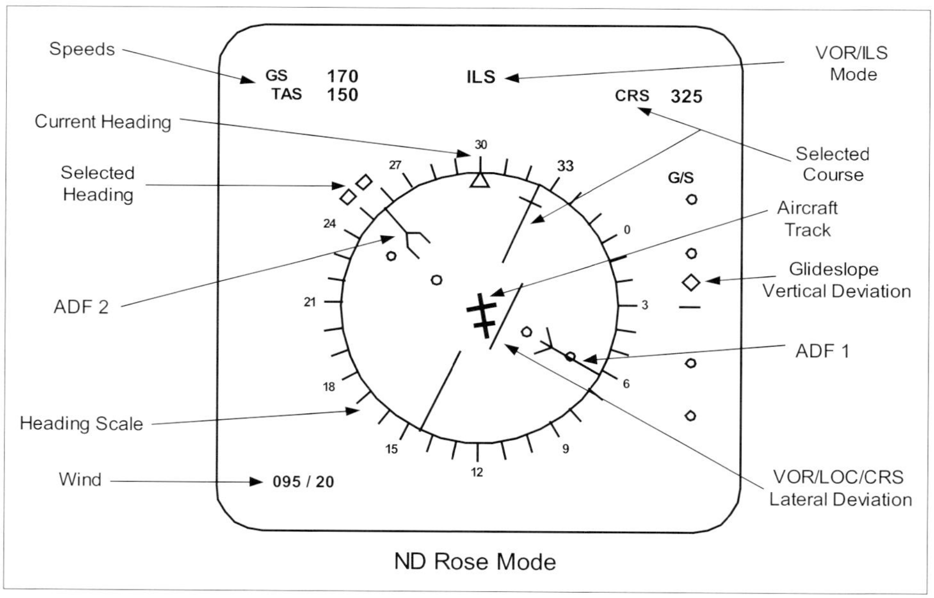

Fig. 5.27 *Typical navigation display – plan and rose formats*

Current wind velocity information, derived from the inertial sensors, is also displayed diagrammatically and numerically. In addition to these selectable basic modes there are various display options available that can be superimposed one at a time. These include weather radar, constraints, additional waypoints, VOR/DME, NDB, and airports.

Speed trend and flight path total energy vector information on the PFD enables the pilot to anticipate the short-term behaviour of the aircraft more precisely. The

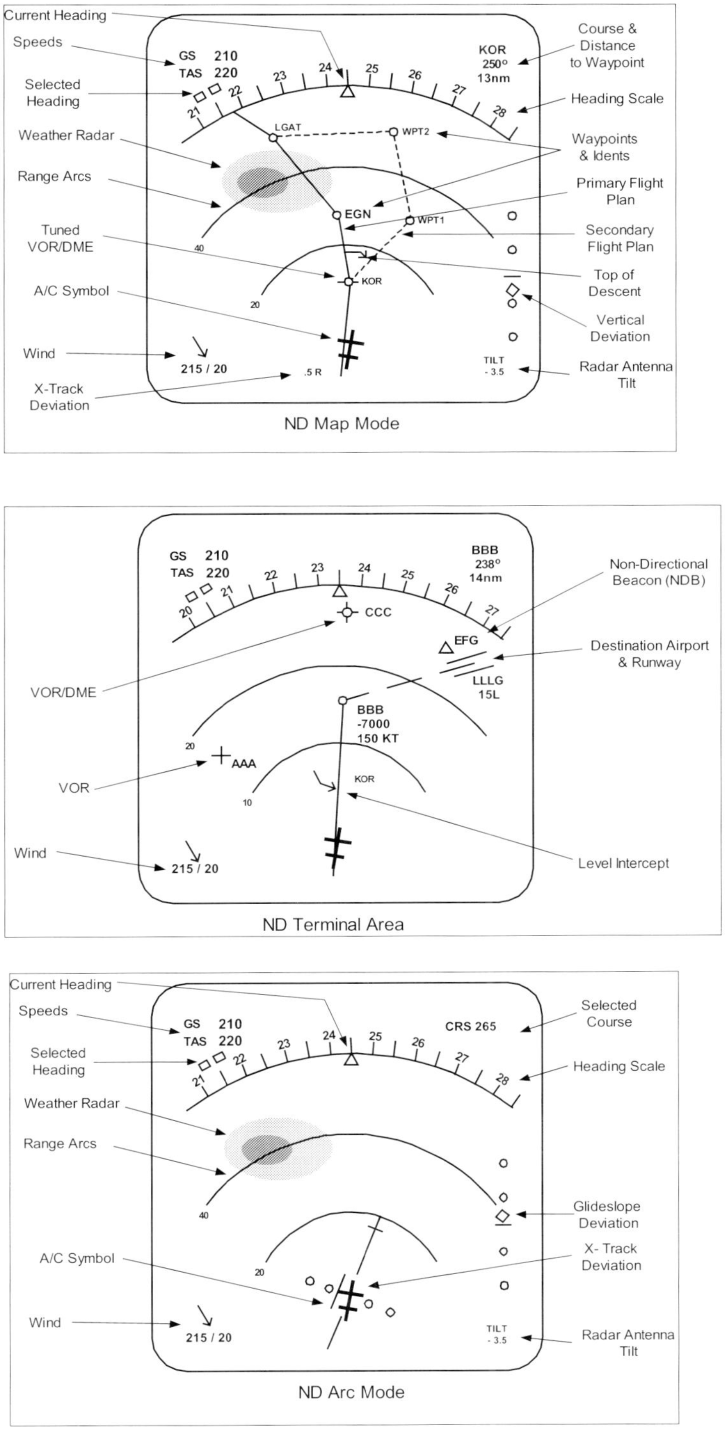

Fig. 5.28 *Typical navigation display – map formats*

ND displays the point ahead of the aircraft at which, during climb or descent, the selected flight level will be reached.

Information and management systems

The Electronic Centralized Aircraft Monitoring (ECAM) system sensors throughout the aircraft are continuously monitoring systems behaviour and reporting back to the central computer. The 'need to know' logic ensures that the data displayed are matched to the need for information; never less than necessary, never more than required. Data are immediately presented when needed.

The ECAM display consists of two CRTs, identical to the flight instrument CRTs, mounted side by side on the flight deck centre panel. Both pilots see the displays clearly, without moving from their normal seating positions. Below and between the CRTs is the warnings lights display panel, also in full view and within easy reach of both pilots. The ECAM control panel is on the forward centre console, again easily accessible.

In normal 'routine' operation, the left-hand CRT continuously displays total fuel state, outside air temperature, and 'MEMO' items which remind the crew whenever systems and functions are in operation that are only used temporarily and could be forgotten. There is no need for the crew to scan the overhead panel, the information they need is within their look-forward scan. As the display is always active, there is no likelihood of a dormant failure. The left-hand CRT also clearly describes any malfunction in plain language/checklist form and prescribes the necessary corrective action. When this has been taken, any resultant operating limitations are displayed as a reminder. The use of paper checklists is thus eliminated during the handling of an emergency.

The right-hand CRT normally displays synoptic diagrams appropriate to the particular flight phase. Should an alert occur, the relevant system synoptic diagram is automatically displayed.

The ECAM system has four display modes: the automatic flight phase, advisory, and failure-related modes, and the manual mode. In normal operation, the flight phase mode automatically presents on the right-hand displays appropriate to the current phase of operation: preflight, take-off, climb, cruise, descent, approach, and after landing. The following examples show typical arrangements:

Before engine and APU start	DOOR page
When APU master SW is selected ON	APU page (maintained for 30 s after APU reaches 95%)
When ENG START selection is made	Secondary ENGINE parameter page
When PARKING BRAKE is released	Brakes page
After slat retraction	Cruise page
After slat extension	Pressurisation page

At any time, the crew can manually select any desired aircraft system for routine checking. However, the ECAM system is itself a continuous monitor, by checking that all systems are operating within normal limits. The automatic advisory mode ensures that, should any parameter drift out of the normal operating range, the relevant data are brought clearly to the crew's notice, well before a warning level is reached. In this case, the parameter value is flashing.

Systems trends are thus automatically monitored, and the crew is warned of possible trouble well in advance, without having to refer to the overhead panel. ECAM provides the foresight that ensures efficient command.

The failure-related mode has precedence over the two other automatic modes and the manual mode. As soon as any warning is triggered, the appropriate system synoptic diagram is displayed on the right-hand CRT, while the left-hand CRT displays warning and corrective action guidance information. ECAM provides the crew with an immediate analysis of systems state in the failure case as well as during corrective action, and eliminates the need to diagnose failures and their implications from the overhead panel indications.

Pilot experience has shown that ECAM reduces the time required to react to an emergency, thus increasing the time available to deal with it and manage the flight. More intimately in the man/machine communication loop and always in full practice in systems operation, the pilot is more efficient – a definite contribution to improved flight safety.

The ECAM CRTs clearly indicate, in colour, whether there is a problem, how serious it is, how to deal with it, and its consequences. When an alert occurs, and simultaneously with the audio warning, the left-hand CRT shows the affected system in red or amber, depending on the degree of urgency, in plain language, with the corrective action required in blue. The change in display format and colour is an immediate attention-getter. The ECAM logic identifies, from a number of apparently simultaneous failures, the primary, or 'generating', failure which appears first on the CRT in a special 'boxed' format. The crew therefore deals with it first. On the warning lights display panel, the appropriate caution/warning light flashes, while the lights for services or functions affected by secondary or 'consequential' failures are identified by an asterisk. However, they are not displayed unless the attempt to correct the generating failure is unsuccessful, and further crew response to the situation is needed.

In conjunction with the blue instructions on the left-hand CRT, the appropriate push-button selector/indicators light up on the overhead panel. The crew simply presses each button to set the component it controls in the required configuration. As each action is performed, the blue instruction on the left-hand CRT turns white, and the synoptic display on the right-hand CRT is reconfigured accordingly.

Actions performed automatically, and thus needing no crew action, are also displayed in white. Interactive operation ensures that the information displayed is never less than necessary, never more than required. The crew are thus never swamped by information and instructions necessary at some other stage but at that moment irrelevant. At all times they are able to check and if necessary override the automation.

ECAM operation

Preflight

SGUs	ON
FAULT and OFF lights	OUT
CRT brightness	ADJUST
Mode pushbuttons	EXTINGUISHED

In-flight	
Without failures	Monitor both CRTs as required
With failures	Perform all corrective action (cyan colour)
	Push CLEAR until MEMO page is again displayed

The ECAM control panel incorporates a 'CLEAR' button which is pressed to remove the message on the left-hand CRT after an action. After making all necessary selections on the overhead panel, the crew check that the left-hand CRT does not call for any further action – 'no more blue'.

Pressing the 'CLEAR' button calls up on the left-hand CRT the systems in which there is a consequential failure, accompanied with the appropriate system synoptic on the right-hand CRT, in which all sections operating normally are shown in green and those no longer available in amber. When the least important system affected has been dealt with, a press of the 'CLEAR' button calls up a 'STATUS' display on the left-hand CRT, which provides a summary of the aircraft's condition and any resultant flight limitations. A further press of the 'CLEAR' button recalls the 'Memo' display on the left-hand CRT, together with the appropriate flight phase diagram on the right-hand CRT.

The crew can revert to the STATUS display at any time by pressing the 'STATUS' button on the ECAM control panel. Limitations will in any case be displayed as a reminder whenever the flight phase so requires. If, during the handling of any malfunction, another and unrelated alert occurs, it is immediately displayed above the existing messages on the left-hand CRT if it is of higher priority. Unrelated alerts of the same or lower priority are added below, according to a predetermined hierarchy, and are dealt with in logical sequence.

References

(1) **Jukes, M.** (1997) Designing Displays for Severe Environment Military Applications, SPIE (The International Society for Optical Engineering), Conference Proceedings Vol. 3057, *Cockpit Displays IV*, Orlando

(2) TSO-C113, Technical Standard Order: Airborne Multipurpose Electronic Displays, Federal Aviation Administration.

(3) TSO-C3d, Technical Standard Order: Turn and Slip Instruments, Federal Aviation Administration.

(4) TSO-C4c, Technical Standard Order: Bank and Pitch Instruments, Federal Aviation Administration.

(5) TSO-C52b, Technical Standard Order: Flight Director Equipment, Federal Aviation Administration.

(6) TSO-C6d, Technical Standard Order: Direction Instrument, Magnetic (Gyro Stabilized Type), Federal Aviation Administration.

(7) **Carr, G.P.** and **Montemerlo, D.** (1984) *Aerospace Crew Station Design*, International Center for Transportation Studies, Elsevier Science Publishers.

CHAPTER 6

The Head-Up Display

Introduction

The Head-Up Display (HUD) has been a vital display in the military cockpit since the first installation in the UK Royal Navy Buccaneer in 1961 and is a mature technology. In more recent years a few HUD installations have become classified as primary flight instruments.

The HUD evolved out of a need for a display that referenced the outside world and could be used for weapon-aiming purposes. At first this consisted of a simple weapon-aiming reticule, but it quickly developed into a more sophisticated projection device through which the user could correlate the position or vector of the airframe and weapons with the outside world.

Although the HUD started its evolution as a weapon-aiming device, it did not take long for the community to realize that a great deal of information could also be displayed to aid with the basic control of the aircraft. The Head-Up Display has been proven over many years to be a means of providing flight navigation, aircraft data, and weapon release parameters. The earliest gunsights provided little more functionality than an aiming system for the guns, but, with the introduction of increasingly sophisticated computational capability, complex manual air-to-air and air-to-ground weapon release became possible with enhanced accuracy. Now the HUD usually forms part of an integrated weapon aiming and release system where automatic modes of release are programmed in for the different stores and release conditions. In recent years it has been possible to introduce sensor video from Forward-Looking InfraRed (FLIR) onto the HUD such that operation by night can be achieved with much the same capability as daytime operation (**1**).

The Head-Up Display provides the key operational feature to present the pilot a collimated image (focused at infinity) in register and overlaying his view of the outside world. He does not need to refocus his eyes to read his instruments since all critical flight and mission data are available Head-Up or Head-Out in 'contact analogue' with the real world.

This chapter describes the optical principles of refraction, reflection, and diffraction employed in a Head-Up Display and the means to generate the displayed image.

By virtue of its operational use, the HUD is installed in an area of 'prime real estate' in the instrument panel, and therefore the HUD is 'designed into' the aircraft to optimize operational capability and performance with cockpit geometry and available space envelope. Seldom is a high-performance HUD reusable from one application to another. Almost always there is a need for a degree of customization. At the very least it is usually necessary to adjust the fold angle of the optical elements to match the cockpit geometry. Often significant electronics repackaging is necessary to shoehorn the HUD into the available space envelope.

As designers became more and more challenged by the advent of new control and display technologies, display designs have become abundant. Every airframer has his or her own version of tactical displays, situational awareness displays, map displays, and HUD symbology. On the one hand, the freedom of the media allows for creativity and invention of new ways to display important information; on the other hand, pilots desire to be able to cross-train from one aircraft to the next. Each new crewstation potentially poses new display formats for the pilot to learn and become proficient with in a short period of time. Because of this dilemma, there has recently been an emphasis on standardizing certain display formats, especially the HUD format, because it is capable of being a primary flight instrument. Standardization of the HUD symbology will allow pilots to maintain familiarity with the symbology regardless of the aircraft they fly. In 1991, the US Air Force (USAF) started a programme to develop and test baseline formats for its electronic displays. The first phase of work led to a published design of HUD symbology for fighter-type aircraft, MIL-STD-1787B (**2**), and work is currently ongoing to standardize head-down displays in transport aircraft also.

The configuration of an HUD traditionally comprises the Pilot's Display Unit (PDU) and a display driver unit. The PDU is rigidly fixed to the airframe and accurately aligned to the aircraft axes by a process known as boresighting. Even with this simplex configuration, the integrity is reasonable and the HUD is now regarded as a primary flight instrument on the basis of proven reliability over many years. Most of the fast jets currently in service still retain a set of reversionary instruments, usually mechanical, which are routinely scanned and act as a natural cross-check on the HUD.

HUD principles

The optical operating principles of the HUD are relatively straightforward, though to integrate these principles in the restricted space confines of a fast jet fighter crewstation results in some ingenious light-bending solutions.

Neither are these principles new. In 1901, Sir Howard Grubb patented a device to improve the sighting for guns. The patent, 12 108, shown in Fig. 6.1, embodies the key elements of an HUD, namely:

- An image source; in this case an illuminated reticule.
- A reflector; in this case a sophisticated prism assembly.
- A collimating lens to produce parallel light.
- A semi-reflecting combining glass.

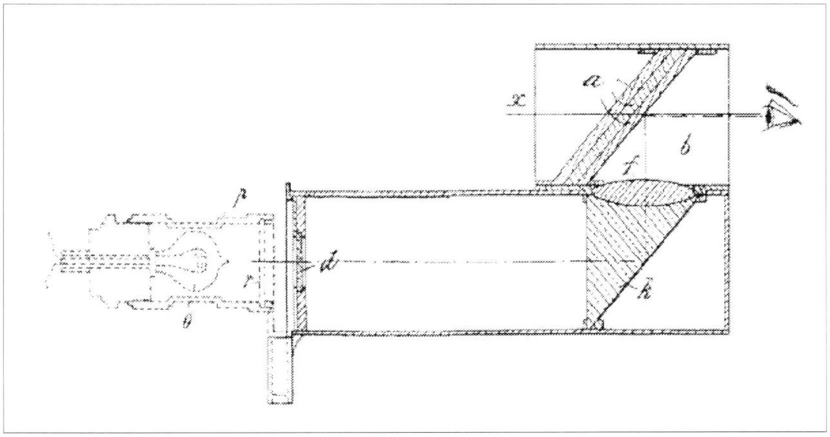

Fig. 6.1 *Sir Howard Grubb's patent of 1901*

The head-up display injects a virtual image of an object (the symbology or sensor video) into the pilot's line of sight with the key attribute of being collimated, or focused, at infinity, so that to the pilot it appears to overlay and be fixed on the outside world scene. Being focused at infinity, the pilot does not need to refocus his eyes to read the data as he has to when he looks inside the cockpit to read his head-down instruments. But absolutely fundamental to the operational capability afforded by the HUD, the collimated symbology appears to be in exact register and in the same focal plane as features of the outside world scene.

Light from distant objects in the outside world scene reaches the observer as parallel (collimated) rays at an angle of subtense relative to the aircraft axes. A point light source in the image plane of the HUD injected into the pilot's line of sight with the same angle of subtense relative to the aircraft axes appears to the observer to be emanating from the same point in space and therefore coincident with it, no matter how the pilot moves his head. As will be described later, this attribute allows the aircraft systems to place symbology in a manner that is often referred to as in 'in contact analogue' with (or conformal to) the real world. Conformal symbology is used to identify the flight path vector, the horizon, and sight lines to targets in the real world. In addition, primary flight data (speed, height, heading, etc.) and flight guidance cues allow the pilot to fly the aircraft without having to look inside the crewstation at his Head-Down instruments during critical phases of the mission.

There are two means of forming virtual collimated images, using the optical principles of refraction or reflection. These are shown in Fig. 6.2.

HUDs using the principle of refraction are often known as collimating HUDs (Fig. 6.2a). The lens system generates a collimated image of display symbology reflected to the eye via a flat combiner. The instantaneous field of view is limited by the size of the collimating lens, and the pilot has to move his head around the eyebox to see the total field of view.

HUDs using the principle of reflection are often known as pupil-forming HUDs (Fig. 6.2b). The lens relay system generates an intermediate image of the display symbology which is then collimated by reflection from a curved combiner. The instantaneous field of view is larger, defined by the size of the curved combiner.

The curved combiner in the pupil-forming HUD may use the optical principle of diffraction provided by a hologram rather than reflection to provide optical

Fig. 6.2 HUD optical arrangements

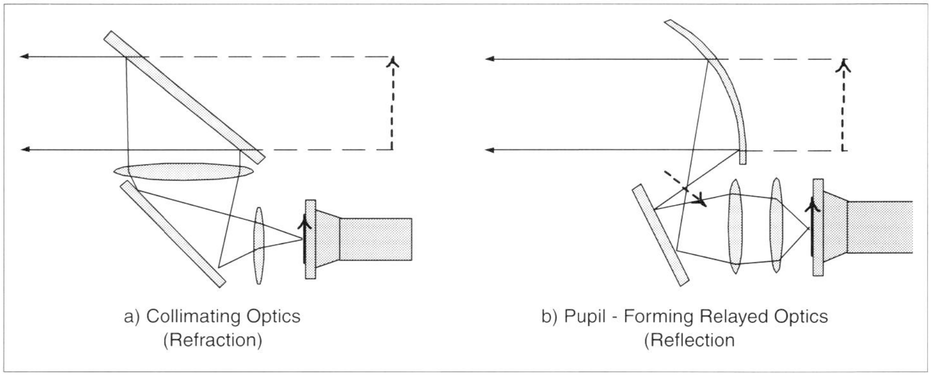

power, and so this arrangement is sometimes also known as a diffractive HUD (or variously DHUD or DOHUD – Diffractive Optics HUD). The spectrally selective feature of a hologram is also used to increase the combiner efficiency.

The gyro gunsight

The first head-up display was an extension to the gunsight. The first fighter aircraft had a pilot to fly the aircraft and a gunner to shoot down the enemy. The gunner had a simple machine gun, turret mounted on the aircraft, which he could aim and shoot using a conventional sight. Soon, the gun became fixed to the aircraft fuselage. The pilot became the gunner too. Now the pilot had to point and aim his aircraft at the target and shoot. Mechanical synchronization ensured bullets from the gun were timed to pass through the plane of the propeller without damage.

However, in complex dogfight manoeuvres the simple fixed gunsight was of limited use. The skill of the fighter ace came in predicting how to phase advance his manoeuvre to the manoeuvre being performed by the target so that bullets fired from his gun would intercept the target at the point in space that the target would occupy during the time of flight of the bullet. However, until the introduction of tracer bullets it was not possible to see bullet trajectories and make adjustments to the manoeuvre to achieve a hit.

A gunsight was developed that could provide guidance to improve the probability of a hit. It is sometimes known as the gyro gunsight, sometimes the Lead Computing Optical Sight (LCOS). The principle of the LCOS is to measure own-aircraft body rates by means of gyroscopes within the sight itself and reposition the aiming reticle to indicate not the sight line of the gun but the sight line to the point in space where bullets will intercept the plane of the target. Of course, this computation requires knowledge of the range to target, which in the absence of a radar was obtained optically. The pilot adjusted a ranging twist grip until the diameter of a circle centred on the gun aim point (piper) coincided with his perception of the target wingspan. From this diameter, with knowledge of the actual wingspan of the target aircraft (which the pilot also dialled in), it was possible to estimate range to target and hence time of flight of the bullet to the plane of the target to feed back into the computations. This method of range estimation is known as stadiometric ranging.

The pilot's job was first to identify the target aircraft and dial in its wingspan; then to manoeuvre his aircraft to track the target by placing the gun-aiming piper

Fig. 6.3 The gyro gunsight (with aiming reticle) (BAE SYSTEMS)

over the target, readjusting the ranging twist grip as necessary so the sight could estimate range and lead angle based on aircraft body rates; hopefully maintain the manoeuvre with low accelerations (if the target is cooperative) to allow the computations to become stable, and then fire the gun.

The first gyro gunsights, an example of which is shown in Fig. 6.3, were electromechanical devices using the simple refractive optical arrangement described earlier to project the gunsight image into the pilot's line of sight.

The collimating (refractive optics) Head-Up Displays

The optical principle of refraction is shown in Fig. 6.4. The image source (the object) is placed just inside the focal length, *f*, of the collimating lens, shown in the figure as a simple convex lens. The focal point, at distance *f* from the centre of the lens, is the point at which rays of light parallel to the lens axis will be focused by the lens to a single point. A ray of light from a point on the object will pass

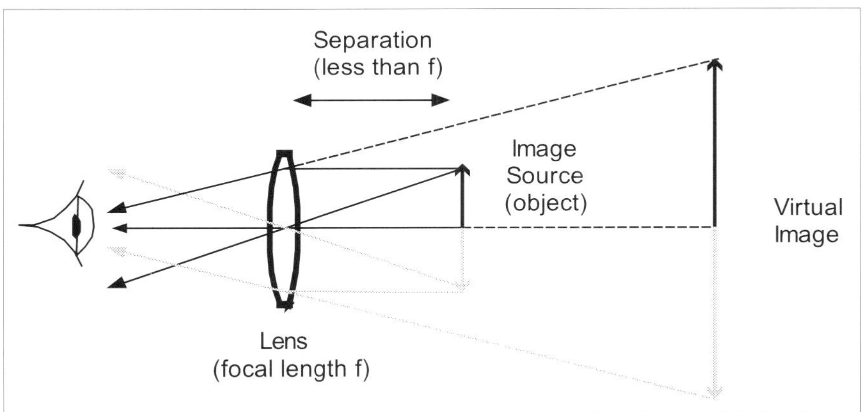

Fig. 6.4 Optical refraction

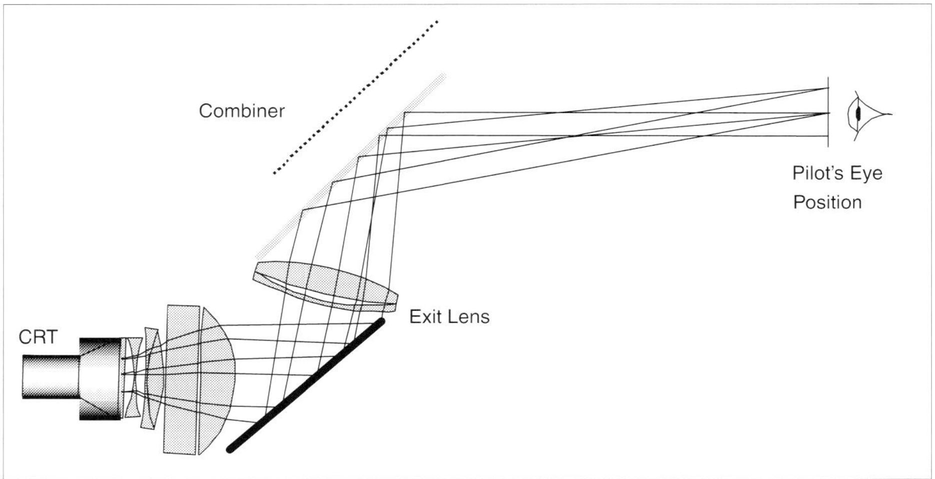

Fig. 6.5 HUD using refractive optics

through the centre of the lens to the observer unrefracted. Other rays of light, from the same point on the object, are refracted by the lens to converge towards the observer. The observer sees a magnified virtual image of the object. This is the principle of the magnifying glass and simple terrestrial telescope. If the object is placed in the focal plane of the lens, then the rays of light refracted by the lens from each point on the object will be parallel. The resultant virtual image is said to be collimated.

To form a practical HUD, as shown in Fig. 6.5, the image source must be placed out of the line of sight. The collimating lens, together with field-flattening lenses, focuses the CRT image at infinity and the image is introduced into the pilot's line of sight by a semi-reflective plane mirror, known as the combiner since the pilot sees the HUD image combined with his normal forward view of the outside world. A second fold mirror or prism is usually introduced between the image source and the collimating lens to produce a compact design for convenience of installation and more importantly to ensure there is minimal obstruction to the pilot should he need to eject from his aircraft.

In practice, the simple convex lens comprises a number of elements to minimize aberrations, together with a number of elements in the image plane to 'flatten' the field. These additional elements are designed to ensure that a collimated image is generated for all points of interest in the image plane from all viewpoints of interest to the observer. This describes a volume in space known as the eyebox. In the confines of a fast jet fighter cockpit, head movement is constrained by the simple expedient that the pilot is strapped into his seat. Even so, maintaining head movement with the HUD eyebox can be problematic during violent manoeuvres.

The development of radar during World War II paved the way for broadcast television using the Cathode Ray Tube (CRT). A new device was now available as the image source to replace the hitherto optical film reticle in the electromagnetic gunsight. Early head-up displays used CRTs developed for tele-cine and cinema projection applications. For that reason the HUD CRT generally has a faceplate with an active area corresponding to 35 mm film, nominally 40–50 mm in diameter. Later, high-performance HUDs use larger-diameter CRTs, typically 75 mm. The CRT is still the device of choice, though, as will be discussed in

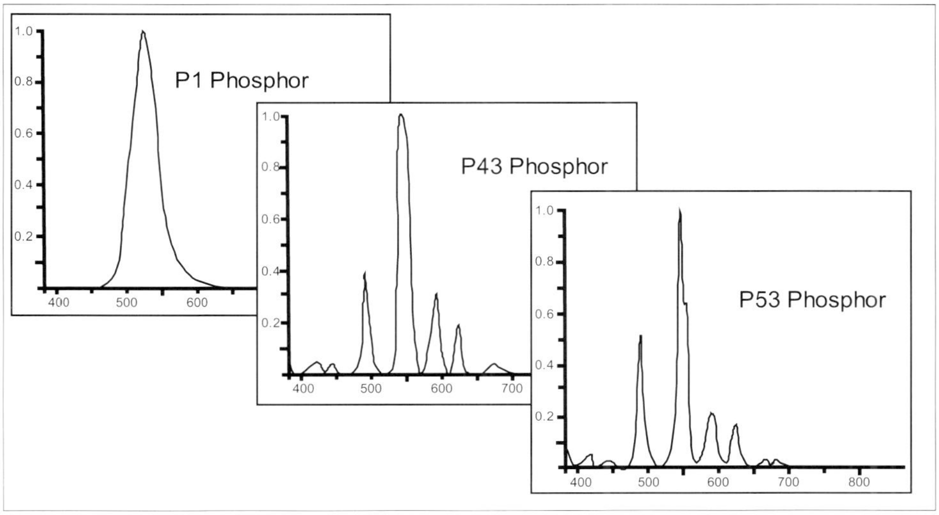

Fig. 6.6 *Typical CRT phosphor emission spectra*

Chapter 9, there are some new technologies emerging that have the potential to supersede the CRT in the HUD application.

The combiner is mostly transmissive to the wide-band daylight visible spectrum so as not to degrade the pilot's view of the outside world in low light conditions. Early HUD combiners were 70 per cent transmissive to daylight, 30 per cent reflective. To achieve maximum efficiency, the CRT phosphor is chosen to have a narrow waveband emission in the green part of the visual spectrum. A number of rare-earth phosphor types are now available. Most commonly, the chemical mix classified as P-53 phosphor is used, which has a spectral emission peak at around 545 nm (see Fig. 6.6). The photopic response of the eye is conveniently also most sensitive to green light (see Fig. 6.7).

The modern HUD combiner is optically tuned selectively to reflect the CRT spectral wavelength with high efficiency. Reflection of this narrow band of green

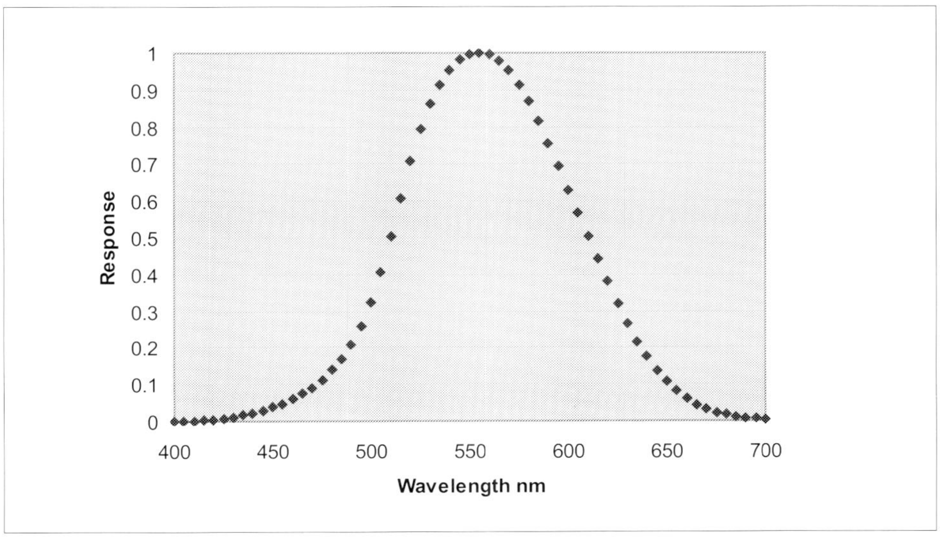

Fig. 6.7 *Human eye photopic response*

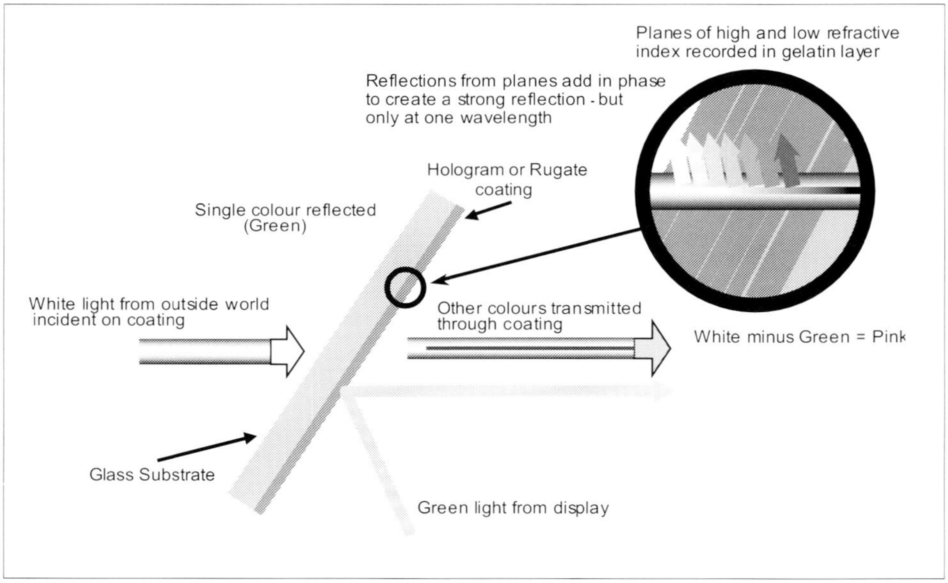

Fig. 6.8 *Selectively coated combiner operation*

light will mean that this component of the outside world scene is suppressed, giving a slight 'straw/pink' tint to the outside world (see Fig. 6.8).

The narrow-band reflection characteristic can be fabricated by a number of processes. The most common process for HUDs of this class is to vacuum deposit a number of discrete coatings of high and low reflective index (known as multilayer coatings). The depth of each layer is carefully controlled to match the phosphor peak emission wavelength. More recently, two alternative technologies have been used, namely rugate coatings and holograms (see Fig. 6.9).

Holograms contain a large number of planes of high and low refractive index material optically recorded in a film that can be arranged to synthesise a mirror. Reflection efficiencies close to 80 per cent can be obtained over the narrow band of frequencies of interest while at the same time only removing a very narrow band of frequencies from the outside world scene and hence achieving close to 80 per cent transmission to daylight also.

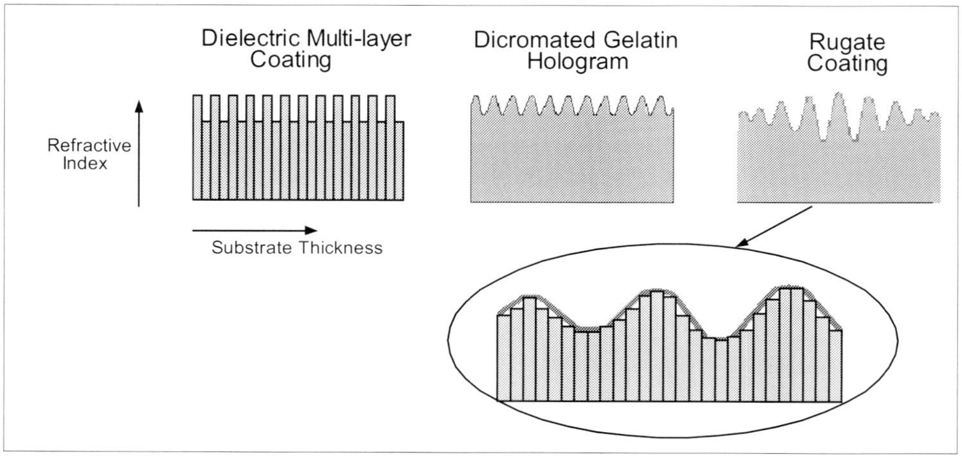

Fig. 6.9 *Multilayer coatings, holograms, and rugates*

Rugate coatings are sophisticated vacuum-deposited dielectric materials deposited in discrete layers to synthesise a sinusoidal variation in refractive index. Apodization facilitates reduction in sidebands.

Rugates are now sufficiently mature to be used in most applications where an unpowered, wavelength-selective mirror is required. Rugates offer better efficiency than multilayer coatings. They are:

- More efficient than a discrete multilayer coating.
- More durable than a hologram.
- Less expensive than a hologram (but more expensive than a discrete multilayer coating).
- Lower weight than a hologram since they can be deposited on a single piece of glass whereas a hologram must be sandwiched between two pieces of glass for environmental protection.

Field of view

The Total Field of View (TFoV) of any HUD is the total angle subtended by the display symbology seen with head movement from any location.

The Instantaneous Field of View (IFoV) of a refractive HUD is the angle of display seen from one head position using one eye and can be simply derived from the size of the collimator exit lens that is seen by the observer, as reflected in the combiner. A typical arrangement, shown in Fig. 6.10, for an HUD with a 150 mm (6 in) exit lens viewed 450 mm (18 in) from the combiner yields an IFoV of about 15°. If the pilot moves his head forward towards the combiner, his instantaneous field of view will increase. Similarly, if he moves his head laterally, he will see angles further off-axis.

In practice, of course, the pilot has two eyes and so he sees two portholes from one head position, one with each eye. This describes the Binocular Field of View (BFoV).

It is possible to increase the instantaneous field of view in the vertical axis by adding a second combiner parallel to and vertically above the first. This has the

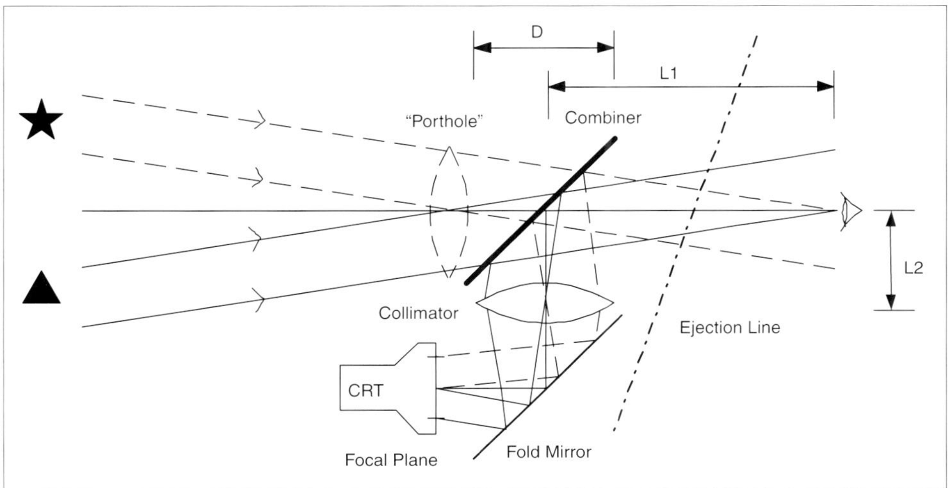

Fig. 6.10 Refractive HUD instantaneous field of view

effect of allowing the observer to view the CRT image through a second porthole vertically above and superimposed over the first.

Provided the quality of the optics is good, all rays of light emerge from the collimator in parallel, and provided the two combiners are exactly parallel, the two images seen by the pilot will exactly register with one another and fuse as a single image. The combiner coatings are graded progressively to fade out one image and fade in the other at increasing elevation angles. Great care has to be taken to chamfer the lower edge of the upper combiner to ensure that the pilot is unaware of its presence and to grade the coatings to ensure that there is no discontinuity at any point which might potentially be interpreted by the pilot as a false horizon. This arrangement is shown in Fig. 6.11, which also shows the relationship between the TFoV, the IFoV, and the BFoV.

Some other terms used to define the performance of a HUD system include:

- *Eye motion box* – the allowable movement of the head within which the display is visible.
- *Cockpit design eye* – the nominal operational position of the pilot's eye (usually taken as the 'cyclops' eye – the central position between the two eyes).
- *HUD design eye* – the nominal head position for optimum viewing of the HUD.
- *Parallax* – a measure of how well the optics provides identical images to both eyes.

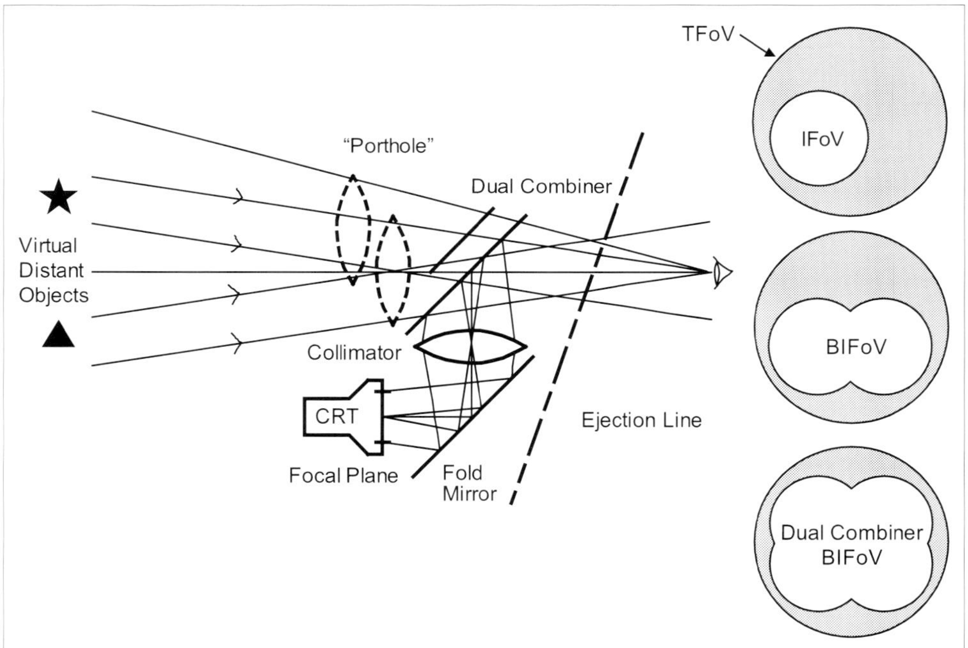

Fig. 6.11 *Dual combiner (including FoV)*

Collimating (refractive) HUD – examples

The Blackburn Buccaneer

The Blackburn Buccaneer was the first UK operational aircraft to have a CRT-based HUD. The Buccaneer was a twin-seat, low-level strike aircraft that saw service with the Royal Navy and the RAF from the late 1950s through to the late 1970s.

The Pilot's Display Unit (PDU), shown in Fig. 6.12, was a small field of view refractive design with a 3 in collimating lens. The single combiner could be folded down for installation and to clean the windshield. Note the large rubber bumper to protect the pilot's head in a crash landing. The remote electronic unit housed a set of circuit cards arranged in a radial fashion around a central core. The HUD provided rudimentary flight and weapon-aiming symbology for the air–surface attack role of the aircraft.

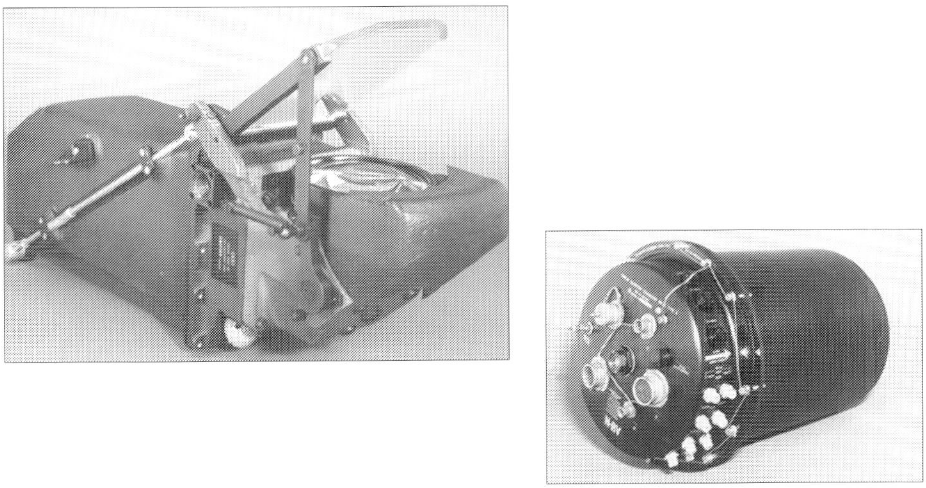

Fig. 6.12 The Buccaneer HUD (PDU and electronic unit) (BAE SYSTEMS)

The British Aerospace Harrier GR.Mk1

Subsequently, the vertical take-off Harrier and the Anglo-French Jaguar were the first UK aircraft to enter service with a designed-in HUD in the mid-1960s. The HUD was now fully accepted as an essential feature for fast jet fighter aircraft.

The head-up display for the Jaguar and Harrier, shown diagrammatically in Fig. 6.13, contained only those optical and mechanical supporting structures described above, together with the CRT and its electron beam deflecting yoke and focus magnet. All other electronics were contained remotely. The high-voltage power supplies required for the CRT were contained in a cockpit-mounted unit. The remaining electronics were packaged in a conventional LRU in the equipment bay. This architecture minimized the size and weight of the PDU itself to facilitate its installation into the cockpit. The PDU had a 4 in collimating lens and provided an IFoV of about 16° and a TFoV of 25°. The single combiner was servo-assisted to travel in the fore–aft direction to position the HUD nominal eye position in elevation about the aiming vector.

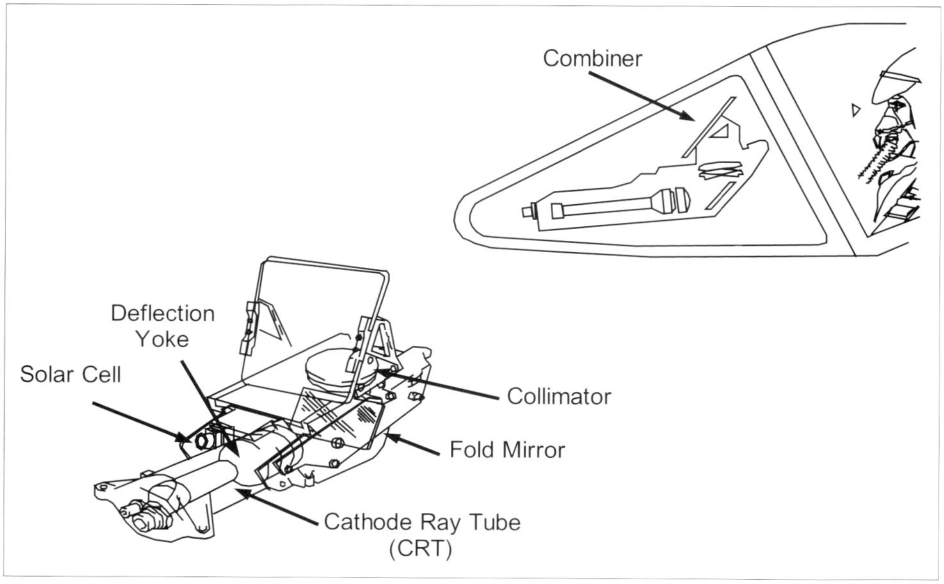

Fig. 6.13 Jaguar/Harrier HUD installation (Smiths Aerospace)

McDonnell Douglas/British Aerospace Night Attack Harrier II (GR-7 and AV-8B)

The Night Attack Harrier II entered service in 1989 and incorporates a high-performance, wide field of view, refractive HUD capable of day/night operation. The HUD shown in Fig. 6.14 incorporates the CRT, the deflection amplifier, the high-voltage power supplies, and all the services to make the HUD a stand-alone line-replaceable unit.

The HUD employs a dual combiner and, in order further to increase the field of view, the collimating exit lens is truncated. That is, instead of being circular, the exit lens is sliced horizontally in the fore and aft axes. This allows the optical axis of an effectively larger-diameter exit lens to be installed nearer the pilot without encroaching upon the ejection envelope. Together with a dual combiner, it is possible for the 6.5 in truncated optics arrangement to achieve an effective 8 in optical field of view. In the Harrier II, the binocular IFoV is 20° azimuth × 15° elevation. The TFoV is 22°.

The HUD presents daytime images in cursive (stroke) mode and night-time images from the FLIR sensor as raster with stroke symbology overlay during the field retrace period. The HUD is compatible with Gen III 'Cats Eyes'™ night vision goggles.

An up-front control panel provides means to control the aircraft major system moding, to enter waypoint and navigation data, and to set and control the radios. A video camera mounted on the side of the HUD provides the means to record the mission outside world scene and HUD symbology through a combining prism assembly which minimizes obscuration to the pilot.

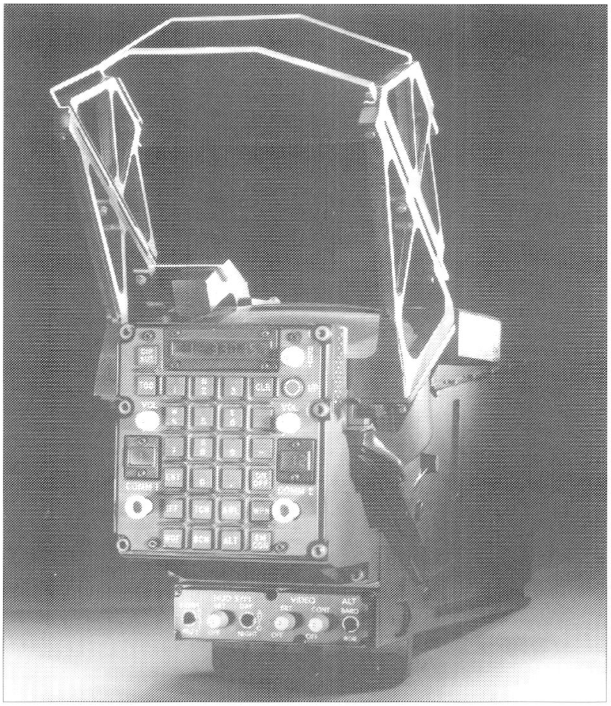

Fig. 6.14 AV8B/Harrier GR7 night attack HUD (Smiths Aerospace)

The Tornado GR.Mk4

The Tornado GR.Mk4 upgrade, which entered service in 2001, incorporates a new HUD, shown in Fig. 6.15, with day/night capability, compatible with the use of a FLIR vision sensor and NVGs. It was the first HUD to employ a rugate-coated dual combiner to give the photometric performance of a holographic HUD.

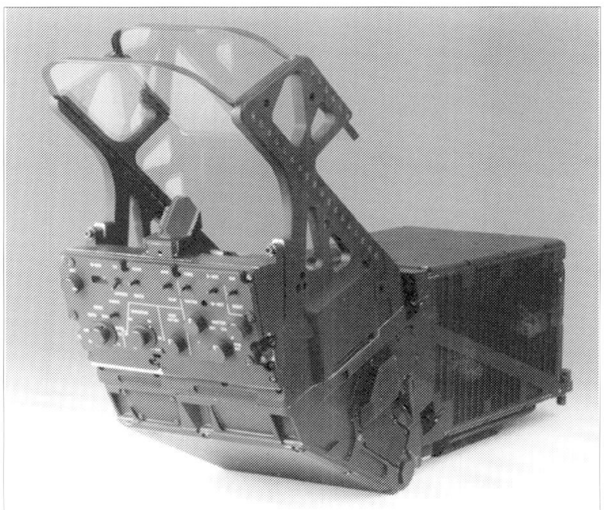

Fig. 6.15 Tornado GR.Mk4 head-up display (BAE SYSTEMS)

The colour-selective, graded, multilayer rugate coating is vacuum deposited and designed to have a high reflectance at the peak of the P53 (green) emission 545 nm. This enables the HUD to display a raster image of sufficient brightness to be daylight viewable and usable with FLIR sensor vision in cloudy/hazy conditions. The graded coating was a particular challenge because it is important that the pilot does not see any horizontal banding in the combiners, which could be mistaken for a horizon line when in cloud. Colour-selective combiners need to match both luminance and colour from a range of eye positions. The colour-selective combiner makes the outside world look a little pink. The HUD has a 6.5 in diameter collimating lens, truncated in the fore–aft axis to provide a 25° TFoV and a 21° azimuth × 19° elevation binocular IFoV.

Pupil-forming (reflective/defractive) Head-Up Displays

The stealth tactic of flying low and fast can be effective in countering all but the most sophisticated air and surface defences and would be further enhanced, because of reduced defence effectiveness, if such missions could also be accomplished with reasonable safety at night and in poor weather conditions. This possibility can be enhanced by providing the single-seat day fighter pilot with pseudovisual conditions, projecting a forward view in a one-to-one relationship with the outside world derived from an imaging sensor (LLTV or FLIR). By making the image large enough, the benefits of perspective and peripheral vision can be derived to give the pilot the sensations of flying under visual (VMC) conditions, enabling him to use familiar visual cues to judge speed and terrain clearances, to navigate, and to identify targets.

Existing collimating HUDs of the type described so far have been used to project FLIR night vision scenes into the pilot's forward view. Experience has shown that a minimum instantaneous FoV, i.e. that part of the total scene visible from one head position, of 20° azimuth by 15° elevation is adequate, but that a larger Instantaneous FoV (IFoV) is essential to support more aggressive manoeuvring or improved off-track waypoint/target detection. The largest practicable FoV is the critical parameter governing the pilot's ability to fly safely at low level, and this requirement has driven the design of a new generation of HUDs using different optical techniques.

The sensor scene, received by the HUD as a standard television video signal, is projected with the necessary one-to-one angular correspondence with the real world. Textural, perspective, and peripheral cues are available to the pilot to judge speed and terrain clearances. The greatest practicable 'up' vision is required to allow the pilot to 'look into turns', while a very large lateral view enables him to detect off-track targets. The sensor scene is overlaid with the normal daytime symbology to provide the pilot with VMC equivalent conditions. Indeed, experience has indicated that pilots usually elect to reduce symbol brightness against the scene background, flying on normal visual cues with occasional scan checks of symbolic HUD flight data, just as they would fly head-up at low level in daytime.

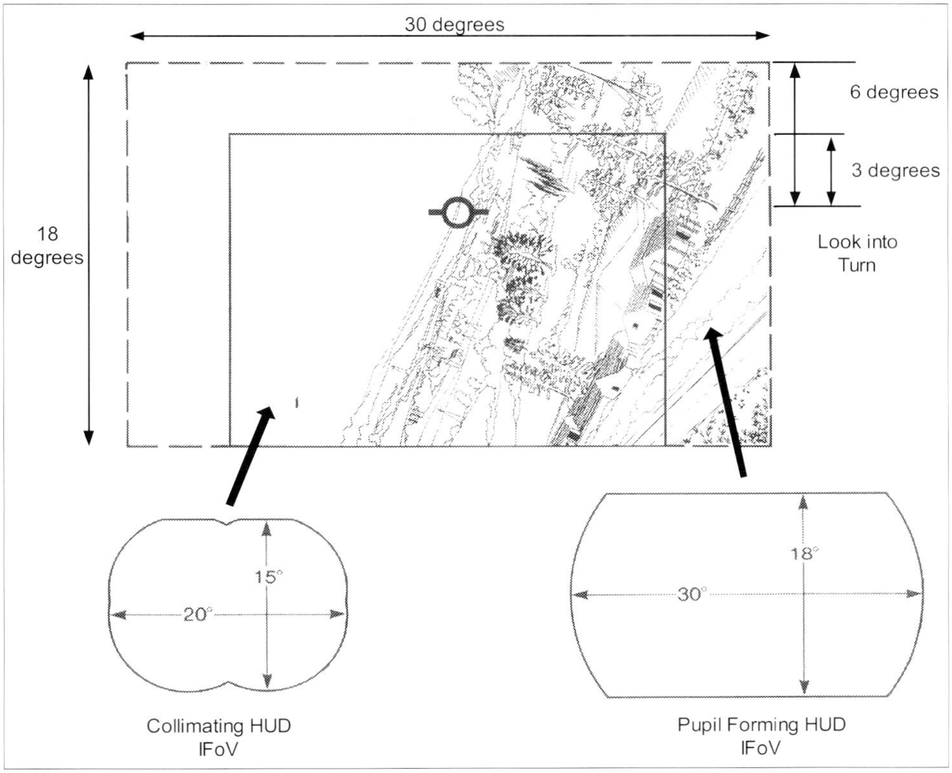

Fig. 6.16 *Operational improvement of increased IFoV (BAE SYSTEMS)*

Figure 6.16 illustrates the significant operational improvement that would be achieved if the HUD IFoV were to be able to be increased from the $20° \times 15°$ practical limit of a refractive collimating HUD to an IFoV of $30° \times 18°$.

The IFoV of a collimating HUD is limited by two critical dimensions: the size of the collimating exit optic which, when reflected by the combiner, provides a virtual image of the display superimposed on the pilot's forward view of the real world, and the distance of this image from him. Large exit lenses are also heavy and liable to crack under extreme mechanical and thermal stress. The most advantageous HUD IFoV would be achieved if the collimating element were to be above the instrument panel where its size would be limited only by canopy clearance, and if it were as close to the pilot as possible. This is achievable using the other means to generate a collimated virtual image, namely the optical principle of reflection, shown in Fig. 6.17.

Here, the object is placed within the focal length of a concave mirror. In this case the rays of light are reflected by the mirror towards the observer who sees a magnified virtual image of the object. This is the principle of the shaving mirror and astronomical telescope. Again, by placing the object in the focal plane of the mirror, the rays of light from each point on the object will be reflected by the mirror to emerge in parallel, i.e. collimated.

As with refraction, it is necessary to design an optical path that allows the CRT image to be introduced into the pilot's outside world line of sight without interfering with it. This is more complex than with the refractive arrangement.

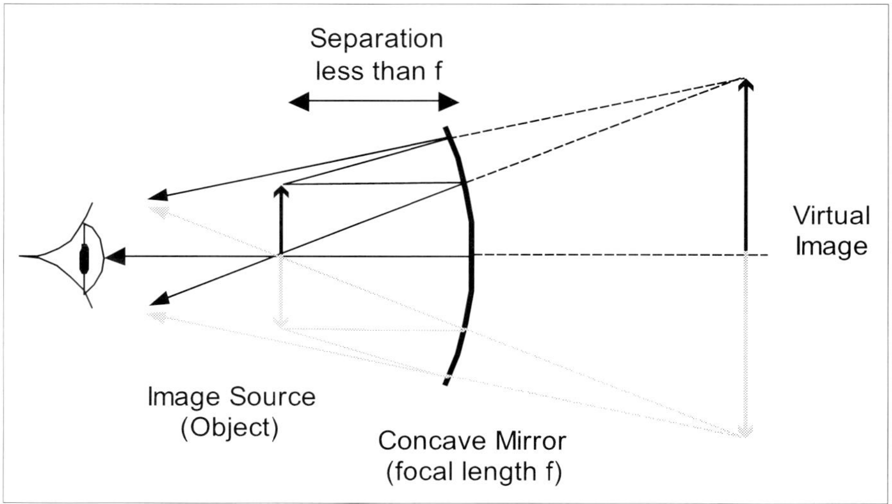

Fig. 6.17 Optical reflection

The most advantageous HUD IFoV will be achieved if the collimating element is placed as shown in Fig. 6.18. This particular arrangement has a number of positive attractions: the large spherical collimating surface provides the largest practicable IFoV; the radius of curvature is large, allowing it to be sandwiched between two planar elements to provide an undistorted view of the real world; the relatively small angle of incidence and long focal length dramatically reduce the aberrations that are present in more off-axis systems. These minimal distortions allow significant simplifications in the design and manufacturing process (**3**).

Of course, the image is in quite the wrong place, and so some plane mirrors are required to fold the optical path and allow the CRT and a relay lens system to be buried in the conventional location, behind the instrument panel. Some of the possible optical configurations are shown in Fig. 6.19. On the right (c) is a class of diffraction optics designated 'off-axis'. It represents one of the earliest configur-

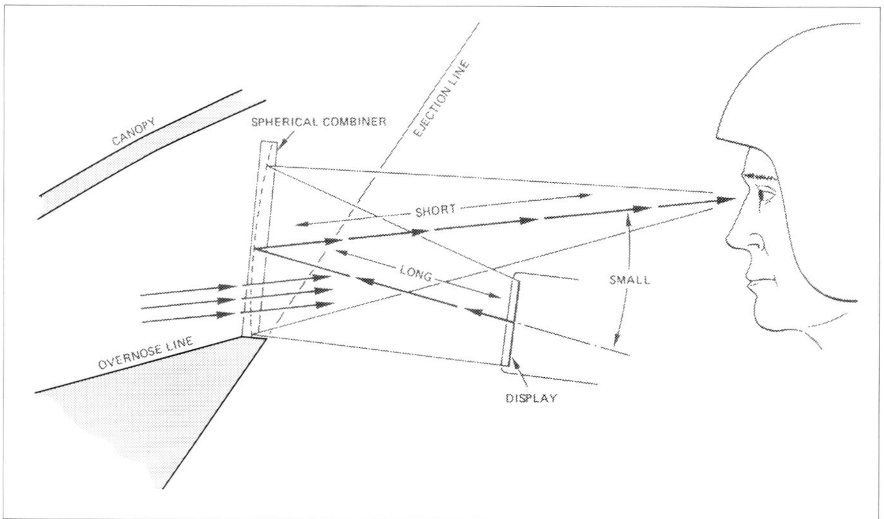

Fig. 6.18 Ideal combiner position (BAE SYSTEMS)

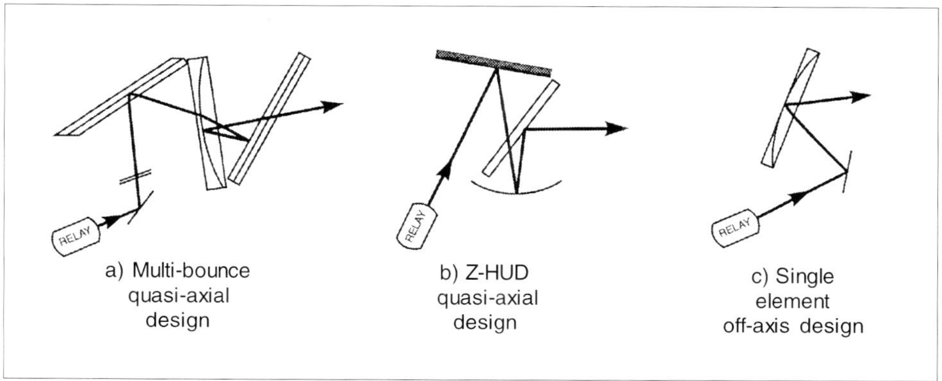

Fig. 6.19 *Some possible reflective optic HUD configurations*

ations to apply the technology, and was pioneered by Hughes Aircraft Co, subsequently Kaiser Electronics on the F.15 and Grippen aircraft. It suffers from a number of difficulties. The curvature of the combiner in such a system provides the principal collimating function but is too great to allow the use of a planar doublet to sandwich the reflective coating. The necessary protection has to be provided by two pieces of curved glass which reduce the thickness of the element and thus its weight to an acceptable level. The curved combiner is more complex to manufacture than a plane combiner, and care must be taken to ensure that the apparent distortion of the real world seen through it is minimal. The large off-axis angle causes significant optical aberrations which must be corrected by introducing compensating aberrations in the reflective combiner itself and in a complex relay lens.

By comparison with the off-axis systems, the other class, termed quasi-axial, have a much reduced critical angle of incidence for reflection. Aberrations are minimal and do not require compensating aberrations in the reflective element itself: the limited corrections necessary can be implemented in a simple relay lens. In the two versions shown, the Z-HUD (so called because the optical path makes the letter Z) separates the reflecting function from the combining function. While this is obviously attractive, it probably places the collimating mirror over the ejection line.

Each of these configurations has been realized in the product designs described below, but, before describing each, it is worth considering the means of fabricating the optical elements.

Diffractive optics (holograms)

Although the main driver for wide-angle pupil-forming HUDs is the night-time sensor video raster role, an essential requirement is that, in daytime use, there should not be any degraded performance in comparison with conventional collimating HUDs. Indeed, it would be beneficial if the raster sensor video image were to be viewable in daytime under conditions of poor visibility also. This drives the need for a highly efficient, spectrally selective, powered reflector. The technology best able to provide all these characteristics relies on the optical principle of diffraction and is best fabricated as a hologram.

This is a specialist subject in its own right, and beyond the scope of this book. However, the following overview describes how the holographic technique is used to create the diffraction surface.

A thin film of photosensitive dichromated gelatin is exposed to two beams of coherent laser light. Owing to the coherent nature of the incident beams, a series of interference fringes are formed throughout the depth of the gelatin film. During the developing process, these fringes are converted to planes of high and low refractive index parallel to the film surface. To a first approximation the refractive index change between adjacent planes is sinusoidal.

To produce a curved (spherical) mirror, one of the lasers is arranged to produce a beam having a plane wavefront, and the other laser is arranged to produce a beam with a spherical wavefront. The construction geometry is set to emulate the HUD installation in which the laser with the plane wavefront represents the outside scene, while the laser with the spherical wavefront represents the desired focal point of the mirror. The glass plates with the gelatin film are placed in the desired location for the combiner. After exposure and processing, the hologram formed in the gelatin film will act as a collimating mirror to light emanating from the chosen focal point at the specific wavelength to which the hologram is tuned. Light from other angles and at other frequencies will pass through the hologram. These principles are shown in Fig. 6.20.

During the developing process, the gelatin swells, producing an increase in the tuned wavelength. Retuning the hologram is achieved by baking the film which reduces the thickness and hence the spacing between planes of constant refractive index. The hologram can therefore be adjusted to reflect at a wavelength that corresponds precisely to the peak emission wavelength of the HUD phosphor. The bandwidth of the angular reflection range is determined by the magnitude of the change in refractive index. This variable can be controlled during the developing process and is specified as the hologram modulation.

Fig. 6.20 Hologram production of a spherical mirror

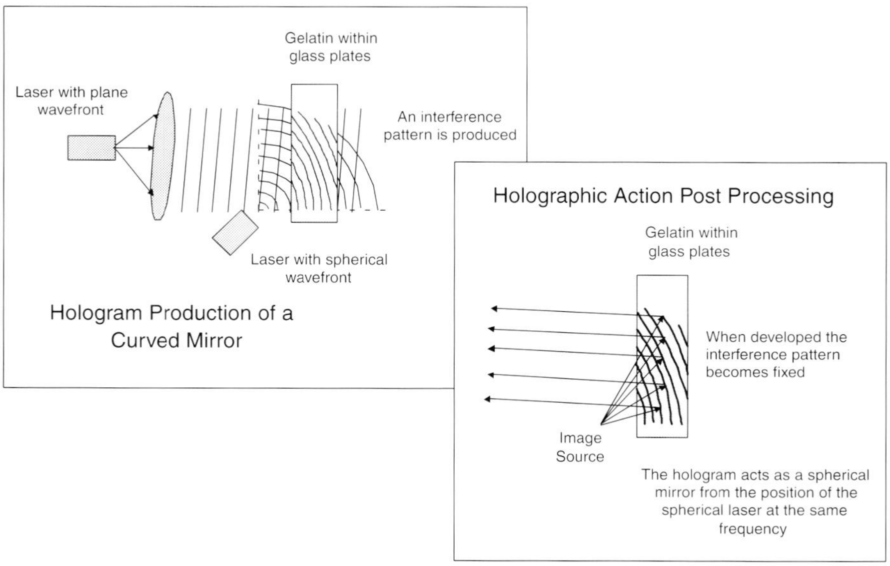

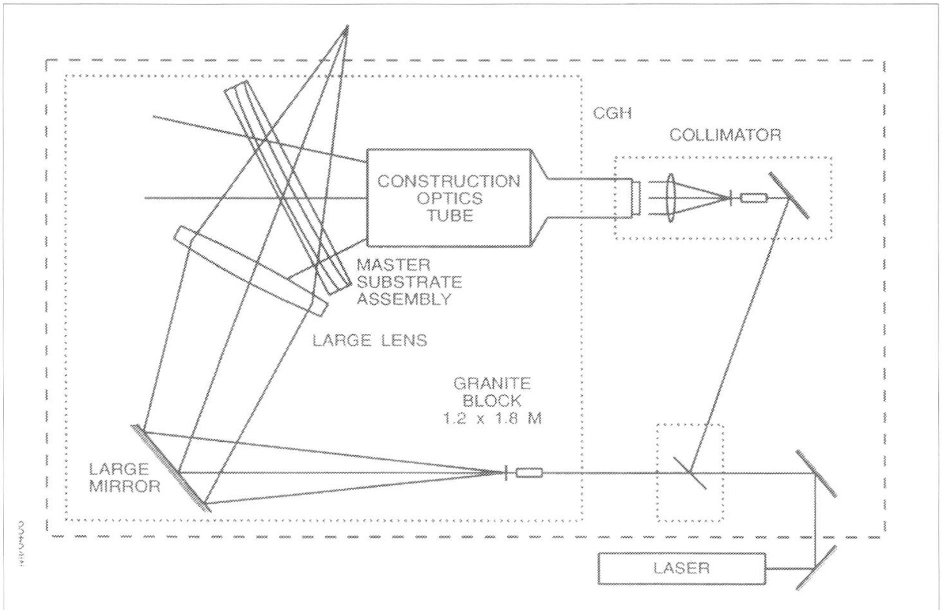

Fig. 6.21 Diffraction element exposure (BAE SYSTEMS)

In a real design, the desired critical angle of reflection varies over the hologram surface and the required diffraction characteristics are largely determined by such factors as the pilot's eye position, head motion, the field of view required, and the geometry. An interactive iterative design process using a suite of CAD programs is used to optimize the optical system design and to meet the particular installation constraints. The construction points can then be determined, and the construction optics designed and fabricated.

Figure 6.21 shows a typical fabrication method used to create the two interfering beams using a beam splitter. Such a technique is necessary in order to introduce the aspherical optical elements into the path of the laser light to create deliberate aberrations as required by the off-axis configuration. However, path lengths are necessarily long, creating exceptional stability requirements of fractions of a wavelength over a period of exposure lasting several minutes. The aberrating elements themselves can also create problems, tending to cause unwanted secondary holograms to be recorded in the gelatin.

Straightforward plane-reflecting mirror surfaces can be produced by a simplified back-reflection hologram construction technique shown in Fig. 6.22. Fringes are created by the interference of incident and reflected rays. As a result, the separated path length can be measured in microns and stability and exposure time requirements can be reduced to a level where they no longer create manufacturing yield problems.

The processing steps to fabricate a hologram are shown in Fig. 6.23 and are:

- Procure the glass elements – substrate and cover glass.
- Deposit dichromated gelatin onto the glass substrate.
- Stabilize the gelatin by drying.
- Expose the gelatin using the hologram construction arrangements described earlier.

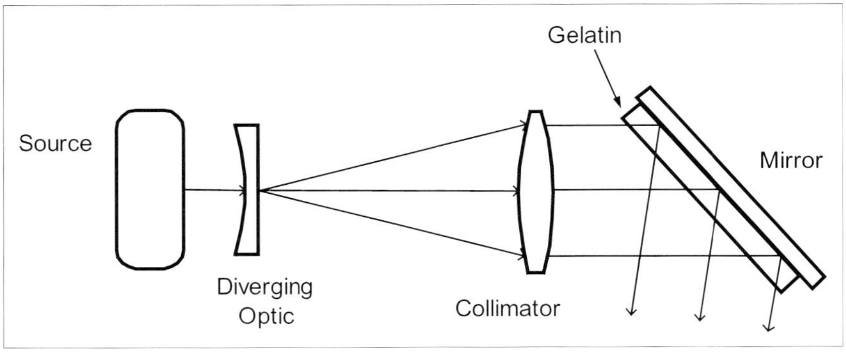

Fig. 6.22 Semi-reflecting mirror hologram construction

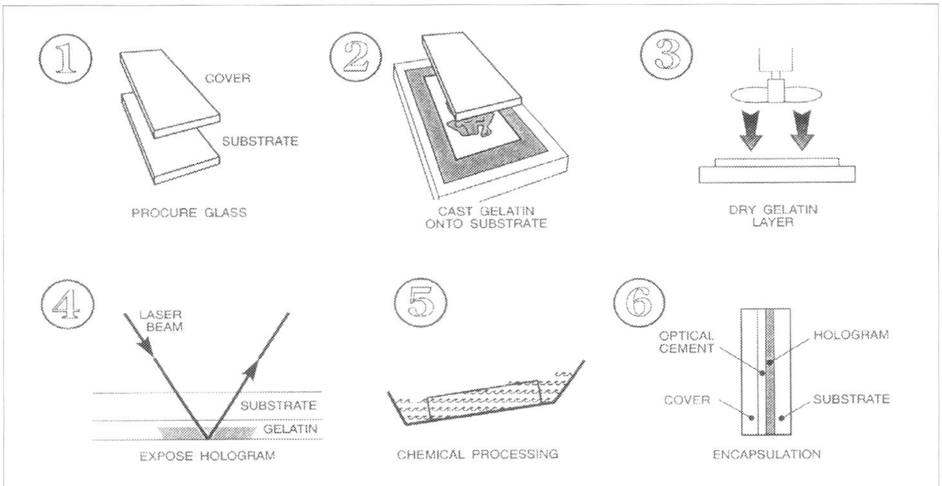

Fig. 6.23 Hologram manufacturing process steps (BAE SYSTEMS)

- Process the gelatin to develop the holographic diffraction fringes and bake to tune the hologram to the desired reflection wavelength.
- Encapsulate the hologram.

In a production environment it is usual to create a master hologram using the process described above and then replicate the master through a copy exposure process.

Other consequences of a pupil-forming holographic HUD

A number of related HUD functions need to be treated differently in the diffractive pupilforming HUD from their implementation in a collimating HUD. A frequent HUD requirement is to provide a totally independent collimated depressible standby sight, available in the event of electronic system failure, although the case for such a sight is weakening as improved technology extends equipment reliability. The most successful method of injecting the, usually red, standby reticle into the optical path has been via a dichroic beam splitter. It is particularly efficient because it reflects most of the red light without significant attenuation of the green CRT light. A red standby sight is not possible with a diffractive HUD since the optics can only operate at a very narrow band of green wavelengths. A different approach is necessary. Therefore, the diffractive HUD must provide an electronically generated standby sight that is completely independent of the separate main symbol generator unit, but of course this does depend on the continued functioning of a large part of the HUD, fortunately now a relatively high-reliability item.

The HUD camera recorder presents a severe problem to a pupil firming HUD. The HUD image is virtual and can only be seen in the pilot's eyebox, and therefore a recorder mounted in the conventional position, looking through the combiner glass, cannot see the HUD image. The only really viable alternative is to provide an electronically scan-converted version of the display symbology which is then mixed electronically with outside world scene video which can now be sourced from a camera mounted forward of the combiner.

Pupil-forming (reflective/defractive) HUD – examples

F16 LANTIRN HUD – multibounce quasi-axial configuration (Fig. 6.19a)

The first application of the multibounce quasi-axial pupil-forming diffractive HUD configuration was in the F-16 as part of the USAF Low-Altitude Navigation and Targeting Infrared for Night (LANTIRN) system. The development contract was awarded in July 1980, and the product commenced production in 1984 (**4**).

The optical arrangement is shown in Fig. 6.24. Obviously, all glass elements above the glare shield must be transparent to allow the pilot an uninterrupted view of the real world. They must also sometimes be transparent to CRT light and sometimes reflect it strongly to allow the HUD image to reach the pilot. This apparent paradox is resolved by the use of highly angularly selective optical

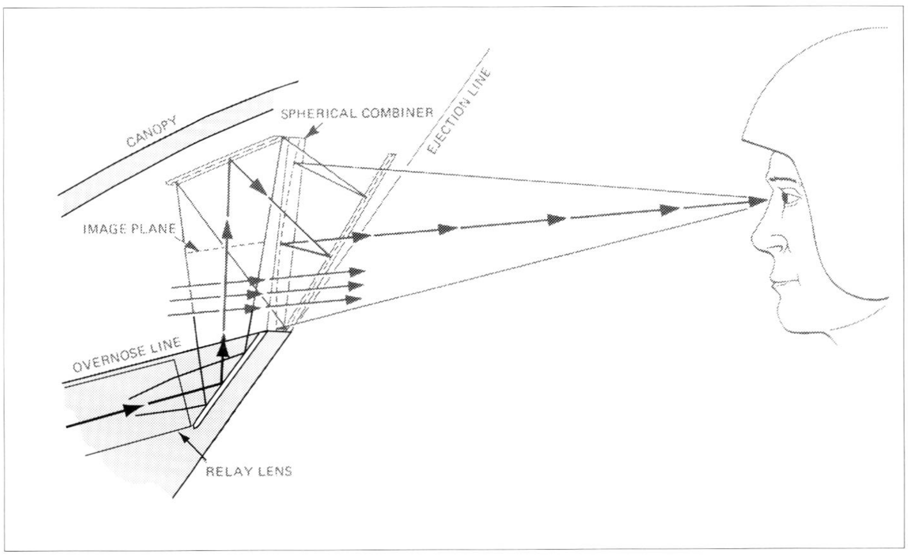

Fig. 6.24 *F-16 Optical configuration (BAE SYSTEMS)*

coatings which will only reflect a narrow band of wavelengths incident at a particular critical angle.

Although the optical geometry would work in theory with multilayer reflection coatings or even neutral-density ones, optical efficiency would be so degraded that the HUD would be unusable. The only coatings offering sufficient selectivity are diffraction surfaces created by the photographic technique used to make holograms. The coatings are designed to act at the very narrow band of wavelengths emitted by the CRT phosphor (P43) and can approach 100 per cent efficiency as individual reflectors. However, because of the number of surfaces involved and because of the inevitable slight mismatches between phosphor bandwidth and coating reflection characteristics and other losses, the light of the CRT actually seen by the pilot is reduced considerably. Indeed, some deliberate detuning is sometimes necessary to ensure that the pilot may continue to view the display while moving his head away from the design eye position. However, the remaining efficiency still exceeds that of a conventional HUD optical system where most of the light is transmitted straight through the combiner glass and is lost. The diffraction coatings equally act on light from the real world in that they select from it the same narrow band of green wavelengths and reflect it away. The effect of this subtraction of green is to make the real world seem to the pilot to have a slight pinkish tinge. One of the principal reasons that the coating is designed to be so selective is to ensure that this discolouration is kept to a barely noticeable minimum.

The complete HUD is shown in Fig. 6.25, together with its installation in the F-16 LANTIRN crewstation. The HUD provides an instantaneous field of view (IFoV) of 30° azimuth $\times$ 18° elevation and operates in both raster and stroke modes to provide day/night operation. It is fully compatible with night vision goggles. The multibounce, quasi-axial design provided the largest practicable IFoV without impacting on the cockpit installation. It provides transmission and

Fig. 6.25 *F-16 LANTIRN HUD (BAE SYSTEMS)*

reflection characteristics superior to conventional HUDs, and allows straightforward hologram manufacturing techniques to be used. Interestingly enough, in addition to its other virtues and almost as a side effect, this optical configuration rejects the bulk of sunlight reflections which can cause problems with other HUD designs, both refractive and diffractive.

Z-HUD – quasi-axial configuration (Fig. 6.19b)

The F-16 optical arrangement shown above has been proven to work well. However, the observer views the outside world through several optical components with consequent attenuation of scene luminance to the observer. There are two optical elements in the over-the-nose view, plus a further element in the upper view, with a potential discontinuity between these views. This is obviously undesirable, especially for night-time, night-adapted flying. Furthermore, although the reflected optical power is contained within the holographic combiner and is transparent to all but the narrow-band green wavelength of the CRT phosphor, the curved surfaces of the concave substrate inevitably produce some distortion of the outside world.

The ideal arrangement would be a single glass element, the combiner, and for that to be a plane surface, not curved. The Z-HUD optical arrangement, shown in Fig. 6.26, achieves these objectives, but suffers from one serious drawback which prevented it becoming a production product. The pilot views the outside world through a single-plane semi-transmissive element. As before, the CRT image is reproduced through a relay lens as an intermediate real image that is in the desired position, after folding by a plane mirror, close to the focal plane of the collimating mirror. The collimating mirror is not transmissive, but purely reflective.

As with the F-16 LANTIRN HUD, it is theoretically possible for the optical elements to be produced by traditional means. However, the practical realization

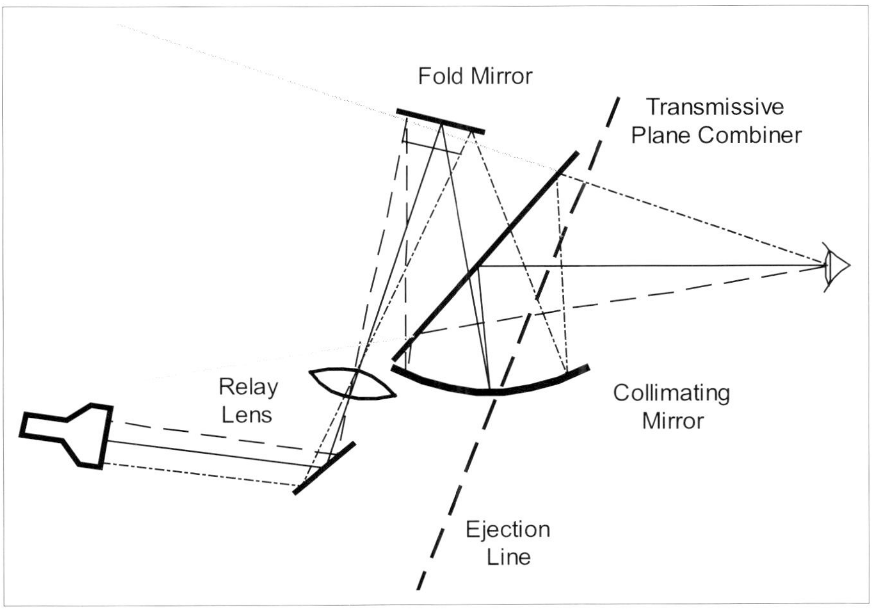

Fig. 6.26 *Z-HUD optical configuration*

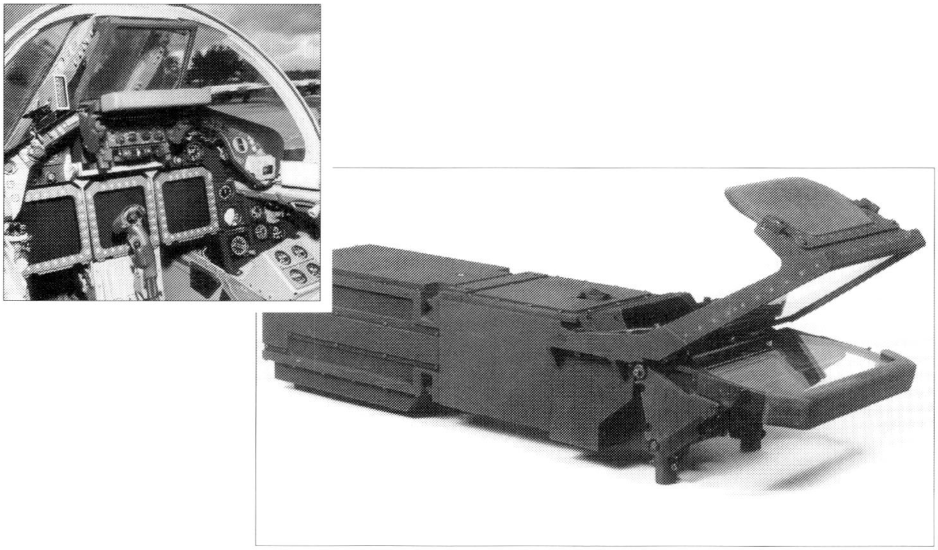

Fig. 6.27 Z-HUD demonstrator (Smiths Aerospace)

uses holographic techniques for the collimating mirror and for the plane combiner. In particular, the performance of the plane combiner is worthy of note. At one and the same time it transmits the outside world scene to the pilot and the intermediate image from the fold mirror to the collimator with low attenuation, but must also reflect the rays of light from the collimator into the pilot's line of sight with high efficiency. This transmit/reflect performance is best obtained by holographic techniques by paying careful attention to the geometry and by tuning the hologram to match the specific narrow-band spectral emission of the CRT phosphor, P-53 in this case.

The optical configuration is known as Z-HUD on account of the folded arrangement of the light path. A technology demonstrator was built (see Fig. 6.27) and has been extensively and successfully flown by the UK Defence Research Establishment DERA in their TIARA Tornado aircraft. The Z-HUD demonstrator provided a 30° azimuth $\times$ 22° elevation IFoV, and operated in stroke, raster, and hybrid modes for day/night operation.

From an optical and electronics perspective the design performed exceptionally well and met all its expectations. However, as can be seen from the optics arrangement, the collimating mirror intrudes into the ejection envelope. Also, the upper fold mirror provides some obscuration to the upper viewing angles.

The fold mirror obscuration can be minimized by arranging it to be parallel to the pilot's line of sight. It is then barely perceptible. To overcome the incursion into the ejection line, the collimating mirror was hinged at the HUD body. A pair of hydraulic pistons (operating in duplex as a fail-safe mechanism) were connected to the ejection seat mechanism. On ejection, the pistons rotate the collimator to its stowed position alongside and parallel to the combiner and outside the ejection clearance envelope. The mechanism was tried and proven to stow the collimator within 2 ms. As a further fail-safe precaution, the mirror could be 'broken out' of the detents holding it in its operation position with a sharp blow, to stow the collimator manually. However good a solution this might have been technically, the psychological barrier was enormous and the design was never pursued to a production product.

Eurofighter Typhoon HUD – single-element off-axis configuration (Fig. 6.19c)

An alternative and much more aesthetically elegant solution is the single-element off-axis configuration shown in Fig. 6.28. This configuration achieves a wide field of view HUD with no incursion into the ejection envelope and with no upper mirror to obstruct the upper field of view. It comprises a single optical element between the pilot and the outside world – the semi-transmissive curved collimating mirror/combiner. As before, the CRT HUD intermediate real image is relayed to be close to the focal point of the collimating concave mirror. The collimated image is combined with the outside world in the same optical element.

The elegant simplicity of this configuration belies its optical complexity. This complexity arises because, by its very nature, the intermediate image subtends a significant 'off-axis' angle to the collimating mirror. This means significant optical correction needs to be applied to correct for image distortion and for distortion of the field.

The collimator can no longer be a simple concave mirror (even if holographic), but must emulate a complex aspheric surface in order to ensure all rays of light from the reflected image emerge in parallel (i.e. are collimated). It is only possible to fabricate this element using holographic techniques in which the hologram itself is not generated by optical means but is computer generated.

The relay lens is complex also. It contains several aspheric elements to provide further compensation for the image distortions produced by the off-axis collimating combiner. Finally, complex geometric distortions have to be applied to the CRT image. These are produced by geometrically distorting the deflection current drive waveforms.

Notwithstanding the above complexities, the clean lines and low forward obscuration make this optical configuration the configuration of choice for high-capability, high-performance, and wide field of view applications. The Eurofighter Typhoon HUD is shown in Fig. 6.29.

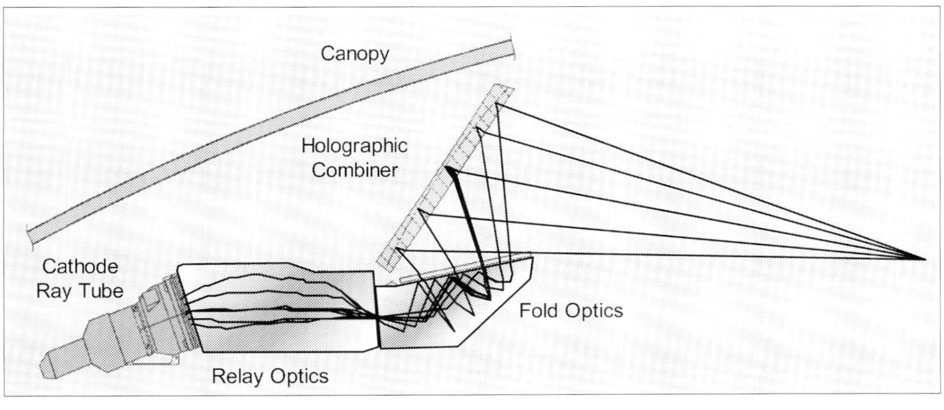

Fig. 6.28 *Eurofighter pupil-forming optical configuration (BAE SYSTEMS)*

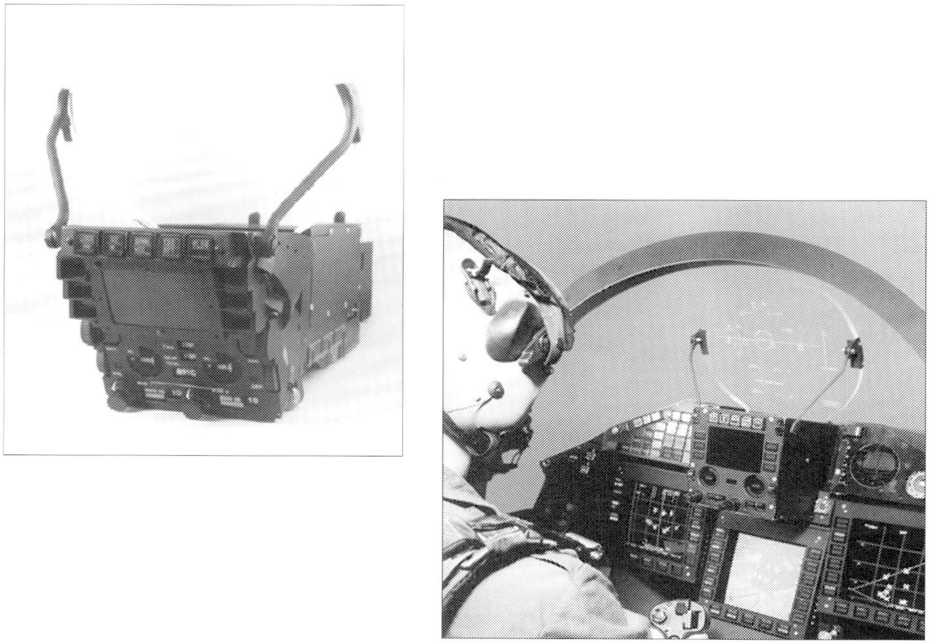

Fig. 6.29 Eurofighter Typhoon HUD (BAE SYSTEMS)

The Eurofighter Typhoon HUD employs advanced computer-generated holographic optics to provide a $30^\circ \times 25^\circ$ elevation Total Field of View (TFoV). The instantaneous field of view is identical to the total field. It provides stroke (cursive), raster, and hybrid modes of operation with outstanding display luminances of 2700 fL (9200 cd/m^2) in stroke (daytime) mode and 1000 fL (3500 cd/m^2) in raster (night-time) mode, this latter being viewable in daytime under cloud and haze. The outside world transmission is 80 per cent.

The HUD also provides a comprehensive up-front control panel with a large-area, daylight-viewable, LED matrix display and programmable keys. The HUD is a high-integrity design and is used as the primary flight display in Eurofighter.

Head-up display functional description

Unlike the Jaguar and Harrier HUD discussed earlier, most modern HUDs contain all the electronic services to support the HUD functions, providing a low-level signal interface to the aircraft computer systems. A typical HUD block diagram is provided in Fig. 6.30. The HUD comprises the following functional elements:

- *Optics assembly*. A set of optical elements comprising the final collimating lens for the collimating HUD or the relay lens for the pupil-forming HUD, image field flattener lenses, and the fold mirror with coatings and filters to reduce sunlight reflections and reduce the risk of CRT burns due to the image of the sun being focused onto the CRT image plane.
- *Combiner*. A pair of optically flat parallel glass plates for the collimating HUD or a curved collimator for the pupil-forming HUD, with semi-reflective coatings tuned to the peak spectral emission of the CRT phosphor.

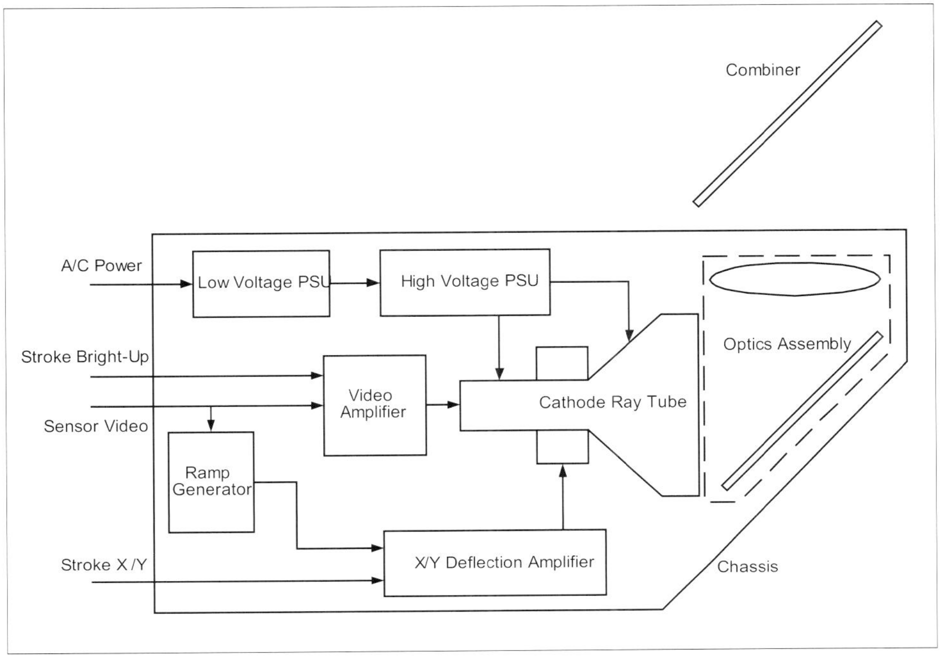

Fig. 6.30 HUD functional block diagram

- *Cathode ray tube*. A high-brightness, high-resolution CRT whose electron beam is usually electrostatically focused and electromagnetically deflected to produce bright, precision, fine-line, stroke-written graphics symbology and raster sensor video imagery.
- *X and Y deflection amplifiers*. Usually dual-mode high-bandwidth, high-precision amplifiers that source current into the CRT *X* and *Y* magnetic deflection yokes to cause the CRT electron beam to trace out the stroke (cursive) symbology and also the raster sensor video scan waveforms.
- *Video amplifier*. Usually a dual-mode amplifier to modulate the CRT beam current by adjusting the CRT cathode bias with respect to the grid electrode. In stroke mode the cathode is driven on or off to brighten symbol strokes and occult inter- and intrasymbol positioning moves. In raster mode the cathode is modulated with the video signal to produce a monochrome sensor image.
- *Ramp generator*. This strips synchronization pulses from the sensor video and generates the raster scan waveforms to align correctly/harmonize the sensor video with the outside world view.
- *High-voltage power supply*. The high-voltage power supply sources the high voltages required by the CRT, the final anode potential (usually around 18 kV), the A1 (focus) potential (usually around 2–4 kV) and the grid potential (usually around 200 V).
- *Low-voltage power supply*. Sometimes combined with the high-voltage power supply, but more usually separated, it conditions aircraft power to provide all the low-voltage rails for the electronic circuits, typically +5 V, ±15 V, ±20 V, etc.
- *Chassis*. Usually a complex precision casting which provides environmental protection to the HUD functional elements and affords a means of fixing the HUD to the aircraft structure.

Fig. 6.31 *Typical HUD subassemblies (BAE SYSTEMS)*

The physical realization of these components into a fully operational HUD tends to be aircraft specific to match the operational capability requirements within the available cockpit envelope. Space in the cockpit is at a premium, no space more so than directly ahead of the pilot in his main line of vision. The HUD must be designed to fit into the smallest space possible, compatible with the space constraints and operational capability of the other crewstation instruments. Therefore, in general the HUD electronic assemblies are tailored to the required capability and packaged so as to make the best possible use of the available space envelope without compromising performance. The electronics assemblies are usually more malleable and are in general designed as complex shapes in order to fit around the major immutable objects such as the CRT, the high-voltage power supply, and the optics. The subassemblies of a typical HUD are shown in Fig. 6.31.

Image generation

As previously discussed, the HUD image is written onto the faceplate of a cathode ray tube (CRT). As with a CRT head-down display, for maximum efficiency, and therefore to achieve maximum display brightness and contrast, the HUD day image (symbology) is written in stroke mode. The information content is therefore in the X and Y beam deflection signals. The X and Y deflection amplifiers require high-bandwidth, high-precision amplifiers accurately and precisely to construct high-quality graphics. The line width of the graphic elements is the CRT electron beam width. For best accuracy and finest spot size, the CRT is usually electromagnetically deflected and electrostatically focused. To avoid flicker, the entire image is redrawn or refreshed at a frequency greater than 50 Hz (20 ms). The absolute brightness range is typically four orders of magnitude (10 000:1) to span the ambient illumination range from bright sunlight to night-time conditions.

Day/night head-up displays also provide the capability to present raster video images in conjunction with Forward-Looking InfraRed (FLIR) sensors for night-

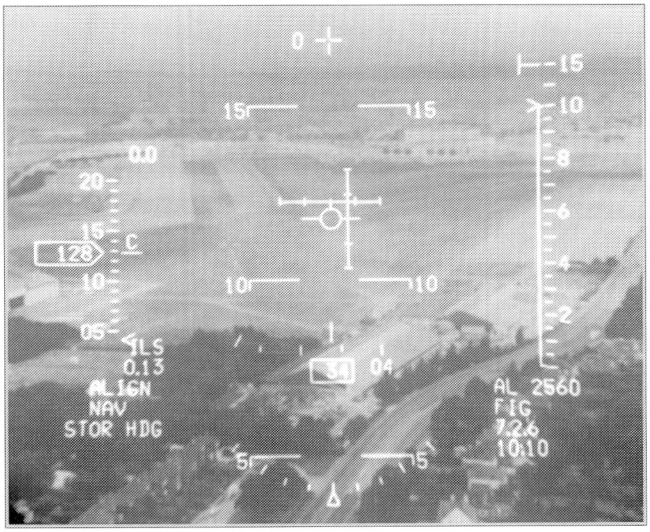

Fig. 6.32 HUD FLIR image plus stroke symbology overlay (BAE SYSTEMS)

time flying. The HUD image is required to be accurately registered with the outside world. It is necessary, therefore, to be able to adjust the size, the position, and sometimes the orientation of the raster scan waveforms precisely to match the field of view of the sensor. The image content is now contained in the video signal, and this video signal modulates the intensity of the CRT electron beam like any conventional video image.

A day/night HUD also provides hybrid operation in which stroke-written symbology overlays the raster sensor image. This interleaved dual raster/stroke procedure is described more fully in Chapter 5. A typical HUD FLIR hybrid (raster video plus stroke overlay) is shown in Fig. 6.32.

Other topics

Boresighting

In order to maintain accurate registration of the HUD image with the outside world, it is necessary to locate the HUD in the aircraft with high precision. In early HUDs this was done by electrically adjusting the position of the HUD image on the CRT with respect to the optical axis and by electrically adjusting the gain (size) of the HUD image also. A telescope was temporarily attached to reference points of the aircraft, through which the HUD image could be viewed, and then by using a special test pattern the aforesaid features were adjusted to align with sighting features on the telescope reticle. This process is known as boresighting.

In modern HUDs with improved optical assemblies and drift-free electronics, it is possible accurately to set up the HUD in the servicing bay to datum points on the HUD chassis. The HUD is then installed into the aircraft onto a mounting tray which accurately and repeatably aligns the HUD datum points to the aircraft. It is necessary to align or boresight the mounting tray to the aircraft only once.

Bird strike

The cockpit of a military fast jet aircraft must be designed to withstand a bird strike (see Fig. 6.33). In cockpits designed up to the 1980s, the cockpit canopy was made of several panes of glass. The front glass was flat, toughened, and rigidly held. There was little deflection on bird strike to interfere with the HUD. The bubble canopy fitted to modern cockpits such as F-16 and Eurofighter Typhoon has no restraining mechanisms. If a bird strikes the canopy, it is designed to flex. The deflections can be several inches, and will move out from the strike point like a series of ripples on a pond.

Adequate clearance must be allowed between the combiner and the canopy to ensure there is little risk of interference and hence of damage either to the canopy or to the HUD combiner/collimator. In the extreme case it may also be necessary to demonstrate that, if in a rare event the canopy were to hit the collimator/combiner, then no broken pieces of the canopy or of the collimator/combiner could become sharp shards which might potentially injure the pilot. Nor might there be any impediment to his safe ejection from the aircraft.

Fig. 6.33 *Bird strike (BAE SYSTEMS)*

Sunlight reflections

Great care has to be taken to suppress unwanted sunlight from entering the HUD optics system, being reflected internally from the various optical surfaces, and then re-emerging, possibly collimated, into the pilot's field of view. This phenomenon has often been described as 'golden balls' which graphically describes the pilot's perception of the resultant images.

The problem can be extremely hard to eliminate. Potential sunlight reflection mechanisms must be thoroughly researched and solutions diligently sought. Usual techniques to minimize the effect is to ensure that all non-functional surfaces are non-reflecting (i.e. black) and install filters that absorb or reflect sunlight but pass the narrow-band spectrum of the HUD CRT phosphor.

A special case worthy of serious attention concerns the situation when the sun is directly overhead and can enter the HUD exit lens on-axis. In this case the sun's orb would be focused onto the CRT faceplate and burn the CRT phosphor in a manner similar to burning paper with a magnifying glass. To reduce this effect, the fold mirror is often made reflective to the HUD phosphor wavelength but transmissive to all other wavelengths especially infrared (IR). Most of the sun's energy is therefore absorbed in the fold mirror mounting structure and not reflected towards the CRT. The fold mirror is a 'cold mirror'.

Up-front control panel

By reason of its existence, the Head-Up Display is located in prime real estate in the cockpit. Every visible surface must be gainfully employed. Therefore, it is usual to find additional controls and indicators attached to the HUD on the surface containing the fold mirror.

These additional controls and displays are usually referred to as the Up-Front Control Panel (UFCP). Typically, the UFCP is an array of switches, keys, and alphanumeric readouts that allow the pilot to:

- Select HUD and major weapon system modes.
- Insert and inspect waypoint and route data.
- Select and control radio and navigation aids.
- Set flying instrument datums (e.g. baroset).

A typical up-front control panel is shown in Fig. 6.34.

The Eurofighter Typhoon UFCP probably represents the most comprehensive capability to date and comprises a set of multifunction keys, each of which has a set of programmable alphanumeric legends and a daylight-readable LED array.

It has been suggested that a range of touch-screen technologies might be appropriate to increase the flexibility of the UFCP. To date, all attempts have been abortive because of the difficulty of being able to meet the conflicting requirements of being able to provide physical support to locate the function to be controlled, provide tactile feedback, and avoid distracting sunlight reflections from smears and scratches.

Fig. 6.34 Typical up-front control panel (Smiths Aerospace)

HUD symbology and principles of use

HUD symbology has evolved significantly from its humble beginnings as an electronic replacement for the gyro gunsight to now including comprehensive flight and navigation information in addition to threat warning and targeting information for a wide range of dumb and smart air-to-air and air-to-surface weapons. In most modern fast jet fighter aircraft the HUD is now the primary flight instrument. In this section we shall discuss some typical HUD symbology in the context of its operational use.

Primary flight symbology

Primary flight data displayed on the HUD usually encompass:

- Flight path vector/marker (also known as velocity vector).
- Attitude (pitch and roll).
- Speed.
- Altitude.
- Heading.
- Vertical speed (rate of climb/dive).
- Angle of attack.

A typical example of HUD primary flight symbology is shown in Fig. 6.35.

The flight path vector is presented in 'contact analogue' or conformal form by the icon known as the flight path marker. This symbol indicates the instantaneous velocity vector, or direction, in which the aircraft is flying (and will continue to fly if the current manoeuvre is maintained) relative to the direction in which the aircraft is pointing, the longitudinal fuselage datum (see Fig. 6.36). This symbol is of vital significance when flying low and fast over non-flat terrain (to avoid radar detection for instance), since it will indicate to the pilot the safe clearance of his current flight path over that terrain. The flight path vector, as indicated by the aircraft symbol, is also most useful for precision landings since it will indicate the current touchdown point.

Attitude (pitch and roll) is also indicated in 'contact analogue' form as a horizon line which will be in register with the real horizon, if visible. Additional lines (the pitch ladder) parallel to the horizon line indicate positive and negative pitch angles, usually at 5° intervals, possibly with a 3° line added during the landing phase to facilitate precision approaches down a 3° glidepath. The pitch ladder is usually centred on the aircraft symbol.

Speed and height are presented as numerical values and a linear tape scale. Sometimes the speed and height presentations are augmented with a circular scale and pointer, one revolution being 1000 ft, emulating the electromechanical altimeter and providing the pilot with a rate-of-change cue. Alternatively (and sometimes additionally), the height scale may be augmented with a vertical speed scale.

Heading is usually presented as a numerical readout on a horizontal scale, emulating a compass card viewed edge-on. Other numeric readouts may also be included if they are generally helpful to the pilot to enable him to prosecute his mission effectively. These include instantaneous *g*, barometric set, waypoint lat/long, time/distance to go to next waypoint, etc.

Fig 2.21 BAC 1-11 advanced civil flight deck (April 1981) (The Royal Aircraft Establishment)

Fig 2.22 BAC 1-11 primary flight display (Smiths Aerospace)

Fig 2.26 Airbus A330 flight deck (March 1988) (Airbus)

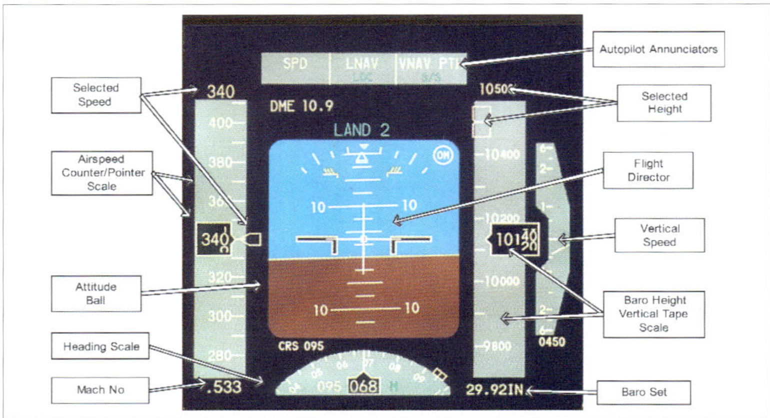

Fig 2.27 *Airbus A320 primary flight display (Airbus)*

Fig 2.28 *Boeing 747-400 flight deck (June 1990) (The Boeing Company)*

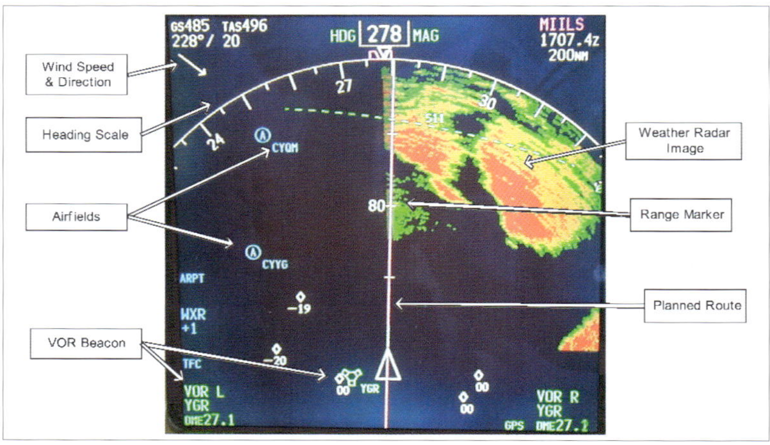

Fig 2.29 *Boeing 777 navigation display with weather radar (The Boeing Company)*

Fig 2.32 KLM 747 – classic upgrade EHSI and engine formats (Smiths Aerospace)

Fig 2.33 KLM 747-classic upgrade – combined display formats (Smiths Aerospace)

Fig 3.11 *McDonnell Douglas F/A-18C cockpit*

(a) Fig 6.25 *F-16 LANTIRN HUD (BAE SYSTEMS)*

(b) Fig 6.29 *Eurofighter Typhoon HUD (BAE SYSTEMS)*

Fig 10.31 *JSF Cockpit (The Boeing Company and Kaiser Electronics)*

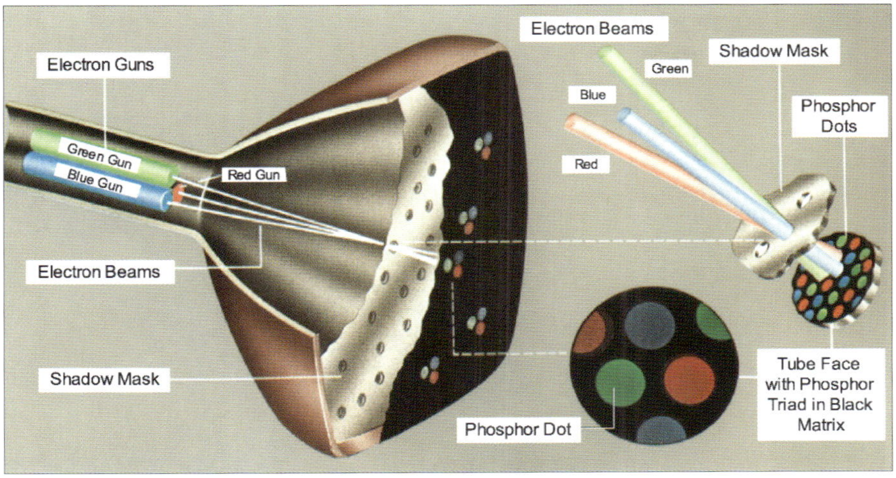

Fig 5.2 The shadow mask CRT

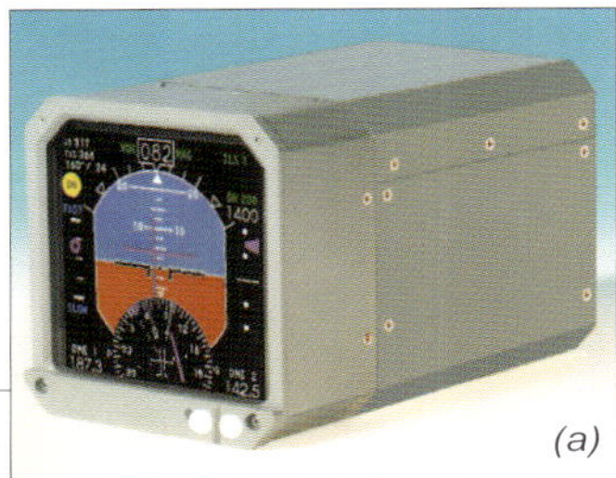

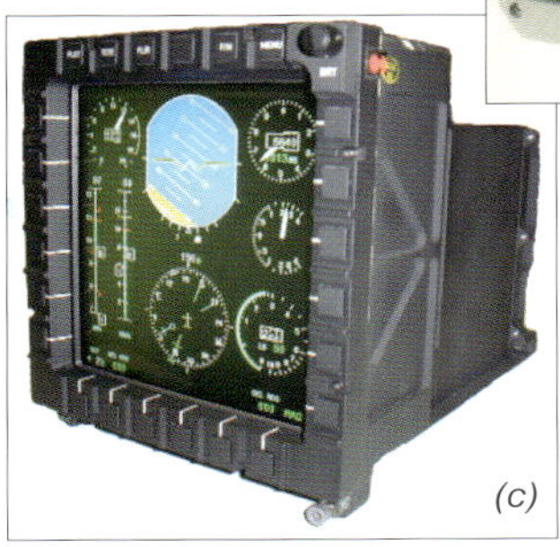

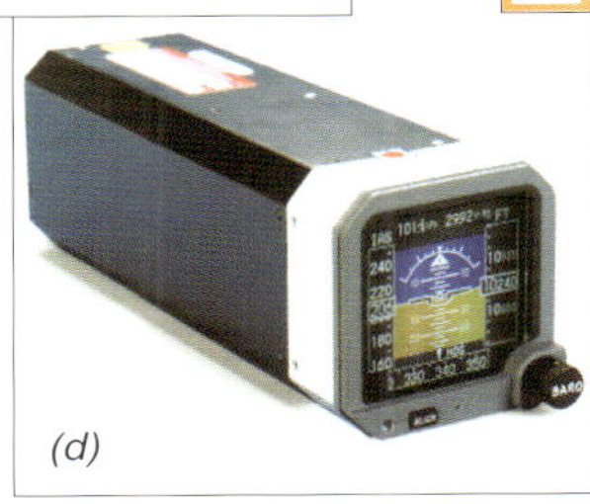

(a) Fig 5.23 *5ATI display unit (Smiths Aerospace)*

(b) Fig 5.7 *F/A-18 and AV-8B multipurpose colour display (MPCD) (Smiths Aerospace)*

(c) Fig 5.22 *Integrated Display Unit (Smiths Aerospace)*

(d) Fig 2.34 *3ATI ISIS with display formats (Smiths Aerospace)*

Fig 3.19 *Super Lynx cockpit (Smiths Aerospace)*

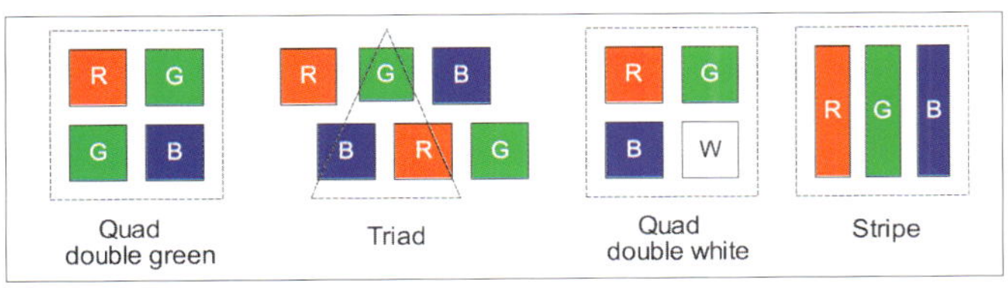

Fig 5.10 *Colour group pixel structures*

(a) Fig 6.14 *AV8B/Harrier GR7 night attack HUD (Smiths Aerospace)*

(b) Fig 6.29 *Eurofighter Typhoon HUD (BAE SYSTEMS)*

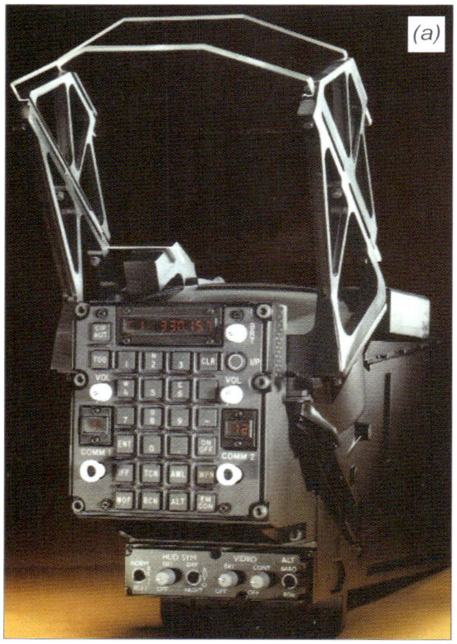

(a) Fig 8.13 *Comanche Helmet Integrated Display Sighting System (HIDSS) (Rockwell-Collins)*

(b) Fig 8.17 *Joint Helmet-Mounted Cueing Systems (JHMCS) (Rockwell-Collins)*

(a) Fig 8.19 *Eurofighter Typhoon Helmet-Mounted Display (BAE SYSTEMS)*

(b) Fig 9.25 *JSF Helmet-Mounted Display (Rockwell-Collins)*

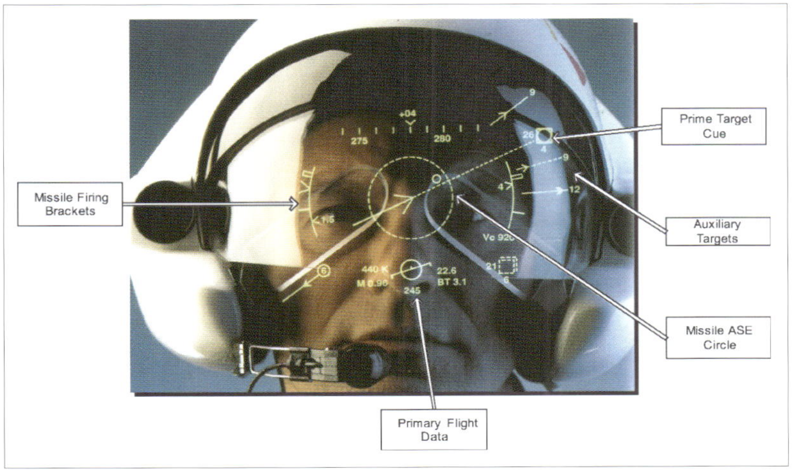

Fig 8.28 *Typical HMD symbology format (Thales)*

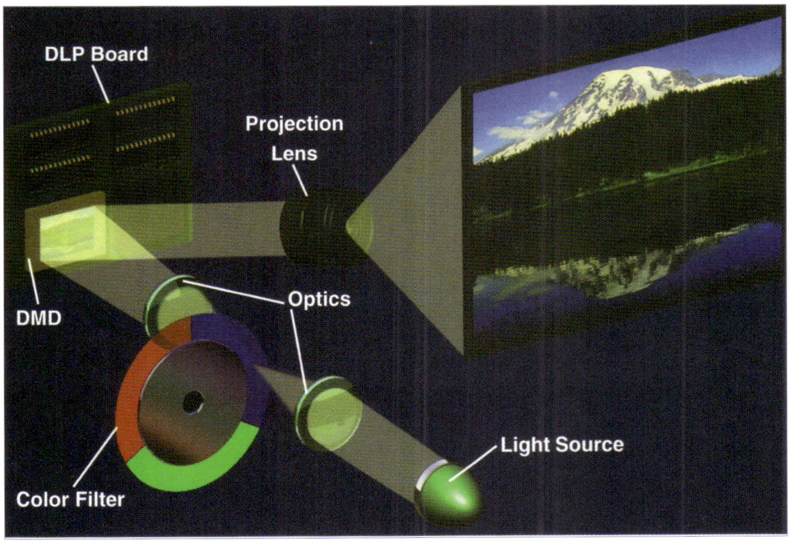

(left) Fig 9.9 *DMD projector (Texas Instruments)*

(above) Fig 9.14 *OLED microdisplay (Cambridge Display Technology)*

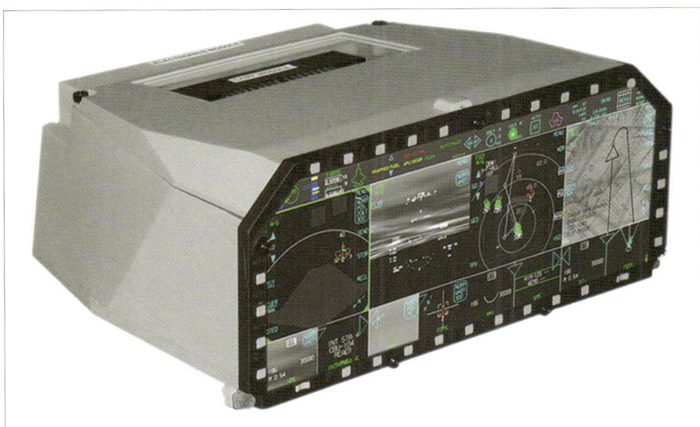

Fig 9.18 *Multiple projector 'big picture' HDD (Kaiser Electronics)*

Fig 10.13 *VSD location on the flight deck (The Boeing Company*)*

(a) Fig 10.14 *Example of a side-looking VSD format (The Boeing Company*)*

(b) Fig 10.15 *Aircraft heading into terrrain (TAWS caution) (The Boeing Company*)*

(c) Fig 10.16 *Final approach (The Boeing Company*)*

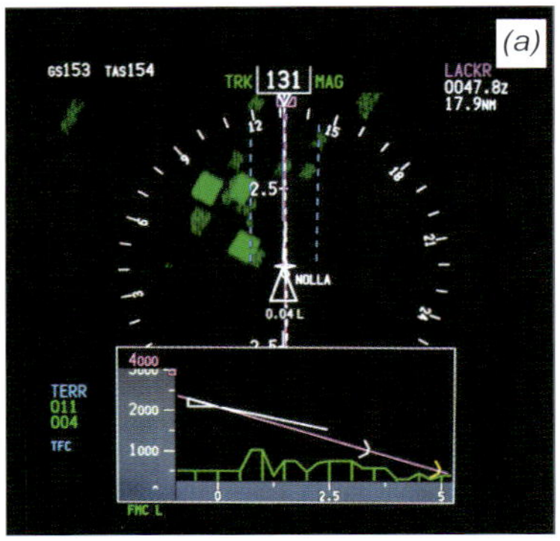

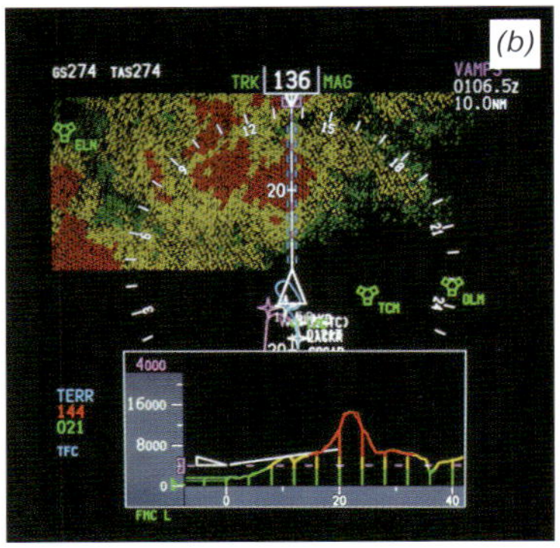

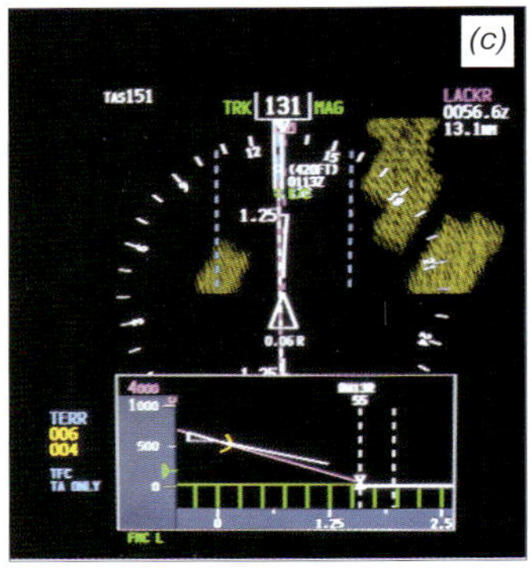

Fig 10.17 *4D primary flight display (VDO-Luftfahrtgerate)*
Fig 10.20 *Perspective PFD – approach (VDO-Luftfahrtgerate)*

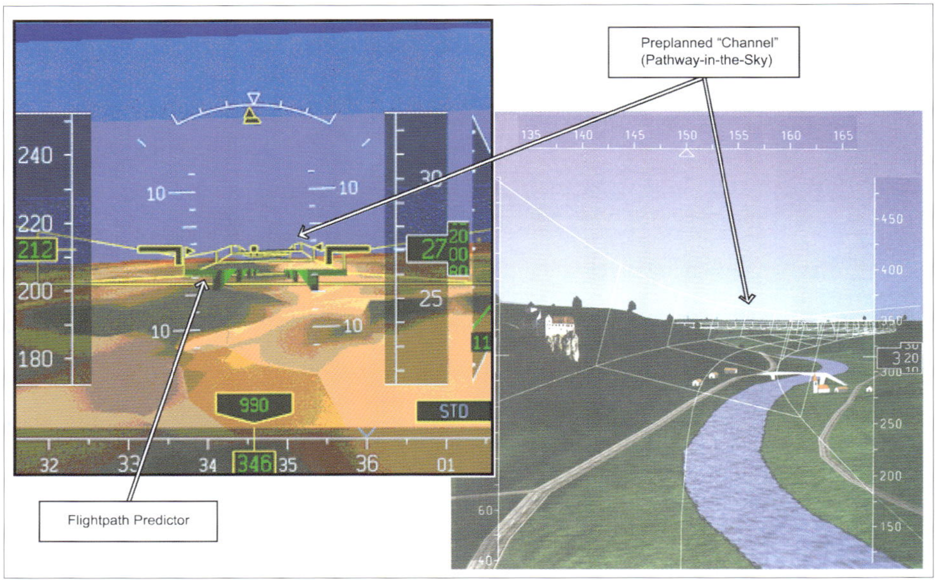

** Reprinted from AERO Magazine by permission of The Boeing Company*

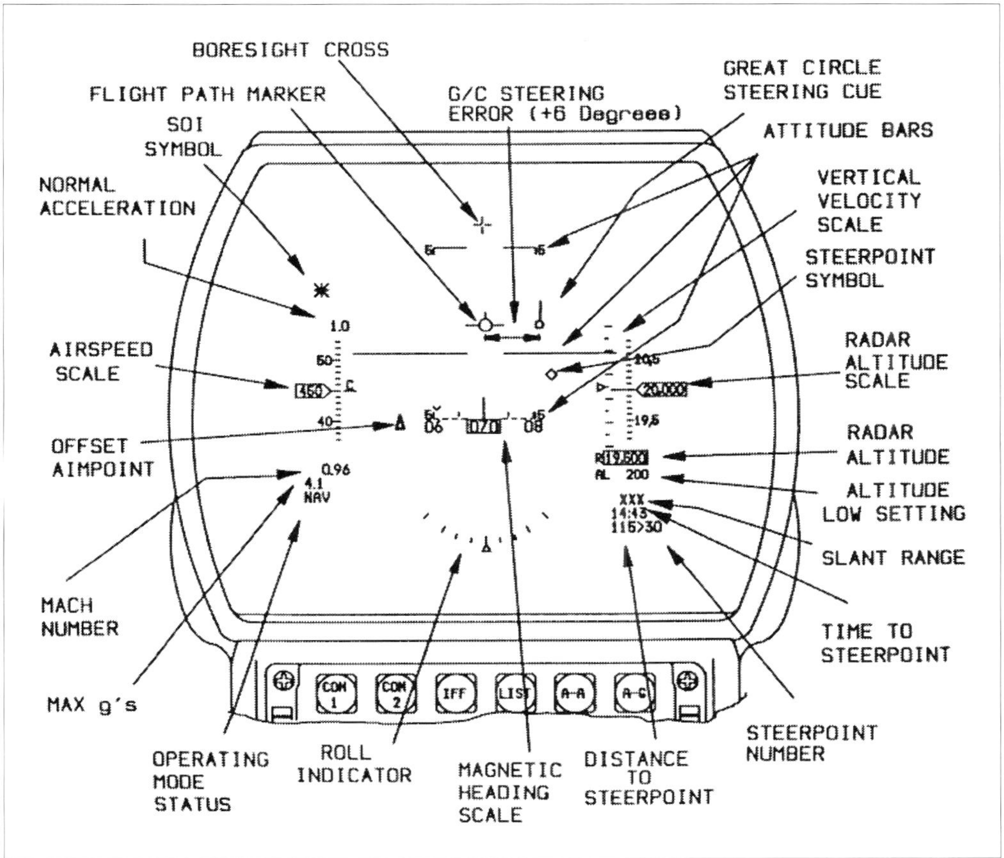

Fig. 6.35 *HUD primary flight symbology (BAE SYSTEMS)*

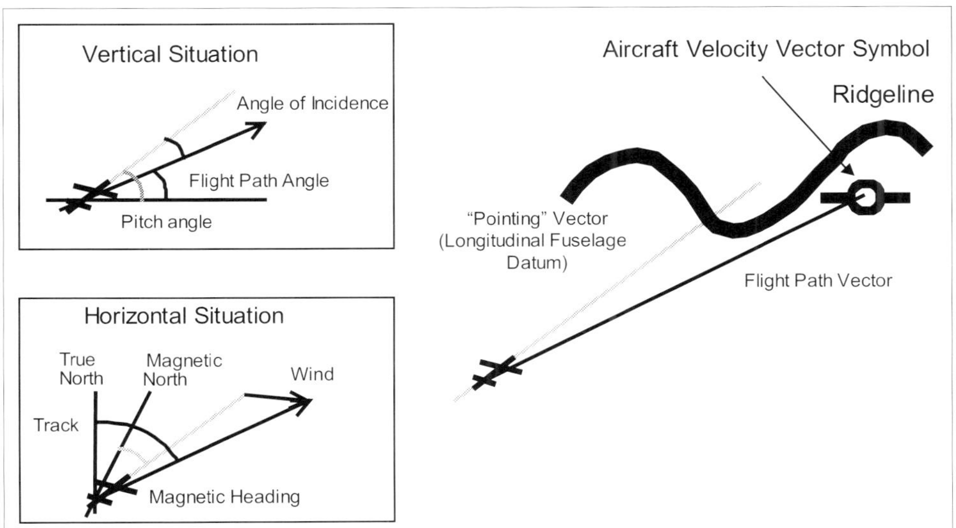

Fig. 6.36 *Flight path vector*

Navigation symbology

Additional symbology is added to the primary flight information to provide en-route navigation to/from the mission target area. This information typically includes:

- Lateral and vertical guidance to maintain preselected flight plan.
- Next waypoint position (latitude and longitude).
- Bearing and range to next waypoint.
- Time to go to next waypoint.

On approaching the next waypoint, a conformal fix cross appears to indicate the computed sight line to the waypoint on the basis of the best estimate of the aircraft's present position. The fix cross should precisely overlay the visually sighted waypoint. Any error represents an error in estimation of present position and can be corrected manually by the pilot.

A typical route to the target area may consist of several legs, comprising a list of waypoints established prior to the mission and entered into the aircraft's navigational system through a mission data loader. To minimize system errors, if at all possible it is highly desirable for the final waypoint to be in close proximity to the target and easily identifiable in order to establish a precise local fix point.

Air-to-surface bombs

On entering the target area, the pilot selects the weapon to be deployed and the HUD symbology changes accordingly. Figure 6.37 shows typical symbology for the release of dumb cast iron, free-fall bombs onto a preplanned target.

The following description is illustrative only. A target designator box replaces the fix cross to indicate the sight line to the target, based on best current estimate of present position. As before, the target designator should overlay the visually sighted target. The pilot should manually correct any errors.

A bombfall line indicates the locus of impact points on the ground of all future bomb releases taking due account of wind. The pilot steers the aircraft so that this line passes through the target.

The solution cue indicates the Continuously Computed Release Point (CCRP) for the selected bomb type to hit the target. This point is computed from a knowledge of the aircraft manoeuvre, height above target, bomb ballistics, and ejection velocities, and an estimation of wind over the target. Height above target can be estimated in a variety of ways. The simplest is based on barometric data and makes the assumption that the ground is flat. A complex method, but with significantly improved accuracy, measures the slant range to the target with a laser.

A safe pass height marker is often associated with the bombfall line to provide an indication of the fragmentation zone of the bomb. The bomb should not be released if the CCRP marker is close to or infringes the safe pass height marker. For a successful release the pilot should:

- Fly the aircraft to position the bombfall line over the target ahead of the CCRP marker.

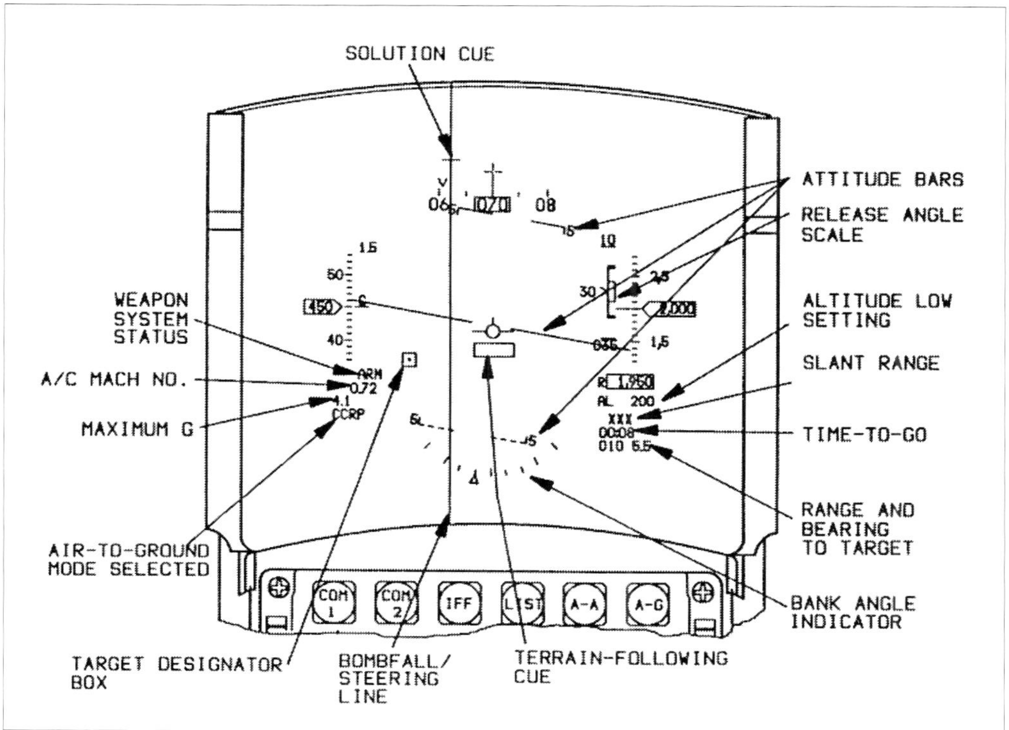

Fig. 6.37 *Typical HUD air–surface bombs symbology (BAE SYSTEMS)*

- Continue to fly towards the target, tracking the target with the bombfall line. The target (and target marker) will move down the bombfall line as the distance to the target decreases.
- When the CCRP marker and the target marker coincide, the bomb should be released, either manually or automatically.

Specific adjustments are made to the air-to-surface symbology for different weapons classes, different manoeuvres, and of course different ballistics.

Air–Air

Should the pilot elect to fire his guns or missiles, then the appropriate aiming symbology appears to enable him accurately to aim and fire his weapons. Typical symbology is shown in Fig. 6.38. A track line indicates the computed trajectory of the gun or missile on the basis of current aircraft manoeuvre and weapon ballistics. A ranging circle is positioned on the line at the range of the target aircraft. A target designator box indicates the radar sight line to the target. Other information derived by the air–air radar about the target is also displayed. This might include, for instance, target motion relative to own-aircraft motion (closure rate), possibly augmented with target slant range and *g*.

Infrared heat-seeking missiles are in general able to be commanded to seek targets at angles significantly 'off-boresight', that is, away from the aircraft longitudinal fuselage datum. As with the radar, a missile target marker indicates the missile sight line. Typically, the missile will be slaved to the radar, that is, be commanded to point in the same direction as the radar. Once the missile detects an

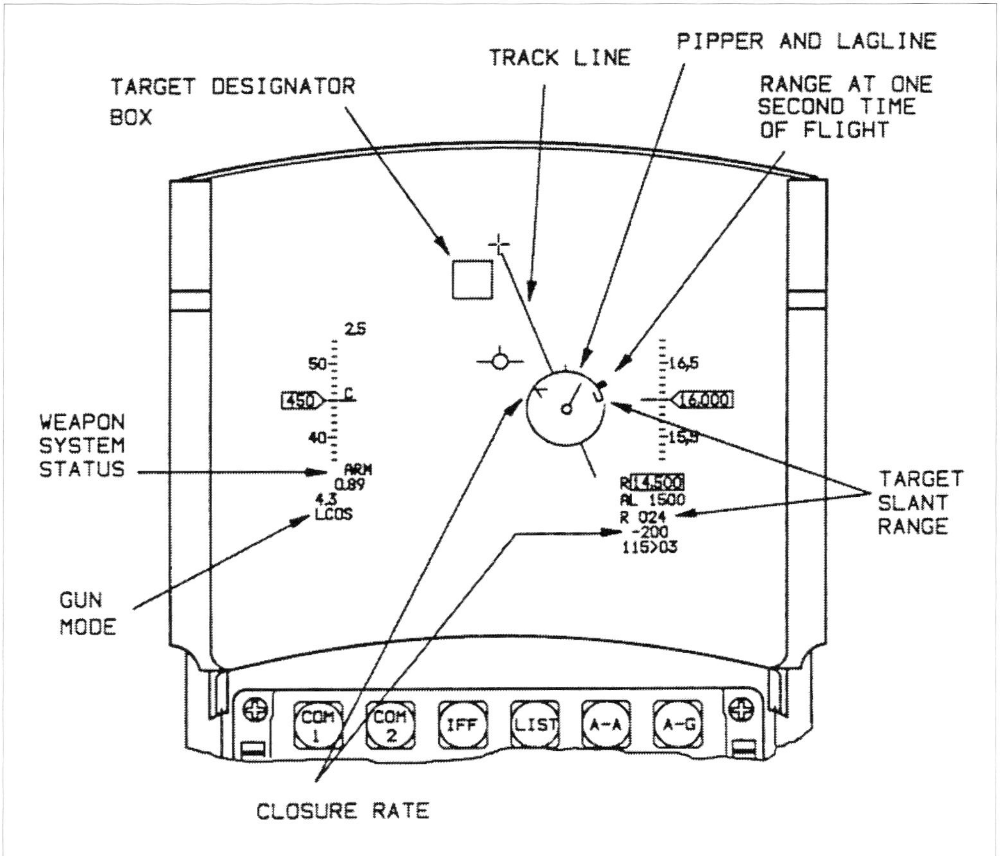

Fig. 6.38 *Typical HUD air–air symbology (BAE SYSTEMS)*

IR source, it will lock to that source and track it independently of the radar. Using the HUD, the pilot can confirm that the missile is locked to the correct source before committing to launch.

A firing solution exists when the radar target designator and the ranging circle are coincident and stable. Usually, the ranging circle will have a number of other bugs indicating the minimum and maximum firing range, together with the probability of success based on the manoeuvres executed by both aircraft and on the manoeuvre required by the missile for a successful intercept post-launch.

References

(1) **Quaranta, P.**, The Evolution of Visionics Systems, Military Technology MILTECH 7/2002.

(2) **Garland, D. J., Wise, J. A.**, and **Hopkin, V. D.** *Handbook of Aviation Human Factors.*

(3) **Fisher, J. F.** *The Operation Of Wide Angle Raster HUDs*, Marconi Avionics, Rochester, Kent, UK.

(4) **Hussey, D. W.** (1981) Wide Angle Raster HUD Design And Application to Future Single Seat Fighters, AGARD 42nd Symposium, Andreas, Greece, October.

CHAPTER 7

Civil Head-Up Displays

Introduction

Chapter 6 described the application of the Head-Up Display (HUD) to fast jet fighter aircraft. This was their first and is still their primary application, though that position is being challenged by helmet-mounted displays. This chapter discusses the application of the HUD to civil and military transport aircraft.

The principle attribute of the HUD is to be able to provide the pilot with flight- or mission-critical information in a form that is conformal, that is, in 'contact analogue' with his real-world view, without the need for him to take his gaze away from the real world and refocus his eyes onto his head-down instruments inside the flight deck. For the transport aircraft pilot, these attributes are likely to be of most benefit during bad weather take-off and landings and for other aspects of the mission that require precision flying such as cargo drops into remote areas or formation flying and refuelling.

In the future it is probable that additional sensors will be installed in the aircraft to augment the visual acquisition of terrain in poor visibility. These systems are already in military use for the obvious reasons. Systems such as Forward-Looking Infrared (FLIR) and Millimetric Wave Imaging Radar (MWIR) can 'see' through fog and rain and present a picture of the outside world on the HUD. The pilot can fly the aircraft 'visually' with the aid of these sensors, acquiring the real world when finally he pierces the gloom. Experimental approaches have been made using sensors in this way, with much success.

Commercial aircraft HUDs came into prominence about 20 years ago as a way of improving the poor visibility capability of aircraft by offering Cat IIIa operations where it was not considered economically viable to develop and retrofit a Cat IIIa autoland to the aircraft. Pilots perceive flight deck instrument data projected into their field of view as they look ahead through the windshield for the runway. This is a great benefit when approaching to land in poor visibility, because the pilot no longer loses critical seconds between looking down at his instruments and then looking up through the windshield for the runway. In addition, with the projected data focused at infinity on the HUD, it is no longer necessary for the pilot constantly to refocus his vision.

HUDs are currently available as optional equipment on several civil aircraft models but are standard equipment on only a few, including the Boeing Business Jet and several Gulfstream models. That is changing. More than 30 airlines worldwide have equipped at least portions of their fleets with HUDs. Some business airplane manufacturers offer HUDs as standard equipment; Airbus soon will follow suit in its commercial airplanes. Boeing, which has a HUD as standard equipment in the Boeing Business Jet, is considering doing the same in its commercial airplanes. Manufacturers and operators are embracing or keeping a close eye on emerging technologies, including Enhanced Vision Systems (EVSs) and Synthetic Vision Systems (SVSs), which increase the capability of HUDs.

In 1990, the Flight Safety Foundation (FSF) completed a contracted study to determine the potential of HUDs to improve civil jet transport safety. Analysis of more than 1000 accidents that occurred between 1959 and 1989 indicated that a properly functioning HUD operated by a correctly trained flight crew might have prevented or positively influenced the outcome of 31 per cent of the accidents. In 1998, the FSF Approach and Landing Accident Reduction (ALAR) task force identified unstable approaches as a major factor in approach and landing accidents. FSF studies have pointed to HUDs as a potent and available tool for safety enhancement.

The International Federation of Airline Pilots' Associations (IFALPA) has recommended that all jet transport airplanes be equipped with HUDs. HUDs should be installed on both the captain's and the first officer's sides of the cockpit in order to provide both pilots with the best possible instrumentation. Dual HUD installation is also desired for monitoring purposes and redundancy.

Currently, all HUD installations in civil aircraft are single installations; that is, the installations comprise one set of equipment with the combiner on the captain's side of the flight deck. Some military aircraft have twin installations, with combiners on both sides of the flight deck but with a single computer. Only two military jet transports – the Boeing C-17A and the Lockheed Martin C-130J – have dual installations, which comprise two complete and independent sets of equipment. The dual installation in the C-130J and the C-17A also serve as the primary flight display.

The head-up visual guidance system

Head-up display is a common term. Civil HUD manufacturers use other terms. BAE SYSTEMS and Honeywell are partners in Visual Guidance Systems (VGSs). Rockwell Collins Flight Dynamics markets Head-Up Guidance Systems (HGSs). Thales Avionics markets Head-Up Flight Display Systems (HFDSs). This chapter will use HVGS as the generic title.

The Head-Up Visual Guidance System (HVGS) adds additional situational awareness and thus safety to the landing approach in the following areas:

- Conformal symbols such as glideslope reference line and flight path vector provide the pilot with a clear touchdown aiming point on runways that lack a precision approach system and, sometimes, adequate approach lighting.

- Instantaneous acceleration/deceleration cues (derived from the aircraft's inertial reference system and not normally presented on the conventional flight instruments) assist the pilot in total energy management during the approach.
- Deviations above or below, and to the left or right of, the aircraft's ideal runway approach path alert the pilot fractionally sooner than the conventional instruments, allowing earlier, smaller flight path corrections to be made while flying ILS, MLS, or GPS-guided precision approaches.
- The HVGS can assist in avoiding accidents and incidents, and wheel and tyre damage and wear, because the system allows a pilot to fly more accurately and improves situation awareness.
- The HVGS can also be used to provide very precise runway centre-line tracking guidance during take-offs in low visibility, with resultant reductions in take-off minima.
- The HVGS can also provide optimum guidance in recovery from severe unusual attitudes following jet upsets.

The sum of these benefits results in a safer, more stabilized approach and a smoother touchdown (**1**).

Principles of operation

A typical civil HVGS system, shown in Fig. 7.1, comprises similar elements to those described earlier for any display suite architecture, namely a display guidance computer (data collection, display management, and graphics generation) and an overhead projector (display device, drive electronics, and projection optics), plus the combiner. The fundamental element of the HVGS is a conformal head-up display that presents flight information and guidance in the pilot's forward field of view.

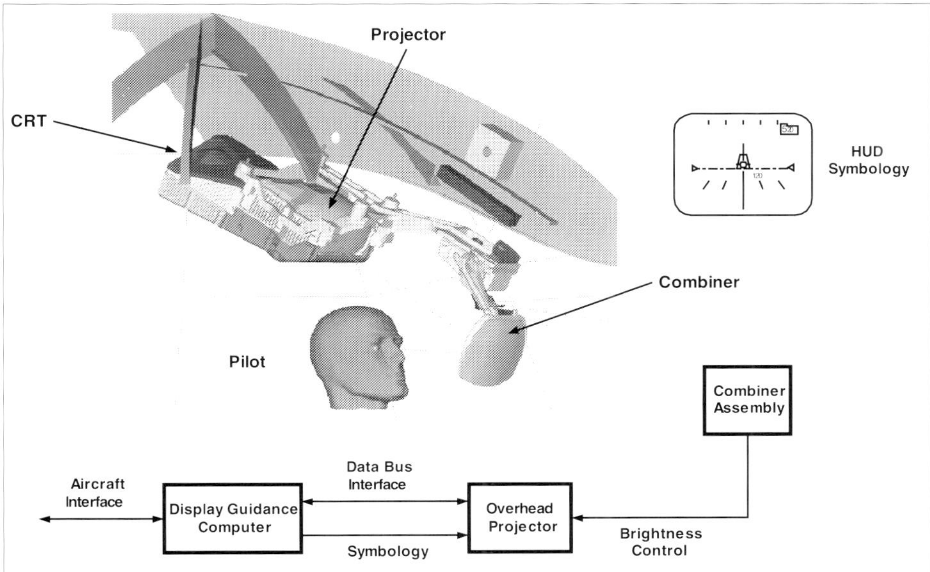

Fig. 7.1 *Typical civil HVGS architecture*

A typical HVGS installation comprises the following Line-Replaceable Units (LRUs):

- *Display guidance computer*, which receives data from aircraft sensors and generates display symbology.
- *Overhead projector*, which includes a cathode ray tube that projects the image onto the combiner.
- *Combiner*, an optical element (glass plate or plastic plate), mounted behind the windshield, which reflects the projected image towards the pilot's eyes, while allowing ambient light to pass through (see Fig. 7.2).
- *Control panel*, which is used by the flight crew to select HVGS modes and features, and to enter data that the computer does not receive from aircraft sensors.
- *Annunciator panel*, which provides HVGS status information and warning information.

The computer in the avionics bay interfaces to the aircraft's sensors and generates the symbology requirements. This is passed to a projector unit mounted above and behind the pilot's head, which generates the symbology and projects it onto the curved combiner positioned in the pilot's forward field of vision. The symbology is collimated, that is to say, effectively focused at infinity such that it appears to the pilot to be 'drawn out on the clouds'. The symbology is also conformal, i.e. the symbols overlay the real-world objects that they are intended to represent.

The overhead projector and combiner are mounted on the cabin roof as shown in Fig. 7.2. The combiner can be stowed (folded) when not required.

The optical arrangement is usually a quasi-axial pupil-forming reflective arrangement. The combiner is a powered reflector, and usually fabricated using holographic techniques or rugate coatings for optimum efficiency (refer to Chapter 6). Figure 7.3 shows the BAE SYSTEMS HUD 2020™, with its optical ray-trace diagram. The HUD 2020 provides a 30° azimuth × 25° elevation field of view (FoV) and operates in stroke (cursive), raster, and hybrid modes to provide a day/night capability.

Fig. 7.2 *Typical HVGS installation (BAE SYSTEMS)*

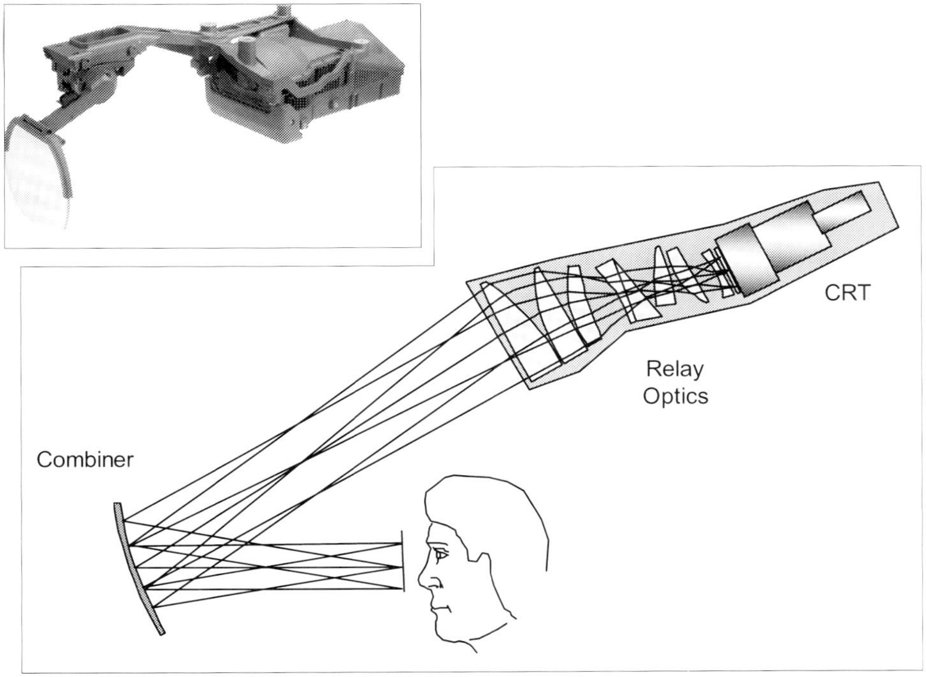

Fig. 7.3 *HUD 2020 optical configuration (BAE SYSTEMS)*

Symbology

The symbology varies depending on the mode and phase of flight. The principal benefits of the system are experienced during take-off, where guidance is provided to allow operation in reduced visual range conditions and during landing where the equipment permits approaches to be hand flown in category III conditions, in some cases at category I equipped airports.

HVGS symbology, shown in Fig. 7.4, provides the pilot with all the flight performance data found Head Down: airspeed, altitude, attitude, heading, etc. Attitude, heading, and directional guidance are conformal, other information is presented in a similar format to that found on the PFD to aid assimilation and transition between the two instruments.

For the civil airline pilot the most important feature of the HVGS is its ability to display the flight path vector superimposed on the outside world. The flight path vector provides an instantaneous and intuitive indication of the direction of flight. Using the flight path vector in conjunction with other cues on the HVGS, the pilot is able precisely to control the aircraft manually, and routinely to maintain desired airspeed to within 2 knots, heading to within 2°, altitude to within 50 ft, and pitch to within 0.5°. This degree of control is difficult to achieve using head-down instruments (**1**).

During approach and landing, the pilot can simply fly the desired glidepath angle and place the flight path vector on the intended touchdown point. The flight path vector in the HVGS enables the pilot to 'spot' land the aircraft and provides important cues if drift or sideslip develop prior to touchdown. In low-visibility approaches, the pilot can fly head-up using visual ILS direction cues. His attention is directed at precisely the right point in space visually to acquire the runway at

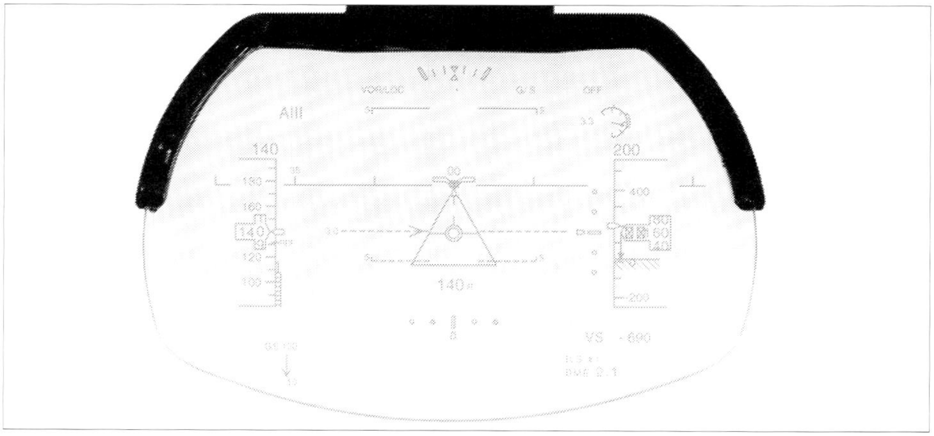

Fig. 7.4 Typical civil HVGS symbology (BAE SYSTEMS)

approach minima. If the minima criteria are not satisfied, he can rapidly and confidently execute a missed approach.

Terrain Avoidance Warning Systems (TAWSs) are a valuable tool to flight crew to avoid terrain. However, if the crew finds itself in a terrain escape manoeuvre where the terrain is visible, then the HVGS provides an unambiguous presentation on whether or not the terrain will be missed. If the flight path vector overlays the terrain, then the aircraft will hit it. The crew must decide on another course of action as opposed to wait and hope.

Traffic Collision Avoidance Systems (TCASs) provide traffic conflict alerts and escape guidance. With an HVGS, both the escape manoeuvre and visual search for traffic can be accomplished head up.

Accidents and incidents still occur today on manual landings. Some are too short, some too long, and some not aligned with the runway. Tail strikes can occur on take-off or landing and can cause considerable damage to aircraft, with consequent out-of-service time. A tail strike can occur on take-off if the pitch rate is excessive. Landing tail strikes can occur with insufficient approach speed and aggressive last minute flight path corrections. Having primary flight data head-up reduces the possibility that a pilot will focus on the touchdown point and allow airspeed to fall or initiate flare too high.

As can be readily seen, the HVGS symbology has great synergy with conventional Head-Down Displays (HDDs) and formats in that the airspeed scale, bugs, and trend arrows are drawn in a very similar style on the left. The conformal symbology is the principle and unique feature of an HVGS.

The symbology is identified in Fig. 7.5 and comprises:

- The *desired approach speed* is entered by the crew into the system, and is displayed on the top of the airspeed scale.
- The actual *airspeed* appears in the box halfway down the airspeed scale.
- The difference between the desired approach speed and the actual aircraft speed is displayed as a *speed error worm* on the flight path symbol as a tape emanating from the 'left wing' of the flight path symbol (this can be seen in Fig. 7.6). If the speed is below the set speed, the tape extends below the left-hand wing of the flight path symbol, and above it if the speed is too high.

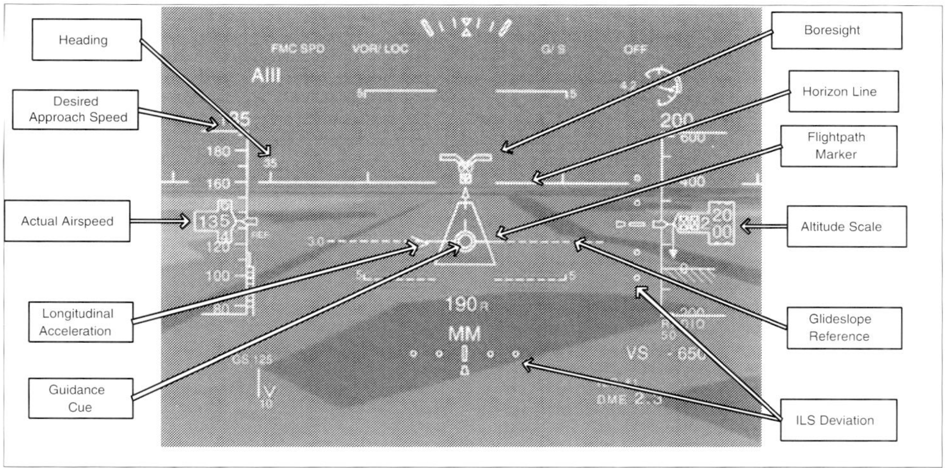

***Fig. 7.5** HVGS conformal symbology subset (typical)*

- The *altitude scale, bugs, and trend arrows* similarly appear on the right.
- The *Instrument Landing System (ILS) scales* appear on the bottom in the case of the localizer, and on the right in the case of the glideslope.
- The *annunciator* boxes appear along the top of the display and indicate the modes both current and armed.

So far, there is not that much difference between the HVGS and HDD symbology, but all this changes with the inner guidance group:

- The *flight path marker* is often referred to as the heart of the HVGS. This is in the form of a small airplane symbol. It provides a conformal display of the longitudinal and lateral instantaneous and inertial flight path. It indicates where the aircraft is going, not where it is pointed. In level flight the flight path symbol overlays the horizon line. On the approach, the flight path symbol can be used to control the actual touchdown point of the aircraft. On a correct approach both the glideslope reference line and flight path symbol overlay the Touch Down Zone (TDZ).
- The *boresight symbol*, which has the appearance of a seagull towards the top of the display, is where the aircraft is physically pointed, but that information is probably of less use to the crew.
- The *longitudinal acceleration caret* (potential flight path angle) is a symbol that aligns with the wing on the flight path symbol when there is no longitudinal acceleration. Inertial acceleration, air mass acceleration, and throttle setting drive this symbol to indicate the correct throttle movement in steady air and wind shear. When the throttles are advanced, giving excess thrust, this thrust can be turned into increased speed or a steeper climb angle.
- The *guidance cue* is generated by the system from the ILS beams and aircraft sensors through the aero model of the aircraft. If the pilot manoeuvres the aircraft such that the flight path marker overlays the guidance cue, the system will accurately take the aircraft down to the runway touchdown zone.
- The *horizon line* represents a level plane through the aircraft's present position. It is always parallel with the outside world horizon, and at moderate heights overlies this horizon.

- *Heading* is marked conformally every 10° on the horizon line. A flagpole on the horizon due east of the aircraft would line up with the 09 heading mark.
- The *glideslope reference line* angle is selected by the pilot. If an ILS is installed, the glideslope reference line will be set to the same value. If the glideslope reference line is set at 3° and this line overlays the visual runway touchdown zone (TDZ), then the aircraft must necessarily be on a 3° glideslope to the TDZ.

Operational aspects

The stabilized approach

During operation into any airport, whether in Instrument Meteorological Conditions (IMC) or Visual Meteorological Conditions (VMC), the use of the HVGS leads to more precise aircraft control. The pilot will touch down at a closely controlled airspeed and sink rate in the ideal portion of the runway.

Non-precision approaches

At facilities where there are no ILS beams and a landing is being made at night, it is a common problem to misjudge the aircraft height on the approach owing to the lack of visual cues. This error is generally in the form of believing that the approach is too high, where in fact the reverse is true. Many serious accidents have resulted in recent years as a direct consequence of this problem.

The operation of a non-precision approach is as follows. Using the HVGS, the pilot selects a 'desired glidepath' angle in the range 2.5–4.5° which appears as a dotted pitch bar at the appropriate conformal position. This may be seen in Fig. 7.6, where a value of 3° has been selected. The pilot then lines up with the runway centre-line and flies the aircraft straight and level until the symbol overlies the touchdown zone of the runway, at which point he commences the descent, and adjusts it such that the flight path marker symbol is positioned on the 3° line. The aircraft will now be descending at the correct glidepath angle to the touchdown point. If in addition, the pilot keeps the power adjusted such that the acceleration

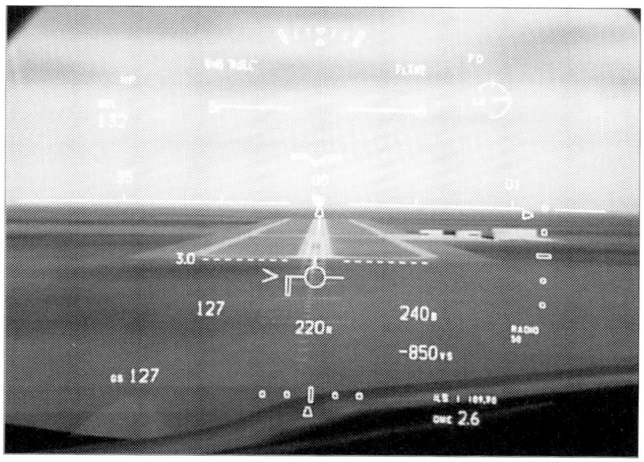

Fig. 7.6 *Approach configuration with desired glidepath symbol (BAE SYSTEMS)*

caret remains on the flight path marker wingtip, then the aircraft will arrive on glidepath and on speed at the desired threshold position (**2**).

If, on arrival over the threshold, the pilot reduces power when about 30 ft is indicated on the radar altimeter, and increases the aircraft pitch attitude such that the flight path marker is just below the horizon line on the HVGS, then the aircraft will be in the ideal flared attitude for touchdown.

Flare and roll out guidance

During Cat III operations, the HVGS will guide the pilot down to the runway threshold at which point the system provides guidance through the flare and advises on power settings. Following touchdown, the system further provides centre-line guidance and a runway remaining indication which enables the aircraft to be kept safely on the runway in poor visibility conditions. Figure 7.7 illustrates the symbology displayed under these conditions.

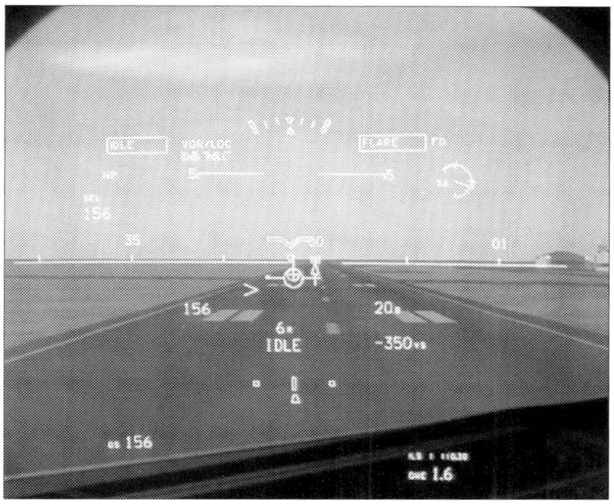

Fig. 7.7 *Landing mode (BAE SYSTEMS)*

Circle to land

The following description illustrates the use of the HVGS to perform one of the most difficult manoeuvres an air transport pilot is required to perform – a night, cloud break, circle to land, visual approach in mountainous terrain. Figure 7.8 illustrates the manoeuvre from the point where the captain has broken cloud, identified the airfield and landing runway, and has manoeuvred visually to intercept the extended runway centre-line.

In Fig. 7.8a the glideslope reference line, set at 3°, is in front of the airport area, showing that it is not yet time to start the descent. This is equivalent to the fly-up ILS signal when flying level to intercept the glideslope from below. Figure 7.8b shows the display at the point where the descent has just been initiated. Throughout the approach, the conformal symbology is used to monitor the aircraft's position and to initiate corrections. Consider the aircraft below the glideslope and to the left. Figure 7.8c shows that the captain has manoeuvred the

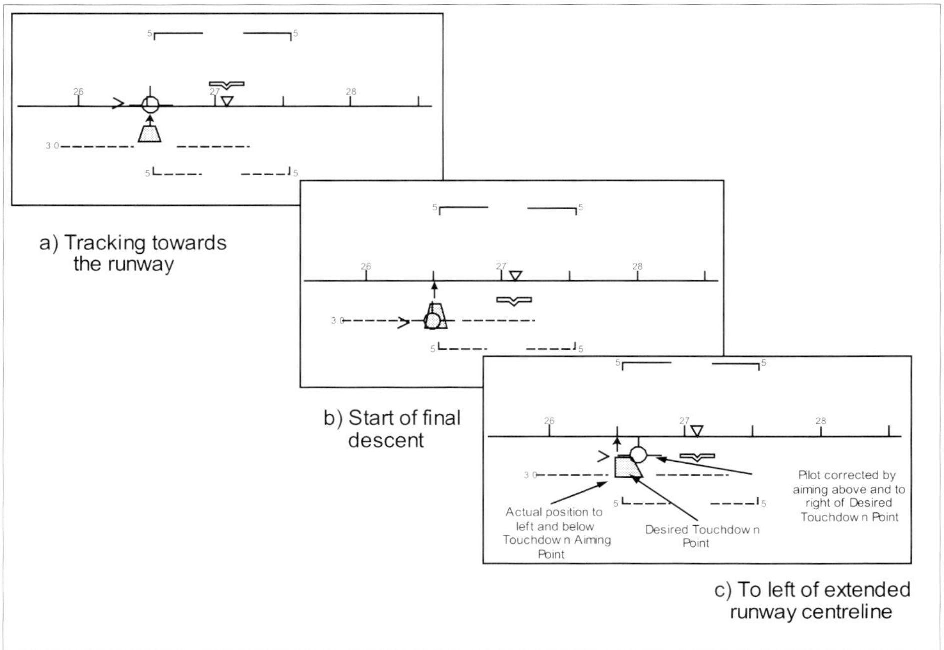

Fig. 7.8 Circle to land visual approach sequence

aircraft so that the flight path symbol is placed above and to the right of the TDZ for correction (**3**).

This guidance information allows the captain to remain Head-Up throughout the approach, avoiding the optical illusions of black-hole approaches and monitoring traffic. It also provides specific guidance cues, even when there is no ILS, to facilitate non-precision and visual approaches to be flown to levels of accuracy similar to precision approaches.

If for any reason the captain has to go around, the HVGS provides two very powerful pieces of information. As soon as the thrust levers are advanced, the acceleration caret moves to indicate the potential flight path. If the caret is above any local terrain obstacles, the captain knows the aircraft will clear them. The second piece of information is the flight path symbol. As the aircraft pitches up, the flight path symbol moves up towards the acceleration caret. The captain can continue to pitch the aircraft until the flight path symbol aligns with the acceleration caret, at which point the aircraft is making a safe, stable climb at the chosen airspeed and thrust setting.

An engine failure is always hazardous, and especially so at low altitude in mountainous terrain. The captain needs to know rapidly whether the current height and speed can be maintained on a single engine. If the flight path acceleration symbol is above the horizon line, there is a positive potential flight path and the aircraft has sufficient energy to climb. As soon as the captain has cleaned up the aircraft and retrimmed to the best engine-out speed, the flight path symbol indicates whether the aircraft can clear any visible ridges, or whether a turn away must be started.

Good energy management is known to be the basis of safe flight, especially in turbulence or wind shear. The flight path acceleration shows immediately whether the aircraft energy is increasing or decreasing. A tail-gust wind shear, for example, will cause a negative rate of change in airspeed which will cause the flight path acceleration to fall below the flight path, indicating the need for more thrust to prevent loss of altitude.

Normal manual throttle control is also simplified by the flight path acceleration symbol. Throttle movements that move the flight path acceleration symbol in the opposite direction to the speed error will correct airspeed errors shown via the speed error tape. All these features lead to a higher percentage of on-speed touchdowns in the TDZ for both visual and ILS approaches.

Take-off guidance

The system provides take-off guidance which comprises runway remaining indications, centre-line guidance, and rotation guidance. The rotation guidance is particularly useful for longer airframes such as the 737–800 in that it reduces the probability of tail strike due to overrotation. The system operates by commanding a 10° pitch up on initial rotation, increasing to 15° when the aircraft has a safe margin of height.

Figure 7.9 illustrates the symbology displayed in this mode. The HVGS provides centre-line guidance in the form of a steering bar, together with a runway remaining indication in the bottom left of the picture. Speed and rotation bugs, etc., are all displayed in a similar manner as they would be head down. The system facilitates take-off in reduced runway visibility conditions.

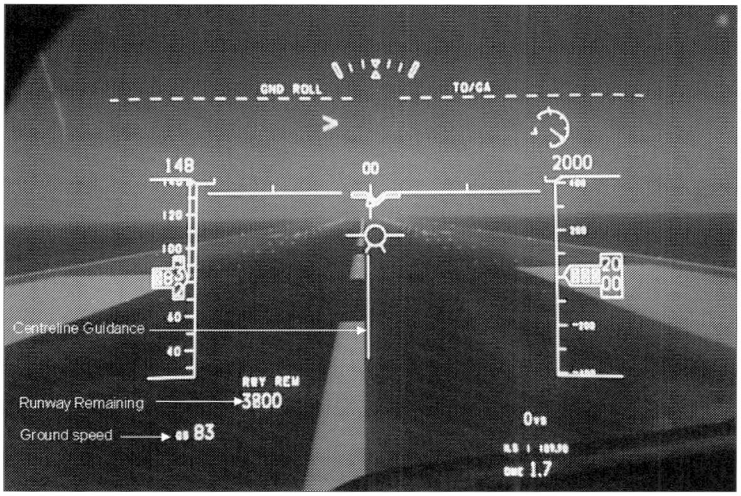

Fig. 7.9 Take-off guidance symbology (BAE SYSTEMS)

Ground guidance

To date, much attention has been expended on the problems of take-off runs in low visibility and in achieving successful landings in poor weather conditions. This is only part of the problem since the aircraft must be able to manoeuvre from the gate to the end of the runway successfully for take-off and likewise must be capable of leaving the runway and finding the gate after a poor weather landing.

NASA has performed some interesting work in pursuit of their low-visibility landing and surface operations programme. A system was tested by a number of pilots at (a simulation of) Chicago Airport which provided taxi guidance both as synthetic 'traffic cones' drawn on a HVGS and also as a situational moving map display. Both forms of navigation aid yielded an increase in taxi speed, and suggested that the system exhibited the potential greatly to increase the efficiency of ground taxi operations under low-visibility conditions.

Enhanced HVGS (EHVGS)

With the safety benefits of the basic HVGS established, the next step is to add enhancements to the system to allow it to look further than the naked eye through rain, fog, and other visibility impairments while taxiing, during take-off, and on the landing approach.

In reduced visibility, pilots encounter attenuation – or, perhaps more accurately, scatter – of the radiation in the visible spectrum which prevents them from seeing the runway. The electromagnetic spectrum shown in Figure 7.10 illustrates the relationship between the visible range, infrared, and microwave, from which it can be seen that the narrow response of the human eye could be greatly assisted if energy in the infrared and microwave regions could effectively be made visible. Longer wavelengths generally penetrate fog better than shorter ones.

Since atmospheric conditions vary, the type of obstruction to vision (fog) will also vary. Fog is generally classified according to its characteristics by such names as advection and radiation fog. In real conditions, the composition of the fog is infinitely variable, hence its effect on electromagnetic radiation will vary, and so the effectiveness of a particular sensor may not be constant from day to day. Having said that, most researchers believe that infrared sensors perform very well at night in clear conditions, have very good resolution, and present the world in a perspective that is familiar to the pilot.

Fig. 7.10 The electromagnetic spectrum

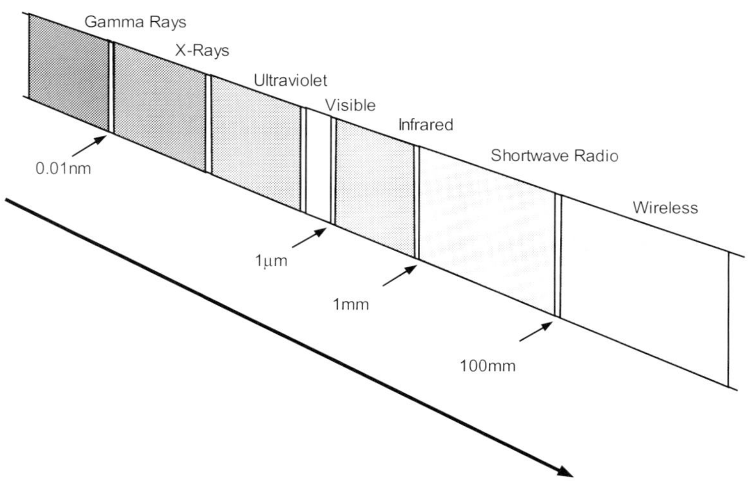

Infrared imaging

Enhanced FLIR vision sensors have been used in military airborne applications for many years. Civil applications of this technology have been longer to come to fruition.

Having recognized the potential, experimental trials have been undertaken for some years to explore the applicability of enhanced vision systems for civil transport applications. Perhaps the most successful of these was the synthetic vision flight trial conducted by a team that included the Federal Aviation Administration, NASA, Department of Defense, and others in the industry using a BAE SYSTEMS holographic HVGS on a Gulfstream GII aircraft. The programme assessed the effectiveness of imaging radars and FLIR cameras in real conditions. The tests showed that the crew was able regularly and routinely to perform visual landings in category III type weather conditions using the sensor image alone on the HVGS.

A small infrared sensor is installed under the aircraft's radome, which detects subtle thermal gradients of the terrain ahead. The pilot sees a realistic portrayal of the airport scene, showing the runway and taxiways standing out clearly against the contrasting surrounding surfaces, including, as he gets closer, images of aircraft holding beside the runway to take off. After landing, he can see the taxiways easily and, more importantly, he can also see other aircraft and vehicles moving around the airport surface beyond normal vision – a valuable benefit at today's crowded airports, especially at night. Pilots can even see animals on the runway (there have been incidents that have caused serious damage to aircraft when struck during take-off or landing).

Two infrared technologies are currently being developed for EHVGS: one operates at a relatively long wavelength, the other at a relatively short wavelength. The more mature technology is an indium antimonide sensor, which operates at a wavelength of 1–5 μm. The other technology is the microbolometer (a device that senses infrared radiation based on the temperature changes induced by the radiation). The microbolometer operates in the range 8–14 μm. The indium antimonide sensor 'sees' through more weather than the microbolometer (**4**).

Lowering minima

While the EHVGS greatly improves the pilot's perception of his surroundings and his situation awareness, it also offers two additional and very valuable operational benefits over the basic HVGS. First, safety is even further increased. Second, and almost equally important, certification authorities are now starting to recognize that EHVGS-equipped aircraft have the ability to descend safely to lower minima on their landing approaches than other aircraft.

This means that, while under the category I standard a conventionally equipped aircraft may normally descend to 200 ft as it nears the runway but then must execute a go-around if the runway is still not visible, an EHVGS-equipped aircraft could be certified to allow further descent to the category II limit of 100 ft above ground before overshooting (**5**).

Millimetric wave imaging radar

The longer-range penetration by FLIR of weather is limited, however, when certain fog densities are encountered. In such cases the Millimetre Wave Imaging Radar (MWIR) often exhibits more effective penetration but at reduced resolution. The ideal solution is to fuse a dual FLIR and MWIR sensor configuration through pyramid processing techniques. This should provide an improved image and also provide a more certifiable system, since there will now be sensor redundancy. The MWIR technique is well proven and has been used in a wide variety of military applications.

During US Air Force evaluation flights in 1998 and 1999 of an HVGS enhanced a BAE SYSTEMS 94 GHz imaging radar, a US Air Force (USAF) Special Forces Lockheed C-130 successfully taxied, took off, and was able readily to fly approach and landings on the image alone, in virtually zero visibility on many occasions, using both conventional runways and tactical dirt strips. According to the project's final report, landing approaches down to 50 ft above the ground – equivalent to today's civil category IIIa standard – were 'easily achievable', along with the observation that 'no special skills were required' to land the aircraft.

Figure 7.11 illustrates the real-world scene and the simultaneous radar image. As may be seen, the radar image exhibits a great deal of detail of the airport runway and taxiways, whereas the naked eye can see little detail at all.

Fig. 7.11 Radar image compared with real-world scene (BAE SYSTEMS)

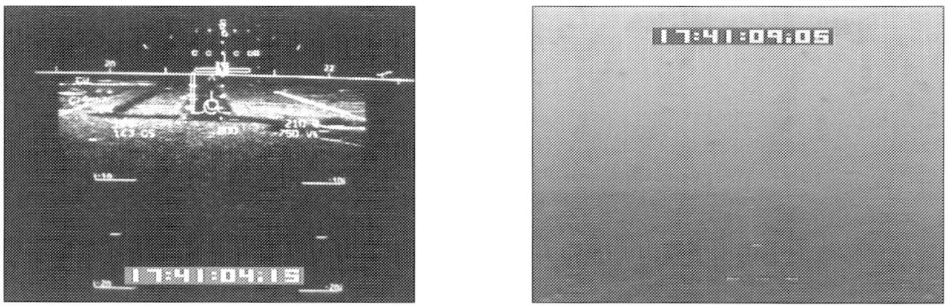

Radar Image with HUD symbology

Real world scene - CATIII conditions

The radar data from the antenna are in the form of range and angle, from which a 'B scope' image may be generated. To process the data into a form that is familiar to pilots and consistent with their normal visual perspective of the world, the information must be further processed. Owing to the nature of the data received from the radar receiver, the B scope to C scope conversion must assume that the target area is flat, a reasonable assumption for most airports. For this reason, buildings, etc., will appear as shadows.

Synthetic vision

Another evolving technology that is being adapted to HVGS is the Synthetic Vision System (SVS) which combines highly accurate terrain data, refined Global Positioning System (GPS) navigation and aircraft attitude, altitude, and heading, to produce a computer-generated image of the outside world for display on the HVGS. The effect when photorealistic images are displayed is that of flying a landing approach on a bright, cloudless day.

NASA has conducted flight tests of an SVS in a Boeing 757 at Vail, Colorado. The test database includes terrain and obstructions, landing and approach patterns, and runway surfaces at the Vail airport, and blends images generated by an SVS with images derived from infrared sensors and millimetre-wave radar. The goal of the project, which will continue into 2005, is to mature the technology sufficiently to reduce the technical risk so that industry can go forward with this kind of system to help prevent CFIT accidents and runway incursions.

Synthetic vision is similar to the terrain awareness and warning system (TAWS) which has a big terrain database. The advantage of SVS is that it is not subject to the limitations of a sensor. The disadvantage is that, if an object is not in the database, it is not going to show up on the display. For example, an airplane on the runway will not be displayed (**6**).

SVS can also provide a Head-Up image after the airplane has touched down to provide guidance to the pilot to get from the active runway to the gate. This is often the most challenging phase of flight. Pilots have fully automatic systems to get the airplane on the ground in low visibility, but then are lost in a sea of blue lights on the tarmac trying to figure out where they are, particularly at an unfamiliar airport.

Traffic collision avoidance system

An HVGS is particularly appropriate for presenting Traffic Collision Avoidance System (TCAS) warnings, providing the pilot with not only avoidance guidance but also potentially a 'threat identifier box'. This system is a variation on the military 'radar target designator box', where a box is placed around the targets (or threats) in the pilot's field of view. If the threat is off the field of view, a limited 'dashed' box is displayed with an arrow pointing in the appropriate direction.

HUD field of view

A typical civil HUD has a horizontal or azimuth field of view of 30°; the vertical field of view is 25°. This is generally deemed to be sufficient; it is usually more of an issue where that field is located with respect to the aircraft waterline (or horizon line), i.e. how many degrees of look-up or look-down.

A recent study (**7**) arose because of a perception that a larger horizontal field of view for a head-up display installed in a civil transport was automatically better. There is a lack of scientific analysis to confirm or deny this view. The study addressed three situations:

- Crosswind landings.
- Curved approaches.
- Runway exit angle.

and each situation was assessed with three HUD FoVs: 30°, 36°, and 40°. The 30° HUD is currently in service. A 36° design was chosen because it was considered that this size is an achievable next-generation HUD. The 40° HUD represents a challenge to the state of the art.

It was found that the aircraft landing restrictions, not the HUD field of view, limited the crosswind situation. The curved approaches gave a more mixed conclusion, but in general the time that the glideslope was within the FoV was not significantly enhanced with a 40° FoV HUD compared with the nominal 36°. The runway exit angle analysis showed that alternative means must be found to show the turn-off point rather than reliance on the FoV.

While it was acknowledged that the tools produced during the investigation require further work, the preliminary results do not support the view that HUDs with larger fields of view than available today would be beneficial. In the specific examples analysed within the referenced paper, an increase in HUD FoV from 30 to 40° did not generally deliver a practical improvement in visibility. Larger optical systems incur weight and cost penalties which are significantly above the pro rata increase in field of view.

The business case

As at September 2002, HVGSs are currently available as optional equipment on several civil aircraft models but are standard equipment on only a few (**6**):

- Boeing offers the HVGS as optional equipment on current B-737 models and is considering offering HVGSs as optional equipment in other models in production.
- The HVGS has been standard equipment in the Boeing Business Jet since the aircraft was introduced in 1998.
- Gulfstream currently offers an EVS as standard equipment on the G550 and as an option for other Gulfstream models.
- Airbus say they plan to begin installing HVGSs as standard equipment on all of its aircraft from 2003.

In spite of the benefits of the HVGSs, 'it is a hard sell'. One reason civil aircraft operators have lagged behind the military in using the HVGS is the price. It is expensive. A typical installation costs about US$500 000. Category III capability is a quantifiable benefit that has value in terms of schedule reliability. Some of the other features are just as valuable but are more safety related, and it is harder to make a business case based on the safety features.

An airline operation is different from a corporate operation in this regard. The corporate operators are less constrained by the requirements of a business case: if they can make a solid argument that the HVGS improves the safety of their operation, then usually they can justify the investment. The airlines are operating in a tighter financial environment, and there are a lot more things competing for the investment.

Publication of an 'ARINC Characteristic', which comprises standards for the design and manufacture of avionics equipment, is likely to accelerate the installation of HVGSs in civil aircraft. The ARINC HVGS Working Group is likely to have a 'mature draft' of ARINC Characteristic 764 in mid-2003 to submit to the airlines for approval.

References

(1) **Will, B.** (1998) HUDs get ahead, *Flight Deck International*.
(2) **Wisely, P. L.** and **Bartlett, C. T.** *Gate to Gate Operation of a Civil Transport Head Up Display*, BAE SYSTEMS Avionics Limited Rochester, Kent.
(3) **Aplin, J.** (2001) Precise approach – every time, *Flight Deck International*, September.

(4) **Caera, D. J.** (2001) The view ahead, *Flight Deck International*, February.
(5) **Wisely, P. L.** and **Moore, J.** Enhanced situational awareness through the use of weather-penetrating sensors and an imaging Head-Up Display, SAE WAC Conference, Phoenix, Arizona, November 2002.
(6) Head-Up Displays In Civil Aviation: A Status Report, Flight Safety Foundation, *Flight Safety Digest*, December 2002.
(7) **Bartlett, C. T., Busbridge, M. L.**, and **Horton, O. T.** *Considerations of Head Up Display Field of View*, BAE SYSTEMS, Rochester, Kent.

CHAPTER 8

Helmet-Mounted Displays

Introduction

Chapter 6 described the head-up display and its evolution from the gyro gunsight of World War II to become the primary flight instrument in today's fast jet fighter aircraft, together with its use to acquire, designate track targets, and release air-to-air and air-to-surface weapons. The advantage of the HUD is that it does not require users to bring their eyes into the cockpit. It also provides the real-world correlation mentioned earlier. However, the major limitation of the HUD is its field of view. Pilots can only benefit from HUD information when looking straight ahead. Because of this limitation, there has been a desire to develop the Helmet-Mounted Display (HMD) so pilots can constantly benefit from the primary flight information superimposed on the real world, regardless of where they are looking.

Inspite of advances in technology, the basic geometry of the HUD optics and of the aircraft limit the HUD field of view to at best a 30° cone centred at or a little above (usually some 6°) aircraft longitudinal fuselage datum (or boresight). It is thus not possible for information (or weapon-aiming reticules) to be presented to the pilot outside this limited field of view. This is fine for gun aiming but is a very limited field of view when compared with the total hemisphere or regard afforded to the pilot in the bubble canopy of a modern fast jet fighter aircraft, and limited when compared with the field of regard of a modern radar (typically 120°) and the acquisition cone of modern air-to-air missiles (typically 90°), as graphically shown in Fig. 8.1.

Helmet- or Head-mounted displays projected onto the visor or onto a combining glass attached to the helmet have been developed to overcome this problem. By the use of miniature display technology producing a display for each eye, combined with accurate head and, in some cases, eye pupil tracking, it is theoretically possible to present a stereoscopic, full colour image to the user in any direction. This could be anything from a simple overlay of information on the outside scene to a totally artificial virtual image. Although the technology of

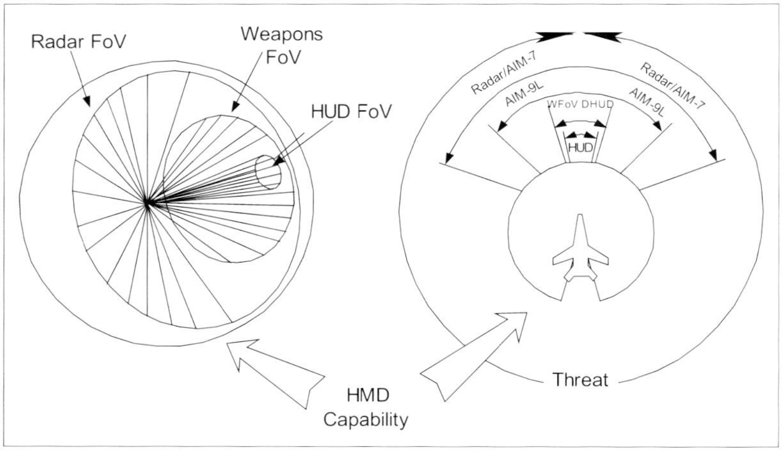

Fig. 8.1 *Field of regard comparisons*

displays and the processing speed of graphics generators have not yet matured to the point where the virtual world image is possible in an aircraft environment, it will undoubtedly materialize in the near future (**1**).

The first military group to embrace HMD technology was the rotorcraft community. Clearly, the field of view limitations of a conventional HUD are raised to a new level of significance when the aircraft is capable of moving sideways and even engaging in reverse! Similarly, it is likely that the new generation of high-agility, fixed-wing fighters, capable of sustaining angles of attack in excess of 70°, will present challenges that may only be satisfactorily addressed by the adoption of HMD technology.

The current rules of engagement require the pilot to steer his aircraft so that he can confirm that the radar and missile systems have acquired the same and the correct target which he can visually identify but only confirm within the HUD field of view. This is fully commensurate with the traditional 'point and shoot' manoeuvre. However, if it were possible to put the HUD onto the pilot's head, even with the same field of view, but now free to move wherever the pilot looked, then it would be possible to cue, acquire, designate, track, and release weapons at targets significantly off-boresight without having to manoeuvre the aircraft. The increase in operational effectiveness would be a significant 'force multiplier'. This is precisely what the helmet-mounted display can achieve.

Figure 8.2 shows some of the possibilities, which include:

- Off-boresight target cueing by directing the pilot/weapon aimer to look in the direction of potential targets or threats detected by his on-board sensors and/or advised by data link from cooperating aircraft and ground stations.
- Designation 'off-boresight' once the target has been recognized and deemed to be a target. Tracking accuracy of the on-board systems and weapons can then be monitored to assure lock is maintained.
- Weapon release can be commanded once a launch solution is reached without having to manoeuvre the aircraft to be within the traditional success cone determined by the HUD field of view.
- Handover is facilitated between crew members and between cooperating aircraft flying the same mission.

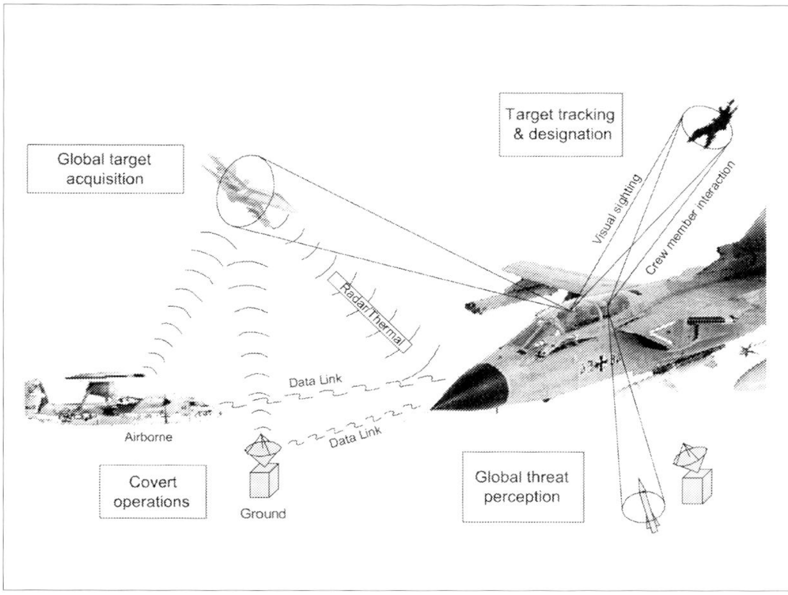

Fig. 8.2 *Possible uses of an HMD to cue, designate, and aim 'off-boresight'*

The advantages of a HMD include:

- Improved reaction time, especially if the weapon sensor can be slaved to the pilots line-of-sight.
- Improved orientation, by presenting flight critical information to the pilot wherever he is looking.
- Improved navigational fixes, by sighting and designating a landmark without the need to overfly it.

The first practical HMD recorded was a cathode ray tube (CRT) based system made by Autonetics in the United States during the early 1960s. By 1970 a system had been flown in a US Navy F-4 which proved the key points. The HMD had a sighting reticle and operated with an optical head tracker that linked the system to allow pointing of radar and missiles.

The most successful rotorcraft system to date is the Honeywell IHADSS fitted to the Apache AH-64 attack helicopter from around 1985. This monocular system operates with an optical tracker and a steered FLIR.

The use of a Helmet-Mounted system for sighting is well established. The Joint Helmet Mounted Cueing System (JHMCS) is installed on the F/A-18 aircraft and similar systems on other fast jets.

The application of a binocular full day and night capability has taken longer to mature. Initial usage has been on helicopters such as the German Tiger. The Eurofighter Typhoon will probably be the first fast jet to install such a device on an operational aircraft. However, the next generation of fast jet fighters, typified by the JSF, are likely to replace the HUD with an HMD, and the HMD is likely to be the primary flight instrument. This represents a significant challenge.

This chapter discusses the operating principles and the technologies employed in the design of a Helmet-Mounted Display, together with an appreciation of the demanding physiological and environmental requirements.

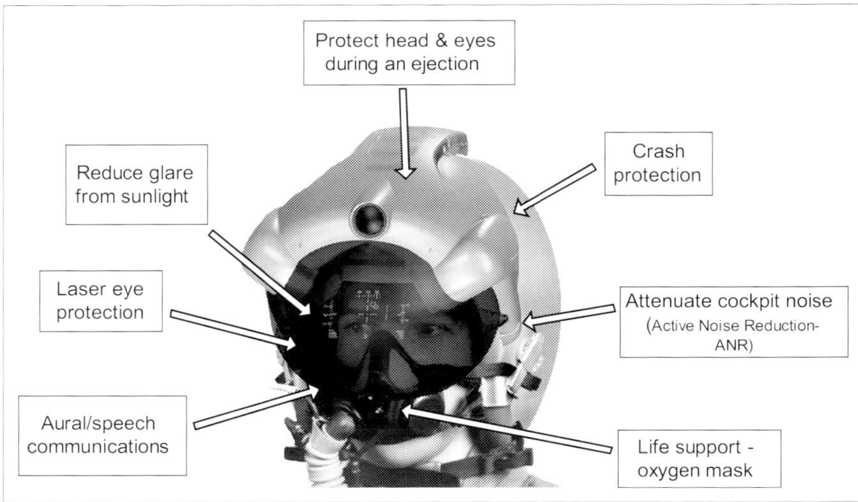

Fig. 8.3 *Normal helmet functions*

Physiological and environmental aspects

The HMD is more than an HUD on the head. The helmet already provides a number of facilities and life-supporting functions with which the display components must be integrated. These are illustrated in Fig. 8.3. The HMD must be a complex system of systems.

The pilot's helmet already provides a number of critical functions into which the helmet-mounted display must be integrated. These functions include:

- Communications microphone.
- Earphones (possibly with active noise reduction).
- Oxygen mask.
- Retractable sun visor.
- Possibly NBC protection.

In addition, the helmet must afford safety protection to the wearer in the event of hazardous events such as:

- Bird strike penetrating the canopy and causing glass shards.
- Ejection.
- Crash landing.

The provision of a helmet-mounted display requires the addition of some significant extra components, namely:

- An image source (e.g. cathode ray tube) or more than one source if a binocular image is required.
- Relay optics to position the HMD image at the focal plane of the collimator.
- Collimating optics to generate a virtual image focused at infinity.
- A partially transparent/partially reflective combiner to introduce the collimated image into the pilot's line of sight.
- Means to adjust the optical elements to match the anthropomorphic features of the pilot.
- Support electronics to the image source.

Humans sight an object by movement of both the head with respect to the torso and of the eye within the eye socket so that the object viewed is focused on the optimal point of the retina. There are systems in the laboratory that can detect eye point of regard with respect to the head (or eye socket), but these are far from being robust in an airborne application. Therefore, all practical head-mounted sighting systems constrain the eye sight line by means of a fixed reticle and measure only the head pointing angle. In the context of a Helmet-Mounted Display, the image area can be considered to be a porthole fixed in space relative to the head as if it were, say, a circle painted on a pair of glasses worn by the pilot with a central cross to represent the boresight aim point. The porthole moves with the head, and the sight line angle of the aim point can be measured with respect to a fixed point relative to the observer, in our case aircraft boresight. Thus, the Helmet-Mounted Display works just like the head-up display except that it is mounted on the pilot's head (or on his helmet head gear), plus the extra set of angles that define the head pointing sight line.

Therefore, the helmet must be a firm fit to the pilot's head so that no movement can occur once the optics has been 'boresighted' to aircraft axes, but at the same time be comfortable to wear and keep the head cool.

Adjustments need to be made to centre the optical image axis onto the pilot's eyeball. Typically, these include:

- The vertical position of the eyeballs with respect to the pilot's head.
- The distance between the eyeballs (the interpupillary distance).
- The eye relief between the eyeball and the first optical element.

The range of anthropomorphic geometry is vast. A one-fits-all helmet is impractical; usually it is necessary to have a range of helmet sizes available. Then the helmet needs to be personalized to the pilot's head shape. Often this is accomplished by using a permanently deformable liner. The liner once deformed personalizes the helmet to the wearer. It may be possible to reuse that liner in another helmet of the same size, but the optical adjustments are unique to an individual pilot and so in practice the helmet becomes personalized to the pilot. With all the HMD optical components added to the helmet, the helmet is a high-value item, certainly not disposable. Some designs allow the HMD optical elements to be removed from the helmet to reduce the helmet intrinsic value.

The provision of the HMD components must not increase the weight of the helmet (a typical helmet before the addition of an HMD weighs less than 2.0 kg (4.5 lb). Indeed, there is a strong desire to minimize the weight of the helmet to reduce fatigue. Nor must it alter the head centre of gravity which needs to be accurately located to make the helmet supportable on the head especially in high-*g* manoeuvres (see Fig. 8.4). The CofG should be in line with the pivotal point of the head on the spine to minimize out-of-balance forces. It is important also to minimize moment of inertia about the yaw and pitch axes since head movements can often be up to 30° per second.

Furthermore, the additional HMD components must not introduce features that may jeopardize the integrity of the helmet to protect the pilot from hazardous events such as:

- Bird strike. It is potentially possible that a bird strike (or other event) may cause the canopy to shatter and release glass shards into the cockpit.

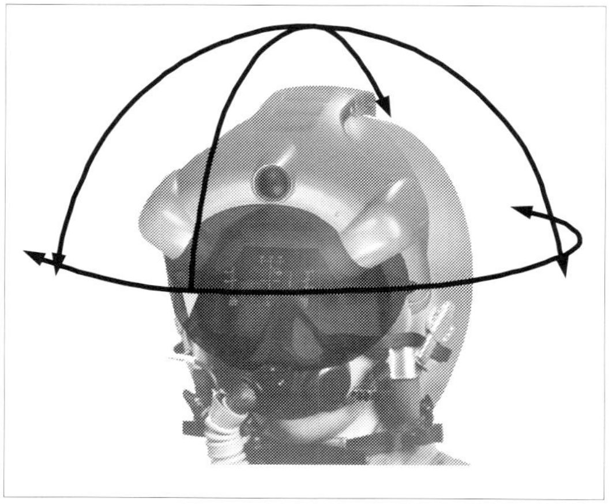

Fig. 8.4 HMD centre of gravity/moment of inertia fundamentals

- Ejection. In the event the pilot has to eject from the aircraft, the helmet must protect him from the acceleration forces of the ejection itself and from the significant windblast, estimated to be in excess of 600 knots.

Achieving all of these requirements represents a significant challenge to the helmet-mounted display designer. The requirements are often conflicting, necessitating inevitable compromises. All these eventualities are addressed by a comprehensive and severe series of tests which are applied to any helmet design by the certification authorities before any helmet can be cleared for operational use.

Head tracker

To be effective as a sight, accurate knowledge is required of the head pointing angles in six degrees of freedom. It can move vertically, laterally, and fore/aft with respect to the body torso and can rotate in all three axes about the pivotal hinge point of the head on the spine.

All practical methods to measure head pointing angles have at some time been tried. None is perfect. The two principle methods are optical and magnetic.

Optical head tracker

A number of schemes have been used. An early scheme used a number of sources positioned around the cockpit, at least 2 and sometimes 3 to generate swathes of scanning beams. Sensors mounted on the helmet detect these scanning beams and provide signals to the head-tracking electronics which decodes the interrelationship between the scanning beams and the time of detection to derive the head pointing angles, see Fig. 8.5. A more recent scheme, and that being used on the Eurofighter Typhoon, uses twin-tracking CCD cameras to sense clusters of LEDs on the surface of the helmet.

In any optical scheme, the location of emitters, the emission wavelength, and the switching frequency must all be carefully chosen so that the tracker will operate satisfactorily in all the ambient lighting conditions to be found in the cockpit,

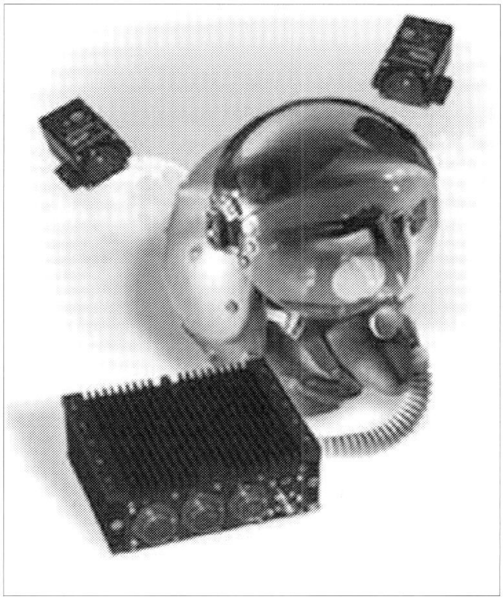

Fig. 8.5 *An optical head tracker (Kentron)*

typically from full sunlight incident on the helmet to night-time operation. Also, the emission and any backscatter from the helmet or cockpit features must not:

- Interfere with the pilot's normal vision under all the operational ambient lighting conditions.
- Interfere with any night vision enhancement devices (NVGs) potentially worn by the pilot to aid night-time operations.
- Adversely impact the covert or stealth aspects of the aircraft.

Electromagnetic head tracker

Again, a number of schemes have been used, both AC and DC electromagnetic fields. In general, a source transmitter is placed just behind the pilot on the bulkhead wall which generates an electromagnetic field in the cockpit. A three-axis orthogonal sensor mounted on the helmet detects the local electromagnetic field and provides signals to the head-tracking electronics which decodes the amplitude and phase signals and derives the head pointing angles.

The electromagnetic field is significantly distorted by the metal structures within the cockpit, and it is necessary to map the cockpit for full characterization of the magnetic field. Sometimes this mapping needs to take account of movable features within the cockpit, for instance seat height position, and several map sets will be required. Usually it is necessary to map only one aircraft type, though the mapping will have to be repeated if there are any subsequent significant changes to the cockpit.

Eye tracker

It is easier to track a target with one's eyes than with the head. It is more natural, more accurate, and does not suffer from buffer or vibration. If the head shakes either voluntarily or involuntarily, it is still possible to hold a gaze on a single

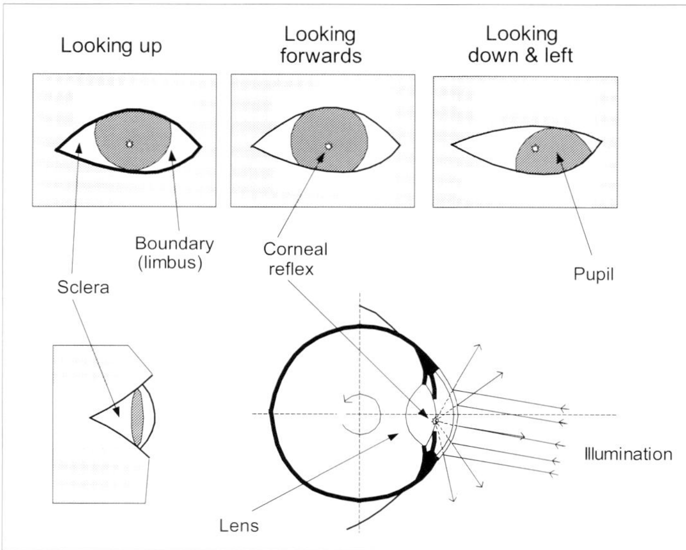

Fig. 8.6 Optical eye tracker principles

point object. It is very difficult to maintain the head pointing in a given direction when under vibration or buffeting. The optimal solution would be to measure eye point of regard with respect to the helmet, head pointing coordinates with respect to the aircraft, and hence deduce eye point of regard with respect to the aircraft. Eye trackers operate by detecting the point of reflection of a light source from the cornea, as depicted in Fig. 8.6.

Working on optical principles, the same issues arise as for the optical head tracker. As yet there is no practical airborne eye tracking system at or near operational readiness.

Boresighting

Since the helmet-mounted sight is not fixed to the aircraft structure, it is necessary to provide a means to 'boresight' the HMD system before each mission. Typically, boresighting comprises sighting a collimated target (e.g. telescope cross-wire) conveniently located and fixed in the cockpit.

Accuracy and dynamic performance

The HUD and HMD have a number of features in common. They can both display raster, cursive, or hybrid information and by definition must be visible over a wide range of ambient lighting. In both cases a FoV of 20–30° is sufficient for general use. The drive electronics and symbol generation are similar, although the HMD requires additional processing and interface for the head tracker.

The differences are, however, more significant. The HUD is very accurate on boresight and the performance is well characterized and stable over the whole FoV. The HMD has reduced accuracy because of helmet instability and tracker errors. The accuracy is not necessarily symmetrical and falls off rapidly at extremes of the head motion box.

For an HUD the error varies radially from boresight generally as the square of the off-axis angle. The HMD errors are much more complex and comprise the radial type of error of the HMD system plus a tracker error which may increase rapidly at extended angles from boresight and at the margins of the head motion box. Typical HUD and HMD system accuracies are shown in Table 8.1.

Table 8.1 Typical HUD and HMD system accuracies

On-axis errors	HUD	HMD
Electrical error	0.8 mr	0.8 mr
Boresighting	0.4 mr	1.0 mr
Tracker error		1.5 mr
Optical error		1.8 mr
RSS	1.3 mr	2.6 mr

A simplified performance comparison is shown in Fig. 8.7. The curve for the HMD represents current state of the art and can be expected to improve with time as technology advances (**2**).

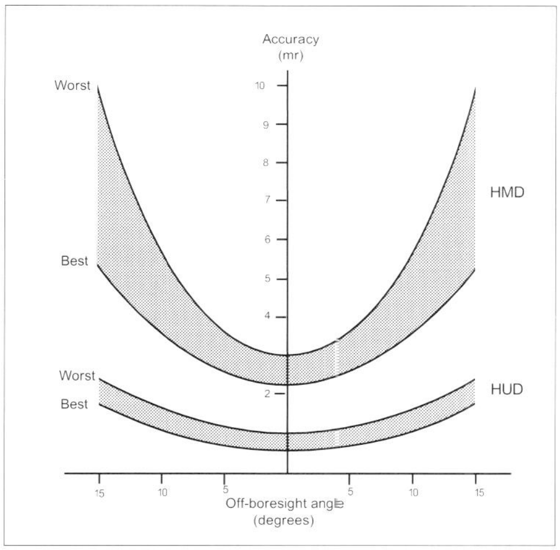

Fig. 8.7 *Comparison of HUD and HMD accuracy (BAE SYSTEMS)*

Accuracy implications

Using a helmet-mounted sight to launch a weapon with 'smart' guidance has been shown to be highly effective, especially since the target can be acquired and designated and the weapon launched 'off-boresight'. The HMD accuracy is adequate for this task. However, using a helmet-mounted sight to target a fixed gun in a dynamic situation such as air–air combat is generally recognized as being ineffective and inappropriate.

The real test of accuracy is with the release of stores such as 'iron' bombs and rockets and the gun. These weapons are still of significance because, in comparison with terminally guided munitions, they are very cheap. They are called 'dumb' weapons because they have no intelligence in the way of guidance and their accuracy depends upon parameters such as:

- Knowledge of the weapon characteristics.
- Knowledge of the environment, e.g. wind speed.
- The precision of the release point.

Total systems analysis (**2**) concludes not unsurprisingly that iron bombs are totally unsuited for high-altitude precision attack, irrespective of the mode of delivery using either a HUD or an HMD. Low-level automatic release of iron bombs and rockets is the only effective mode and, since there are many other factors, there is little appreciable difference between HUD and HMD accuracy.

Dynamic performance

The dynamic performance requires careful consideration since rapid movements of the pilot's head (30 deg/s) are entirely possible. System bandwidth and delays must be commensurate with the operational role in order to avoid disorientation that might be caused by 'swimming' symbology. For this reason it is usual to close-couple the head-tracking electronics with HMD symbol generation and avoid any data latency problems associated with a digital databus.

HMD configurations

The optical configuration should be selected according to operational use. As usual, but more so with an HMD, the choice is made more complex by the many aspects that need to be taken into consideration. The HMD is not just a display but a system of systems, as described earlier. Display performance needs to be balanced with complexity and ease of use in the operational environment. For this reason it is wise not to overspecify the display requirements. Bigger is not necessarily better if it weighs more.

The choices broadly are:

- *Sight or display.* A sight is used simply as a target designator, possibly with the addition of simple direction-finding cues (such as look-up/down, left/right and event annunciators such as weapons lock/fire). A display generally adds other symbology such as primary flight data, weapons data, and imagery at night.
- *Monocular or binocular.* Monocular is suitable for daytime use when the pilot's attention is mostly on the outside world. However, at night, when vision is augmented by other aids such as FLIR or NVGs, a binocular configuration is preferred. Although the AH-64 IHADSS is monocular, and used at night, there have been reported problems of fatigue to the point where the brain rejects the image from one of the eyes, usually the eye with the sight.
- *If binocular, then one image source or two.* Both configurations have been used. With an image single source, the images must precisely overlay one another. If dual sources are used, then it is possible to extend the azimuth field of view with a central area in which vision is stereo with two monocular areas to the sides, as shown in Fig. 8.8. The image sources must still precisely overlay one

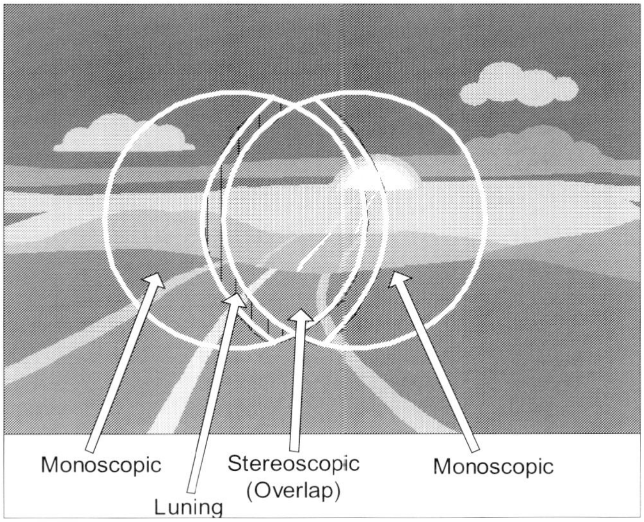

Fig. 8.8 Monocular or binocular configuration

another in the binocular overlay region. Any misregistration is called binocular disparity. The human eye/brain combination will cope with binocular disparity for a time, but small and subtle differences between each image will cause eye fatigue or the inability for the eye/brain to fuse the images, when the brain will again reject the image from one of the eyes. Small vertical displacements are particularly disconcerting, while small horizontal displacements give rise to false depth perception and the image may appear to move in depth or swim with rapid head movement.

- *Off-visor or periscope optical system.* The periscope optical system is in many ways easier to design and manufacture and therefore was the first to be used. Off-visor systems are less obstructive and more technically elegant but demand more complex solutions if wide fields of view are required. Example optical systems will be described later.
- *Day, night or day/night operation.* It is a self-evident requirement that a daytime display must be viewable in direct sunlight. At night the image may be augmented with sensor video and/or night vision goggles. There are many solution choices here, most of which will be discussed later.

Optical configurations

The optical configuration choices for a helmet-mounted display are identical to those described in Chapter 6, using the principles of refraction and/or reflection to introduce a virtual collimated image into the pilot's line of sight possibly with the aid of holographic diffraction in the combiner element. Typical arrangements are shown in Fig. 8.9. These configurations demonstrate the use of the same principles as the refractive collimated optical systems and quasi-axial and off-axis pupil forming systems described in Chapter 6. The lens diameters are of course considerably reduced when compared with those of a head-up display since the effective exit lens is close to the pilot's eyeball, and hence a 30–40° field of view porthole is obtainable with lens diameters around 1 cm. However, the intermediate

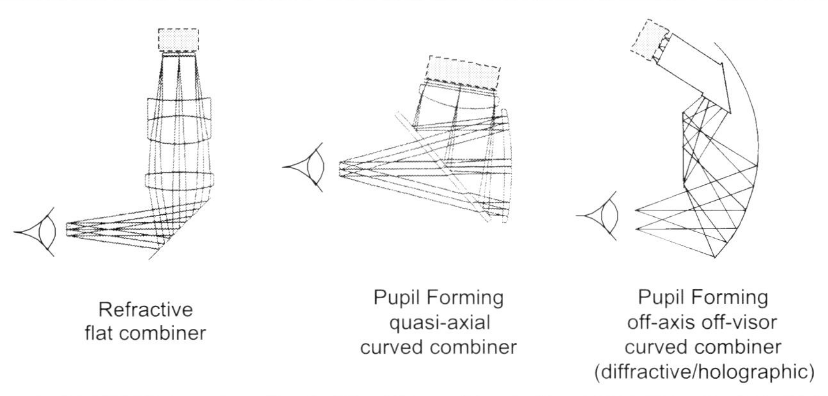

Fig. 8.9 *Typical optical configurations*

optics to relay the image from its source (typically a miniature CRT) to the focal plane of the collimator are often very complex since they need to bend the light path around the head without being obstructive and at the same time maintain the CofG close to the head natural CofG.

In all these optical arrangements, the collimator is a semi-transparent curved surface placed between the outside world and the pilot. In early designs the optical assembly was separate from the visor. However, it is highly desirable and technically elegant if the collimator function can be integral with the visor, achieved by the application of a suitable coating or optical surface. Later designs achieve this, frequently with the use of holographic techniques similar to those used for wide field of view reflective/holographic head-up displays. However, it must be remembered that the visor is subject to a lot of abuse and is likely to need to be replaced from time to time. Therefore, it needs to be a relatively low-cost item. The visor is usually made from a polycarbonate material. Polycarbonate is easily moulded into complex shapes, but has inferior optical quality compared with glass.

Figure 8.10 shows a generic HMD optical system. The EO component elements are:

- Image source. Usually a high-brightness, high-resolution miniature CRT (though early sights used a small LED matrix). Emerging technologies are AMLCD (transmissive and reflective), electroluminescent (EL), organic LED (OLED), and low-power laser scanning direct into the eyeball. These technologies are discussed further in Chapter 9.
- Relay lens and fold mirrors. The relay is a group of lenses or exotic equivalents. The relay prepares the image to be projected into the pilots' line of sight by the combiner. The relay contains mirrors and/or prisms to make the relay lens group fit around the head.
- Combiner. The combiner projects the image from the relay into the pilot's line of sight. It may be flat or have optical power. It is semi-transparent to the outside world, but reflective to the specific wavelength of light from the image source, in the same manner as required for a HUD, as discussed in Chapter 6.

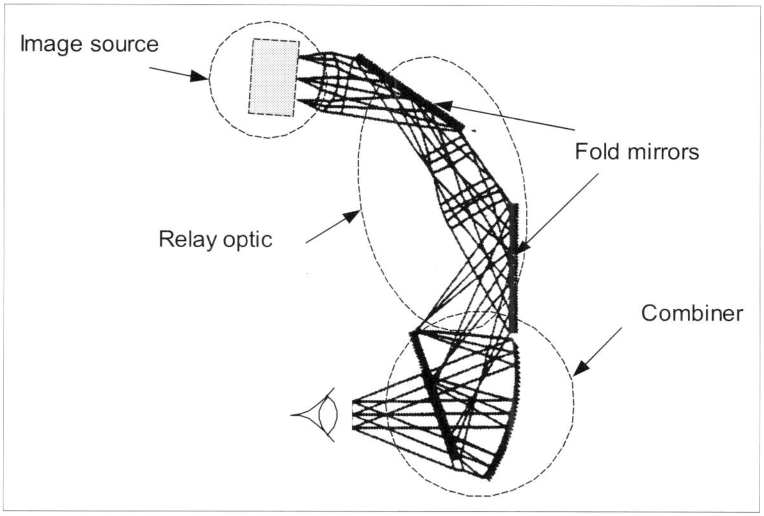

Fig. 8.10 Generic HMD optical system

Helmet-Mounted Displays – examples

The first users of HMD technology in aerospace applications were amongst the rotorcraft community. Clearly, the field of view limitations of a conventional HUD make that technology inappropriate where the air vehicle is capable of moving sideways. The weight issues are less significant in a rotorcraft which is not able to pull as much *g* as a fast jet fighter. Neither is ejection or windblast a concern – ejection from a helicopter is not to be recommended.

Rotorcraft Helmet-Mounted Displays

The Apache AH64

The Apache AH-64 attack helicopter was the first operational rotorcraft to install a Helmet-Mounted Display. The Apache is a tandem two-seat helicopter with advanced crew protection systems, avionics and electrooptics, plus weapon control systems. The copilot/gunner sits in the front seat; the pilot sits behind him and approximately 18 in (0.5 m) higher. Both crew members use various sensor systems to detect and attack targets. The copilot/gunner has primary responsibility for firing both guns and missiles using the Integrated Helmet and Display Sighting System (IHADSS).

Figure 8.11 shows the IHADSS helmet. The optics arrangement produces a collimated image using the optical principle of refraction.

The Honeywell-supplied IHADSS provides a 40° × 30° field of view, monocular (single-eye) collimated image from a miniature Cathode Ray Tube (CRT). The images constructed on the CRT are derived from the nose-mounted Target Acquisition and Designation Sight/Pilot Night Vision Sensor (TADS/PNVS). The TADS/PNVS and the gun are slaved to the copilot/gunner head line of sight. The system has proven itself to be operationally effective. There have been some reports of fatigue after extended use, possibly caused by the contention created in the brain of having significantly different images from each eye.

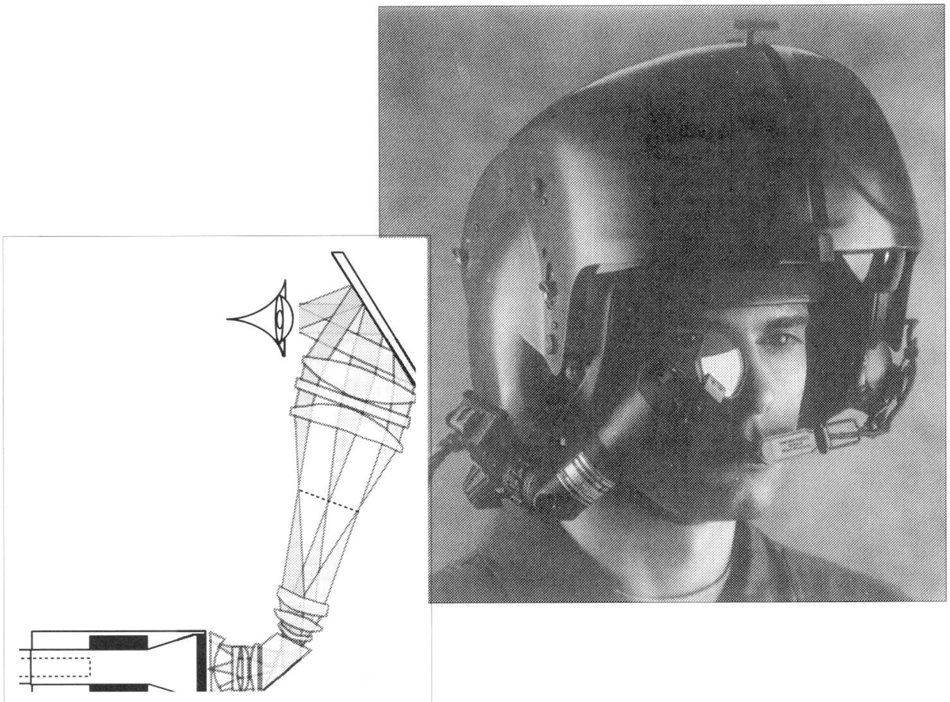

Fig. 8.11 *The IHADSS helmet (Honeywell)*

The Eurocopter Tiger

The Eurocopter Tiger is designed to meet the French and German requirements for an advanced multirole helicopter for battlefield operations in the European theatre. The avionics system employs an MIL-STD-1553B data bus to facilitate subsequent introduction of later generations of sensors and weapons during the course of the helicopter service life. Several variants are planned, depending on the role of the national operational force. Typically, TV, FLIR, and laser rangefinder sensors in the mast-mounted sight system provide data and imagery to the copilot/gunner who cues, designates, and aims the helicopter weapons using his Helmet-Mounted Display. The helicopter may be equipped with a variety of weapons including antitank missiles, short-range air–air missiles, and an undernose turret-mounted cannon which is slaved to the copilot/gunner head line of sight.

Two variants of Helmet-Mounted Display are planned, one designed to meet the requirements of the French operational role, the other the German operational role. The Helmet-Mounted Display fitted to the German variant is shown in Fig. 8.12.

The BAE SYSTEMS Knighthelm HMD for the German Army Tiger helicopter is an integrated day/night Helmet-Mounted Display. It is a two-part helmet design, using a form-fitted inner helmet with an outer display module. It has been refined and enhanced as part of an extensive development programme and is optimized for the attack helicopter application. The design optically mixes the output of an Image Intensifier Tube (IIT) with Cathode Ray Tube (CRT) imagery.

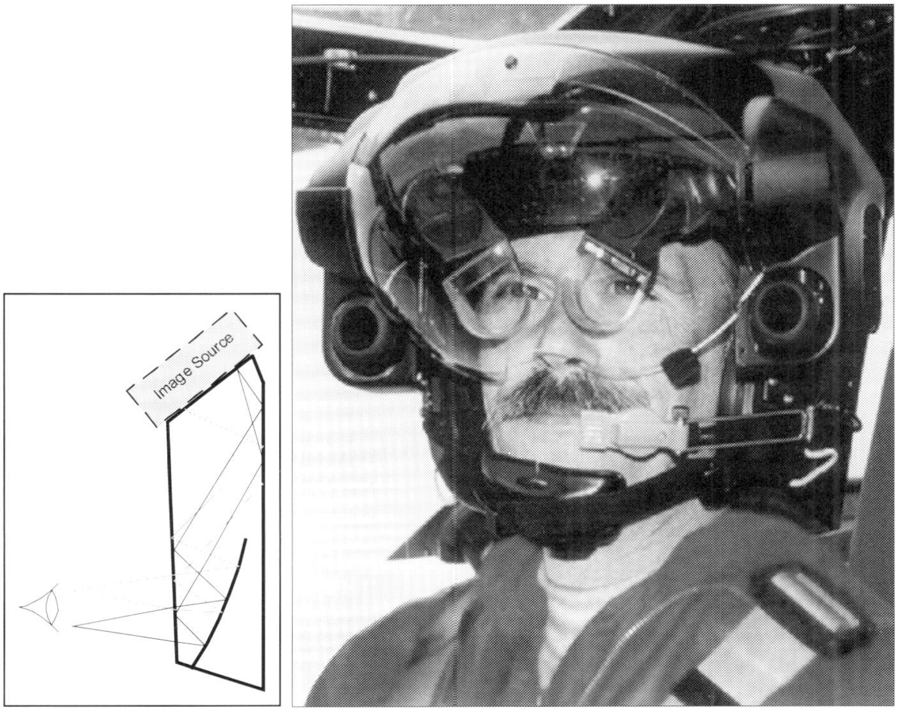

Fig. 8.12 *Eurocopter tiger HMD; German variant (BAE SYSTEMS)*

This provides a flexible display of symbology overlaid on NVG imagery or symbology overlaid on FLIR video viewed as a collimated image in the see-through combiner eyepiece in front of the user's eyes (**3**).

The Knighthelm HMD system provides a high-performance integrated Helmet Display with a binocular presentation of TV sensor, image intensifier, and overlaid symbology information with a high degree of functional flexibility. This flexibility enhances aircrew mission effectiveness in all scenarios and in both day and night conditions.

Key features of the Knighthelm HMDs are:

- Binocular display integrated with a purpose-designed flying helmet:
  - 40° display;
  - binocular presentation, 100 per cent overlapped.
- Dual sensor display capability – night vision Image Intensifier Tube (IIT) or Thermal Imager (TI):
  - integrated, directly viewed GEN 3 IITs for high night vision performance;
  - miniature high-performance CRT display of sensor imagery and symbology;
  - fully rotatable sensor display for image stabilization against head roll.
- Symbology display overlaid over real-world view and sensor image:
  - cursive symbology visible in all ambient lighting conditions, overlaid on IIT or head-steered FLIR imagery;
  - operation in both day and night flying conditions;
  - selectable binocular/monocular CRT symbology presentation.

- High-performance helmet tracking technology for weapon-aiming symbology control, FLIR sensor steering.
- High level of failure survivability by multiple display paths.
- Integral battery back-up for image intensifiers, selected automatically on power failure.

The optical arrangement is an off-axis, multielement combiner with a relayed pupil fabricated into a solid eyepiece using plastic elements to minimize weight.

Comanche HIDDS

The Comanche Helmet Integrated Display Sighting System (HIDSS) is shown in Fig. 8.13. The HIDSS represents the culmination of several years of development and epitomizes the state of the art in rotorcraft HMD design and operational performance for sophisticated rotorcraft applications. The system employs a two-piece design, using modules containing the high-value $35^\circ \times 52^\circ$ FoV binocular optic, high-resolution AMLCDs (1280×1024 – SXGA), magnetic tracking sensor, plus associated drive electronics in an Aircraft Retained Unit (ARU). The ARU is mounted in the aircraft and is considered part of the weapons and electrooptics systems. The optics are stowable and can be used in either a monocular or binocular form. The head borne weight is 3.9 lb and with the optional 40° I^2TV module weighs 4.4 lb. The HIDSS meets US Army flight safety, weight, and CofG criteria. HIDSS has pilot-adjustable Inter-Pupilliary Distance (IPD) and dioptre adjustments required as a result of US Army Aviation Aeromedical Laboratory studies.

Fig. 8.13 Comanche Helmet Integrated Display Sighting System (HIDSS) (Rockwell-Collins)

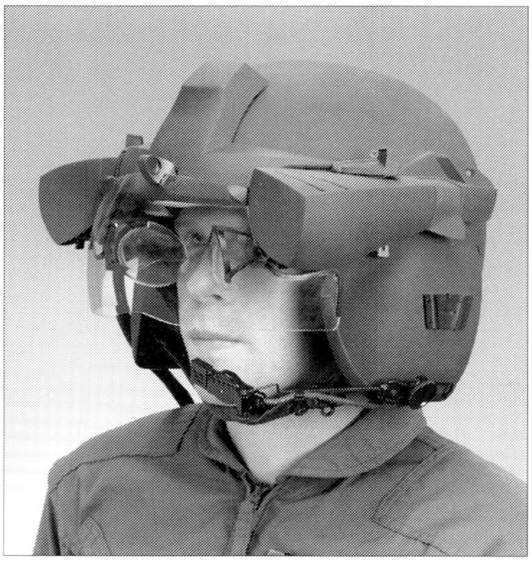

Fast jet fighter Helmet-Mounted Displays

As described earlier, the human factors and environmental aspects surrounding the application of a Helmet-Mounted Display onto a fast jet fighter are severe.

Much time and effort and not insubstantial sums of money have been expended to achieve operational realizations. The experimental aspects have been many and various, and only some of the highlights will be described here.

Helmet-Mounted Sight

The simplest and earliest form of Helmet-Mounted Display used in fast jet applications was the Helmet-Mounted Sight (HMS), used in daytime operations to acquire and designate targets off-boresight. The HMS installed and worn by RAF Jaguar pilots is shown in Fig. 8.14.

Fig. 8.14 *Jaguar Helmet-Mounted Sight (BAE SYSTEMS)*

The HMS comprises a simple fixed aiming reticle which is back-illuminated with a bright light source (such as an array of LEDs) (see Fig. 8.15). Simple optics collimates the image which is projected off the visor into the pilot's line of sight. The pilot is able to use the HMS as a simple sight to 'point' and designate targets to the mission avionics system. With the provision of some additional LEDs, it is possible to direct the pilot to look towards a potential target of interest or a threat.

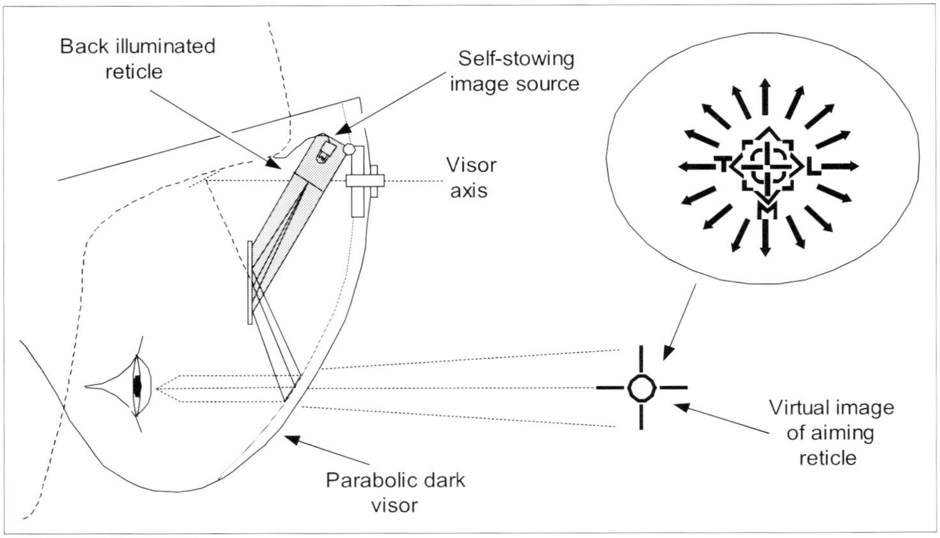

Fig. 8.15 *Simple Helmet-Mounted Sight*

Monocular daytime Helmet-Mounted Sight

The Kaiser Agile Eye™ was flown extensively on a number of aircraft types, most notably the F/A-18, during the late 1980s. The Agile Eye provided a daytime-viewable, monocular 30° circular field of view display. Symbology was generated in cursive (stroke) mode on a high-brightness miniature CRT mounted on the helmet itself. The optical system relayed and collimated the CRT image which was introduced into the pilot's line of sight by a semi-reflective combining patch on the visor itself. The patch had no optical power. The head tracker was electro-magnetic. Figure 8.16 shows the Kaiser Agile Eye HMD and others of the same genre. The CRT high-voltage supplies and services are remote. Although the product itself never progressed to production, much useful data were obtained both on the operational advantages to be gained from the use of an HMD and on product-specific requirements on which to base a production design.

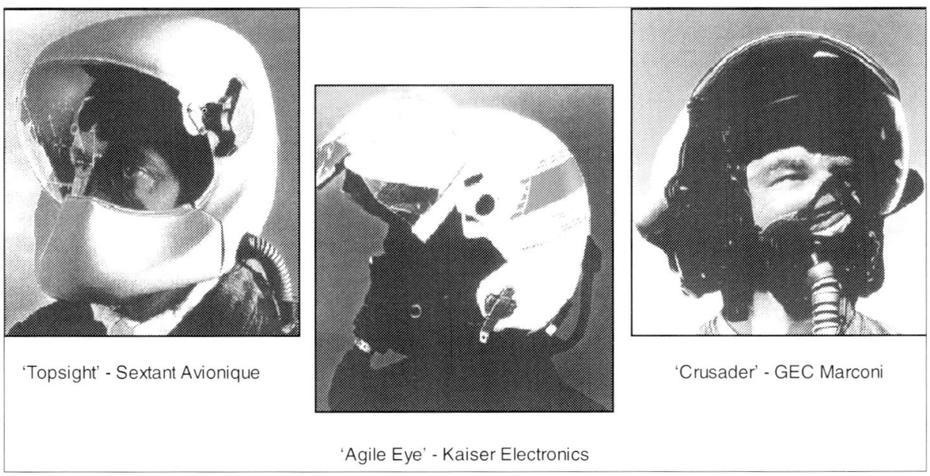

'Topsight' - Sextant Avionique

'Crusader' - GEC Marconi

'Agile Eye' - Kaiser Electronics

Fig. 8.16 *Examples of monocular daytime HMDs*

Fig. 8.17 Joint Helmet-Mounted Cueing Systems (JHMCS) (Rockwell-Collins)

Joint Helmet-Mounted Cueing Systems (JHMCS)

The JHMCS programme is the culmination of the extensive trials evaluations performed using helmets such as Agile Eye. The JHMCS, shown in Fig. 8.17, is designed to provide first-shot, high off-boresight weapon engagement capabilities. JHMCS enables the pilot to direct weapons against enemy aircraft while performing high-*g* aircraft manoeuvres. The pilot needs only to turn his head at the target and weapons will be directed to where the pilot's head is pointing. The system can also be employed accurately to cue the pilot to ground targets. Targeting cues and aircraft parameters are displayed directly on the pilot's visor. The JHMCS is designed to have low HMCS weight, optimized CofG, and in-flight replaceable modules to enhance operational performance – including the ability to be reconfigured in-flight to meet night vision requirements.

The JHMCS Helmet-Mounted Display has a monocular field of view of 20°. The exit pupil is 18 mm with an eye relief of 50 mm. Enhanced cueing is provided by two LED reticles. The display module is compatible with US Air Force HGU-55/P and US Navy HGU-68/P helmets. The weight is 4 lb (1.82 kg) with mask. MTBF is 1000 h.

Night vision goggle HMD

The first night-time Helmet-Mounted Displays were simple Night Vision Goggles (NVGs) attached to the helmet. Born out of necessity, the simple expedient of attaching NVGs to the helmet provided a low-cost extension to the operating role of a daytime aircraft. Of course, there were issues of compatibility to face to make the cockpit lighting, the instruments, and the displays compatible with the goggles. These are discussed in Chapter 12, but are generally soluble by relatively simple techniques and close attention to detail.

Figure 8.18 shows Cat's Eyes™ night vision goggles attached to a conventional helmet. Cat's Eyes are used on a number of fast jet fighters including the night attack F/A-18C/D. The optics is a quasi-axial collimating reflector which introduces the NVG image into the pilot's line of sight. The pilot is able to see the real world and his instruments/displays unaided through the collimator. He sees an intensified image of the real world superimposed on his real-world view. For total compatibility, the instruments and displays must not be 'seen' by the NVG intensifier tubes.

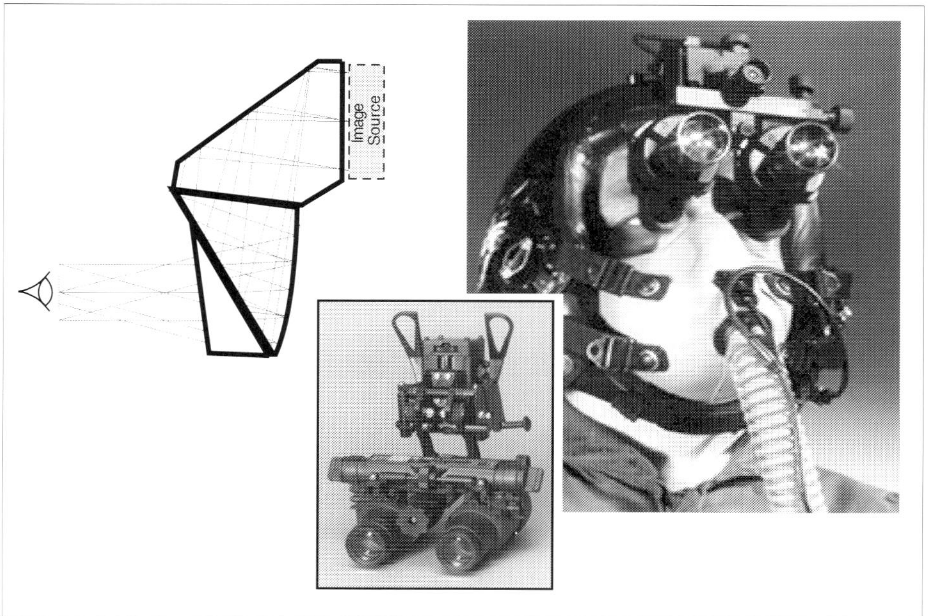

Fig. 8.18 Cat's Eyes night vision goggles (BAE SYSTEMS)

Eurofighter Typhoon binocular, day/night Helmet-Mounted Display

The Helmet-Mounted Display for Eurofighter Typhoon represents the current state of the art in HMD technology. The helmet, shown in Fig. 8.19:

- Provides primary aircrew protection.
- Attaches to the inner helmet.
- Provides a lightweight but stiff platform for the optical components:
  - CRTs/optics/mirror,
  - night vision cameras,
  - blast/display and glare visors,
  - head tracker diodes (infrared).
- Removable night vision cameras autodetach during ejection.
- Mechanical functions:
  - protection features in line with survivability limits,
  - life support up to the limits of human functionality,
  - comfort and stability to support display requirements.
- Display features:
  - $40^\circ \times 30^\circ$ binocular field of view,
  - corresponding $40^\circ \times 30^\circ$ binocular night vision camera,
  - display of sunlight-visible symbology and/or imagery.

The optics arrangement (dual binocular) shown in Fig. 8.20 uses a 1 in high-brightness, high-resolution, monochrome (P53 green) CRT as the image source. A complex relay lens with a brow mirror introduces the relayed image into the focal plane of two spherical diffractive mirrors which are deposited on the visor by holographic techniques. The field of view is 40° and the exit pupil is 15–20 mm. The optical arrangement introduces significant geometric distortion, as shown in Fig. 8.20, which is corrected electronically.

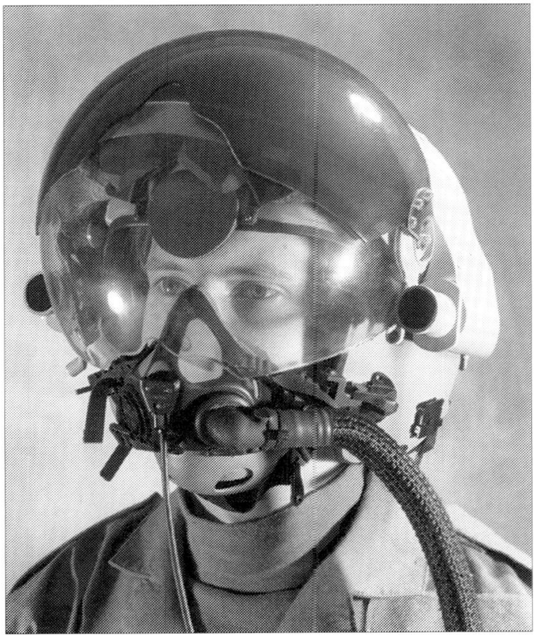

Fig. 8.19 Eurofighter Typhoon Helmet-Mounted Display (BAE SYSTEMS)

The helmet is a two-part arrangement (see Fig. 8.21). The inner helmet fits inside the display outer helmet and can be swapped with a respirator hood version to give NBC protection. The inner helmet:

- Facilitates individual user fitting.
- Maximizes comfort and stability:

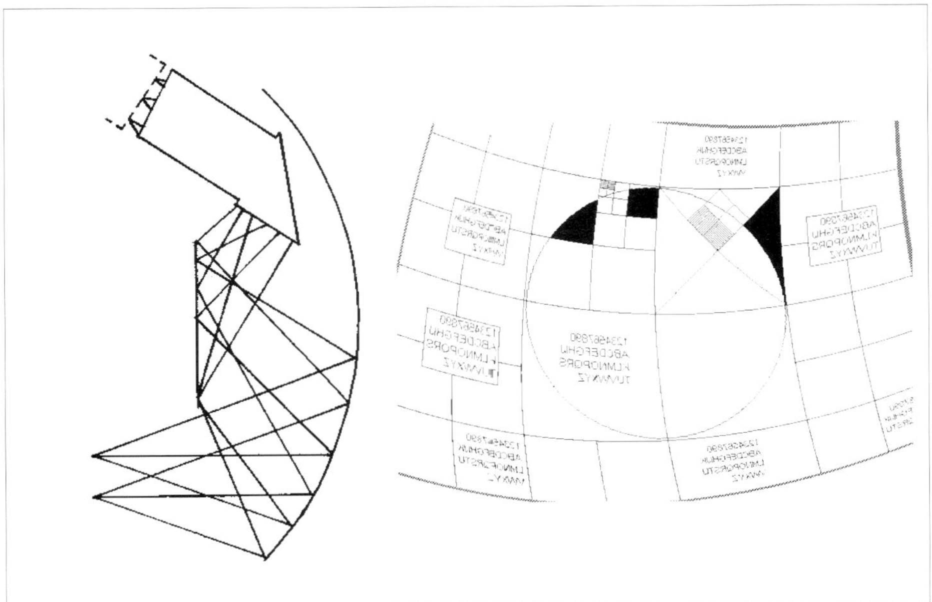

Fig. 8.20 Eurofighter Typhoon optical configuration (BAE SYSTEMS)

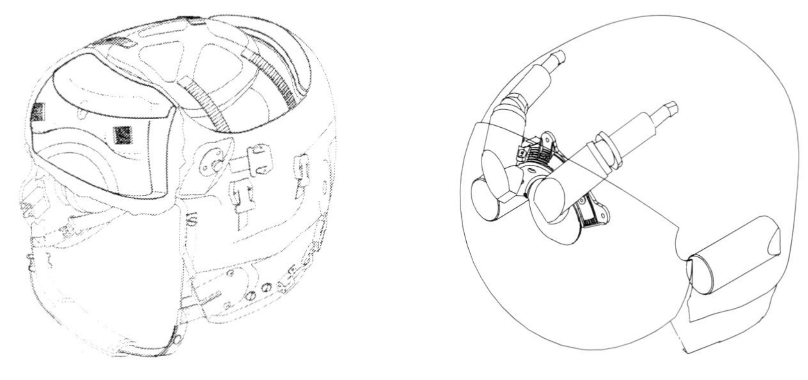

Fig. 8.21 Eurofighter Typhoon two-part helmet (BAE SYSTEMS)

  - the brow pad is moulded to an exact fit,
  - air circulation around the head is assisted.
- Embraces a wide anthropomorphic range:
  - advanced suspension system,
  - lightweight oxygen mask.
- Provides optimum hose/cable routing.

The display outer helmet attaches to the inner helmet. It provides:

- Primary aircrew protection.
- A lightweight but stiff platform for the optical components:
  - CRTs/optics/mirror,
  - NVE cameras,
  - windblast/display visor (clear) and glare visors.
- Head tracker diodes.
- Removable night vision cameras.

Joint strike fighter

The Joint Strike Fighter (JSF) is also planning to have a binocular day/night Helmet-Mounted Display. In the JSF it is intended to take the state of the art one further step and eliminate the Head-Up Display completely. The HMD will provide all the targeting information and sufficient primary flight information to meet the mission objectives.

As discussed in Chapter 10, the instrument panel is planned to provide a large area head-down display with image formats providing enhanced situational awareness to facilitate the easy and intuitive assimilation of flight and mission data.

Figure 8.22 shows a prototype of the JSF HMD. The helmet provides a $30^\circ \times 50^\circ$ total field of view from two overlapping $30^\circ \times 40^\circ$ monocular images. The optics is a pupil-forming arrangement with the relayed image collimated and introduced into the pilot's sight line by a diffractive holographic semi-transparent curved combiner introduced into the visor. The exit pupil is 18 mm. The image

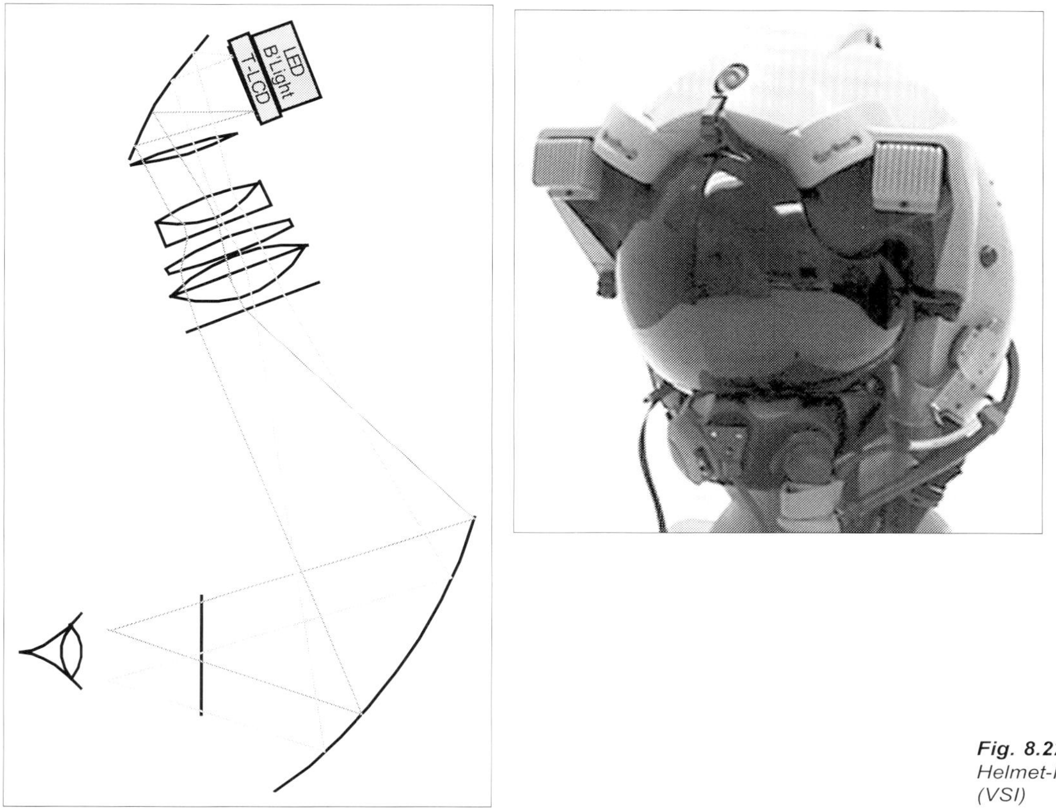

Fig. 8.22 Prototype JSF Helmet-Mounted Display (VSI)

source is a high-resolution (SXGA) transmissive LCD, illuminated with a bright LED backlight. The reverse sides of the image sources are visible in the photograph.

Helmet-Mounted Display functional description

A Helmet-Mounted Display is a heavily distributed system. For obvious reasons, only the minimum of components are carried on the head to minimize weight. A typical HMD system is shown in Fig. 8.23 and comprises the following Line Replaceable Units (LRUs):

- *Helmet shell*. The basic helmet provides a high level of pilot protection and comfort and is fitted with communications equipment, a form-fitting system, and display module-mounting points. The basic helmet is form fitted to an individual pilot and becomes part of his personal equipment.
- *HMD display module*. This often simply clips onto the basic helmet shell and contains the display components in a single-sized lightweight module. Dependent on the functionality, the components located within the display module are dual-image intensifiers, dual CRT displays, optical assemblies, a dual-visor system, a helmet tracker receiver, an auto brilliance sensor, a battery pack, and an umbilical cable.

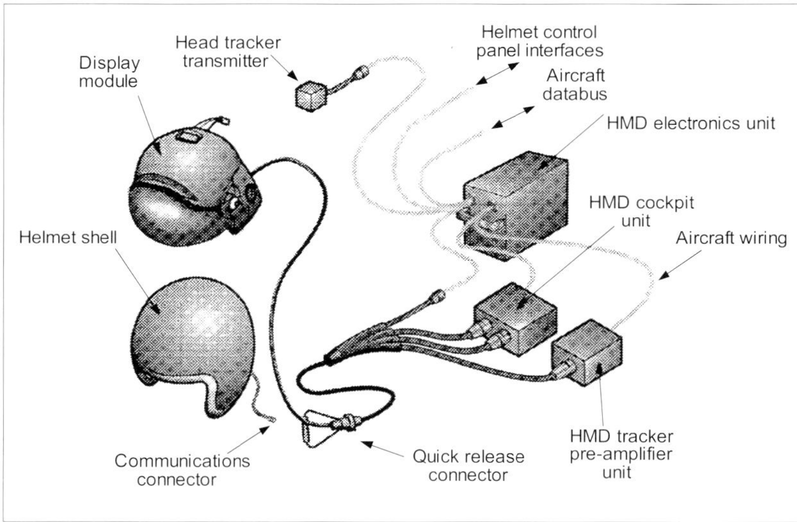

Fig. 8.23 *Typical HMD system components*

- *HMD electronics unit.* Located in the avionics bay, this contains the main HMD system electronics functions including system interfaces, processing, display drive, and helmet tracking.
- *HMD cockpit unit and tracker preamplifier.* These small units are bulkhead mounted in the cockpit close to the aircrew seat and contains the electronic functions that need to be connected by a short cable length to the HMD.
- *Head tracker transmitter.* This small unit is mounted physically close to the HMD, either on the aircrew seat or the canopy, and is part of the helmet tracker function.
- *Boresight reticle unit.* Two of these are mounted in the instrument panel in the cockpit and provide accurate optical references at well-defined lines of sight for boresighting the helmet tracker system.
- *Quick release connector.* This comprises the cable and connectors that run to the HMD quick-release connector.

The basic helmet profile and mechanical features must be optimized to integrate the requirements of the display components in such a way that the fundamental properties of the flying helmet are not compromised:

- *Protection.* The helmet shell (in combination with the display module) must provide the necessary level of impact attenuation and penetration resistance.
- *Weight.* Total head-supported weight is a vital parameter. The addition of display functions must of necessity introduce additional weight, which at the very least may be uncomfortable but may also create unsafe neck loadings. Therefore, it is usual to utilize a purpose-designed helmet shell which, although derived from current state-of-the-art helmet technology, is optimized for the HMD application. This also allows the helmet shape to be developed to enable the display components to be fully integrated into the helmet.
- *Centre of gravity.* The mass distribution on the helmet is such that the centre of gravity of the HMD and head combined must not result in significant out-of-balance moments about the neck pivot point. This is important to avoid

neck strain in both normal use and in 'high-*g*' manoeuvres, to minimize pilot fatigue, and for safety in conditions encountered during forced landings.

- *Helmet sizing*. Anthropomorphic databases are utilized in determining head size, shape, and thus the optimum helmet sizing philosophy to ensure that the helmet design is compatible with all likely pilot populations. Typically, several sizes of helmet shell are required to achieve coverage of aircrew population.
- *Helmet fitting*. In general, all flying helmets are tailored to the individual to achieve a fit, which is comfortable for all operational environments. Adequate helmet stability is desirable to ensure that the helmet does not move in use. The HMD introduces additional constraints, as the helmet fitting procedure must also align the helmet platform with the design eye of the optical system and the user's line of sight. Given the importance of head-supported weight, the helmet suspension system should be light in weight and provide good load distribution to eliminate pressure concentrations or 'hot spots'. Additionally, the liner system design should allow air to circulate around the user's head to avoid thermal stress.
- *Communications equipment*. The helmet-mounted components of the communications system are the headset, which is located within the helmet shell, and a microphone assembly on a support boom.
- *Acoustic noise attenuation*. The properties of the helmet are a product of both the helmet shell and ear capsule design. In practice the ear capsule provides the bulk of the sound attenuation for most helmets. In a fast jet the acoustic features must also be compatible with any Active Noise Reduction (ANR) systems.

Image source

Most current helmet-mounted displays use a miniature CRT as the image source (see Fig. 8.24). The CRT has a faceplate image area of typically 20 mm (0.75 in) and an overall length of around 100 mm (4 in). The CRT image is generated by an electron gun and focused electrostatically. The final anode potential is typically 13 kV and the electron beam current is around 100 μA. The electron beam is electromagnetically deflected by two orthogonal scan coils. To

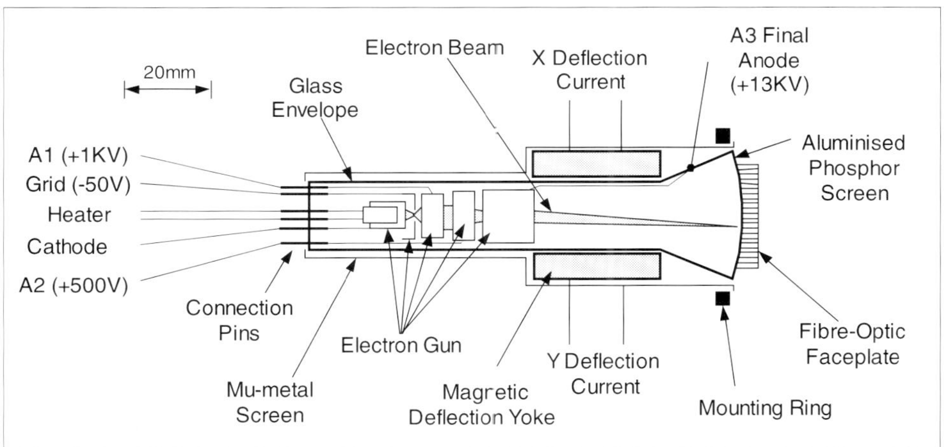

Fig. 8.24 *Typical HMD miniature CRT*

achieve high-brightness daytime viewability, the image is constructed in cursive (stroke) mode. If the HMD is also used for displaying night-time sensor imagery, then the combined sensor plus symbology image is constructed in raster or hybrid (stroke + raster) mode. A fibre-optic faceplate produces a flat image plane and enhances the image contrast.

Only the CRT and essential support electronics are mounted on the helmet to minimize weight. The high-voltage power supplies, video, and deflection amplifiers are mounted off-helmet. In the event of the need to eject, these links must be made electrically safe before disconnect. All high voltages must be removed to avoid any risk of high-voltage corona from the resultant exposed connector.

Next-generation HMDs may employ an LCD as the image source together with a bright backlight. The backlight might be an array of LEDs. This technology would avoid the need for high voltages.

Night vision goggles

To extend the operational use of the HMD, current designs are moving towards binocular systems providing day and night operation. In daytime use, the HMD provides a symbolic image of the type discussed above, generated in cursive (stroke) mode on the CRT. At night-time the symbolic image is augmented with sensor video from a turret-mounted Forward-Looking Infrared (FLIR) and/or Low-Light TV (LLTV) sensor slaved to the head line of sight. In addition, or alternatively, the night-time outside world view may be provided by Night Vision Goggles (NVGs) worn on the helmet.

Pilot-worn NVGs have been used for some time, attached to a standard helmet with suitable mountings and detachable on ejection. The main functional component of NVGs is the image intensifier, the operating principles of which are shown in Fig. 8.25.

The intensifier is sensitive to light in the near-infrared part of the electromagnetic spectrum. An objective lens focuses photons from the outside world scene onto the intensifier photocathode. The intensifier comprises a series of microchannels across which is placed a high-voltage potential of around 5 kV. A s-

Fig. 8.25 NVG image intensifier operating principles

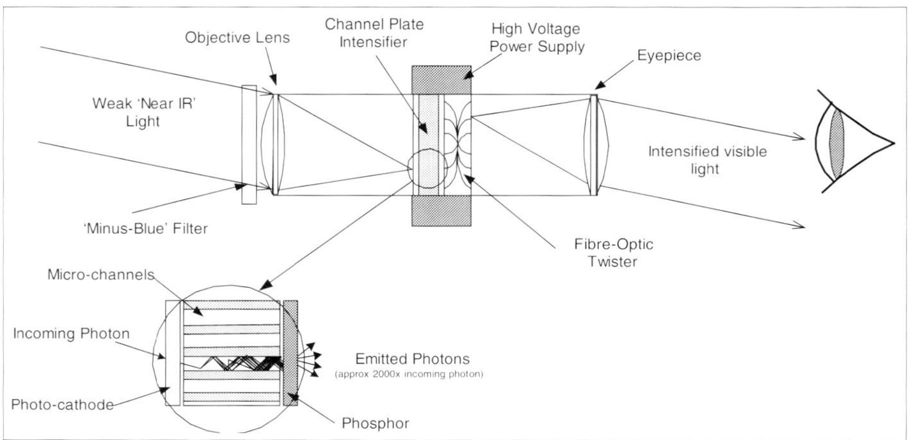

mall battery-powered inverter, which forms an annulus around the intensifier tube, generates the high voltage. An incoming photon generates an avalanche effect in the channel plate, producing multiple electrons from the single incoming photon. The gain is at least 2000:1. The electrons then collide with a phosphor screen which converts the energy into visible (green) light. A twisted fibre-optic bundle inverts the image so that through the eyepiece the observer sees an erect, real, intensified image of the outside world scene.

Binocular day/night HMD architectures

The Eurofighter Typhoon and next-generation HMDs integrate NVGs with the HMD image generation function. There are two means to achieve this, optically and electronically, and these are shown in Figs 8.26 and 8.27.

Optical integration of NVGs with HMD imagery

Figure 8.26 shows the arrangement optically to combine the NVG image intensifier image with the HMD image. The HMD image is generated conventionally by constructing symbology cursively (in stroke) on the CRT. The intensified image of the outside world scene is generated conventionally on the NVG phosphor screen. The two images are combined optically by an arrangement of prisms and a beam combiner. The combined image is the objective for the relay lens system which introduces a combined, collimated image into the pilot's line of sight.

This configuration has the advantage that the two image sources are completely independent. In the event of failure of any part of the HMD image generation system or the head tracker, the NVG image is still available to the pilot. A second benefit is that the NVG components can be carried on a detachable mount and only installed for night-time missions where their use is appropriate, thus minimizing weight and aerodynamic drag when NVGs are not required.

The disadvantage of this configuration is that the mix ratio of NVG to HMD images is fixed by the transmit/reflect characteristic of the beam combiner. Image intensity from both sources is reduced which might impact upon operational performance. At night-time it would be desirable to have a transmit/reflect ratio

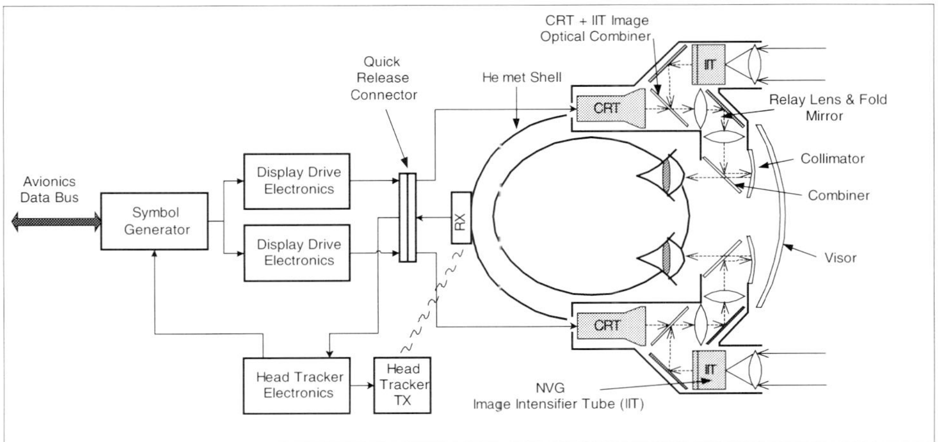

Fig. 8.26 Optical integration of HMD and NVG images

that strongly favoured the NVG-intensified image. However, in daytime, when the NVGs are not used, it would be highly desirable to have a transmit/reflect ratio biased in favour of the CRT image so as not to degrade performance under a high ambient luminous environment.

Electrical integration of NVGs with HMD imagery

Figure 8.27 shows the arrangement electronically to combine the NVG image intensifier image with the HMD image. Now the NVG image intensifier phosphor screen is not viewed directly but is viewed by a video camera. The video output from the camera is transmitted 'off-head' to the HMD electronics drive unit where it is electronically mixed with the HMD symbology. The HMD symbology overlay can be either in video (raster mode) or as cursive symbology in the field retrace period (hybrid stroke + raster mode). The combined electronic image is then transmitted back to the 'head' and written on the HMD CRT. The CRT image is then relayed, collimated and introduced into the pilot's line of sight as before, without any prisms or beam combiner.

This configuration has the advantage that the NVG and HMD mix ratios can be dynamically adjusted to suit the operational need. Furthermore, the NVG aspects are truly add-ons and can be simply detached from the helmet without any impact on the HMD optics should they not be required for the mission. Furthermore, any geometric corrections necessary, for example on account of an off-axis optical design, can be electronically applied to both the CRT and the NVG image.

However, the impact of failure is more severe. Both HMD and NVG images will be lost should the electronics drive unit fail. It is possible to mitigate against this failure mechanism by architecting a duplex system with full segregation between the left and right eye image generation electronics and projection optics. In the event that one lane is lost, the other lane will still continue to provide combined HMD and NVG image intensified images in a monocular manner.

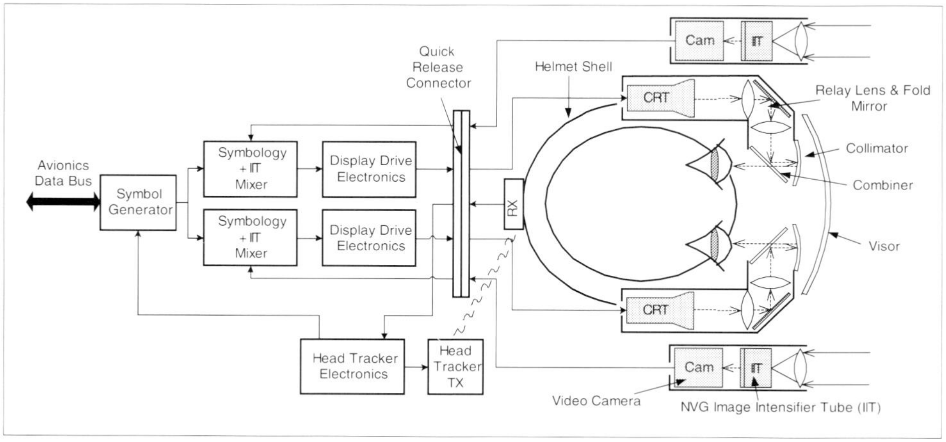

Fig. 8.27 *Electronic integration of HMD and NVG images*

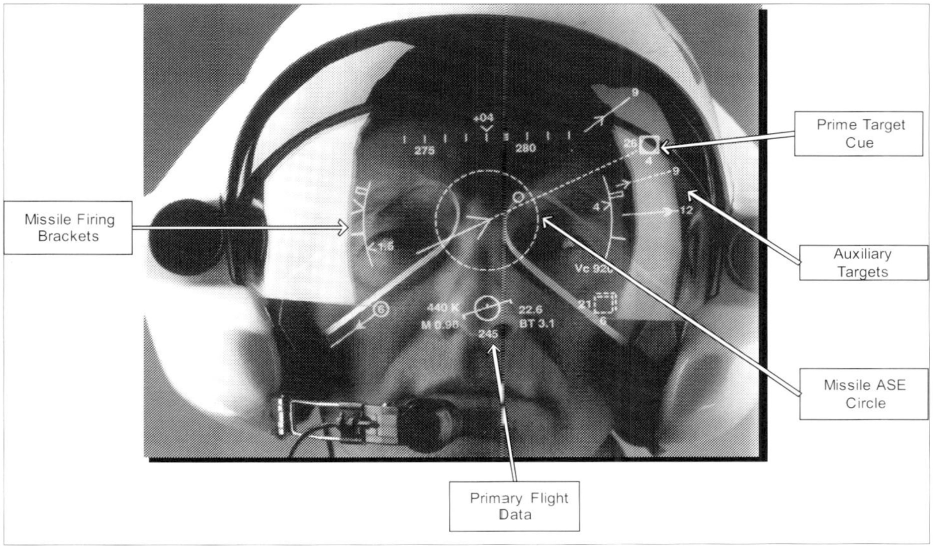

Fig. 8.28 Typical HMD symbology format (Thales)

Symbology

A typical symbology format is shown in Fig. 8.28. The first thing to note is that the primary flight data are subservient to targeting data. In all current HMD installations, a HUD is still fitted. The HUD is generally the primary flight instrument. Primary flight data on the HMD play a secondary role and are only used subliminally for quick orientation. The pilot defers to the HUD if he becomes seriously disoriented. Attitude information is not contact analogue for good reason. Presentation of contact analogue attitude information at any head angle significantly off-boresight would at best be confusing and could be extremely disorienting. Therefore, in this example, attitude information is presented as a simple line indicating pitch and roll against a fixed circle. The extensions at the end of the intersecting line indicate normal/inverted roll. Speed, height, and heading are presented as simple numerical readouts.

The primary aspects of the display format relate to targeting data to several priority targets. If the target is not within the HMD field of view, then a line with an arrowhead indicates the direction in which the pilot should look visually to acquire the radar target. Radar data such as range to target and target *g* level is displayed alongside the target marker. If missiles are selected and have acquired a target, then a target marker also shows the line of sight to the missile target. Additional information indicates probability of launch success as a set of range brackets and allowable steering error, the latter being computed from target and own-manoeuvre dynamics.

Helmet-Mounted Display as a primary flight reference

Primary flight symbology

While HMDs have been used to provide flight reference information in rotary-wing platforms for some time, fixed-wing HMD systems to date have been used primarily for targeting (i.e. JHMCS). It is desirable to replace HUD systems with HMD systems as a primary flight reference. Doing so will free up the significant cockpit real estate occupied by the HUD as well as greatly increase the look-through display field of regard available to the pilot for weapon aiming and sensor display. HMDs provide a virtually unlimited field of regard for off-boresight targeting, weapons employment, and display of sensor information.

The seeming advantages of this increased area of display space must be approached with caution, however. There is a strong potential for targeting sensor information presented off-boresight to draw the pilot's attention away from the traditional on-boresight flight reference information for longer periods of time. During periods of limited visibility, the increased off-boresight time could result in reduced spatial awareness if some sort of flight information is not presented in the off-boresight display. While the presentation of primary flight information in HUDs is well understood and documented, methods for presenting off-boresight information are not as mature. Questions remain as to whether off-boresight flight information should be aircraft referenced or world referenced.

Reference (**4**) describes the ongoing work being undertaken by the USNR Naval Air Warfare Center Aircraft Division, Patuxent River, and by Boeing to design and evaluate operational Helmet-Mounted Display (HMD) symbology formats that fulfil the primary flight reference (PFR) requirements. The objective of the HMD/PFR programme was to develop, demonstrate, and evaluate HMD concepts that can be transitioned into PFR systems for US military aircraft. The study evaluated a number of potential HMD formats, shown in Fig. 8.29, including:

- The Virtual HUD. A baseline HUD configuration that approximated an MIL-STD 1787B configuration for basic flight symbology.
- Modified Arc Segment Attitude Reference (ASAR). ASAR symbology presents pitch information via a circular arc segment. This was modified to include markings for more accurate pitch angle interpretation as well as indications for airspeed, altitude, and heading. The angular relation between a reference symbol and the centre of the arc segment gives roll information.
- Expanded HUD (attitude sphere). The expanded HUD format represents an attempt to create a world-referenced HMD attitude symbol set by expanding the MIL-STD pitch ladder and horizon line throughout the full HMD field of regard. The expanded HUD symbology was designed to provide the pilots with the sensation of being inside a virtual globe formed by the longitudinal lines and the conformal horizon.

The study findings indicated:

- Off-boresight symbology should look distinctly different from conventional up-front (HUD) symbology.
- The symbology should contain airspeed, attitude, heading, and altitude information, as well as rate information for airspeed and altitude.
- The question of whether or not off-boresight flight symbology should be aircraft or world referenced is not easily answered. At this stage the results do not strongly favour one over the other.

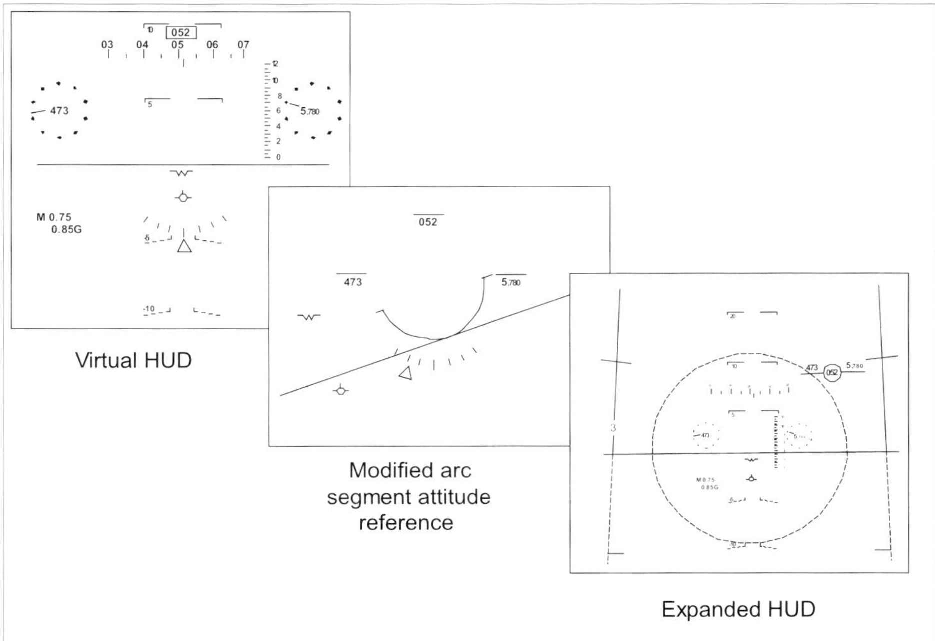

Fig. 8.29 Experimental primary flight symbology

HMD impacts

Expanding the role of the HMD to air to air and air to ground in day and night scenarios, and further as the primary display in the next generation of aircraft, has a significant impact on HMD designs. Most requirements for night use are becoming achievable with potential technologies that promise larger FoVs, greater resolutions, and greater accuracies. Day use has different requirements for display of symbology only over a smaller FoV, and minimal weight for high-*g* combat manoeuvres.

Requiring the HMD to be the primary display significantly increases the accuracy requirement of head tracking to allow precision aiming and delivery of ballistic weapons. Current technologies come close to meeting both day and night needs of the future programmes but fall short of meeting all of the requirements. Table 8.2 outlines a summary of the requirements that are growing out of the understanding for the uses for the HMDs as the primary display for day and night and air and ground sighting.

Table 8.2 HMD requirements for primary display

Field of view	Minimum of 50° horizontal
Weight	Maximum of 4.0 lb
Centre of gravity	Within 2.0 in of occipital condyle
System accuracy	Less than 2 mr r.m.s. in forward cone

Weight, CofG and safety

The major components of helmet-mounted displays for weight and CofG are listed in Table 8.3. The CofG must be within 2 in of the occipital condyle in order to meet the safety requirements. The head and helmet are the major components of the system weight, leaving little design margin for the display module. The weight requirement is less than 4.0 lb for a day/night helmet. This is about the same as most day-only helmets now in production. With the head and helmet driving the CofG and little design margin for the weight, new methods for HMD design must be employed.

Table 8.3 Major component for weight and CofG calculations

Human head	10 lb
Helmet	2 lb
Display module	1.5 lb
Oxygen mask	0.6 lb
Cabling pounds	1.0 lb

Accuracy

Most of the current programmes to advance the HMD technology focus on the display characteristics such as FoV, brightness, and resolution. However, most ignore the need for a fully functional day/night Helmet-Mounted Display that meets the tight accuracy requirements of air-to-air and air-to-ground operations. The major components that make up the accuracy of most Helmet-Mounted Systems can be seen in Table 8.4.

Table 8.4 Contributors to HMD accuracy

Human performance	1–2 mr r.m.s. or more
Boresighting	1–2 mr r.m.s.
Helmet-mounted display	1.8 mr r.m.s.
Tracker algorithms	Less than 1 mr r.m.s.
Aircraft metal	3 mr RMS or more after compensation

The next-generation helmet is likely to use the following technologies to achieve these requirements (**5**):

- Hybrid trackers. Most optical trackers offer a good solution for aircraft environments where large metal content is an issue. However, they are of limited use at night since a light source (usually infrared) is required. Magnetic trackers do not suffer from the limitations of obstruction and IR night use that most optical trackers suffer from. The magnetic trackers can be used successfully in most cockpits where the metal content does not exceed its compensation capability. Hybrid trackers will be developed to offer the best of both technologies. Neither a stand-alone optical or magnetic tracker can fully meet the requirements of a primary helmet-mounted display. However, by carefully combining these two technologies, a high accuracy can be achieved while still maintaining a good dynamic response for the air-to-air scenario.

- Visor system. Reflective coatings on the visor form the optical combiners that allow viewing of the HMD image data. The visor will use a narrow-band coating that provides good see-through visibility and image brightness for daytime operations.
- Relay optics. The optical housing design will use a lightweight material that has proven performance with respect to dimensional stability. Critical components are protected, within the housing, using a low-cost, easy-to-replace window element.
- Image source. Most probably AMLCD and backlight modules, along with associated drive electronics, will provide a complete semiconductor image source. The NVG image will be mixed electronically off-helmet. Semiconductor image sources offer significant advantages over traditional CRT devices in HMD applications. High-voltage power supplies, high-voltage interconnects, sweep generators, and deflection/video amplifiers are all eliminated. The associated power, weight, cost, and reliability problems are also eliminated. The AMLCD plus backlight is capable of achieving the display performance required by the day/night primary HMD application (response, contrast, refresh/update, and grey shades). The backlight will probably be implemented using a miniature LED array. These devices are capable of the light output required to meet the system day viewing requirements.

References

(1) **Garland, D. J., Wise, J. A.** and **Hopkin, V. D.** *Handbook of Aviation Human Factors.*

(2) **Bartlett, C. T.** *Head Up Displays and Helmet Mounted Displays. A Comparison of Accuracy and Integrity.*

(3) **White, J.** and **Cameron, A. A.** *Knighthelm 24 Hour HMD – From Development to Production*, BAE SYSTEMS Avionics limited, Rochester, Kent.

(4) **Christian, L. T.** and **Hart, L.** *Helmet Mounted Display as a Primary Flight Reference*, Naval Air Warfare Centre Aircraft Division, Patuxent River.

(5) **Foote, B.** *Design Impacts to the HMD When Used as the Primary Display*, Vision Systems Int, San Jose, California.

CHAPTER 9

Emerging Display Technologies

Introduction

Chapters 5 to 8 described the operation of the current display technologies used in aircraft display systems, namely monochrome CRT [miniature CRTs for HUDs, and subminiature (or micro) CRTs for HMDs] and colour shadow mask CRTs now superseded by direct-view colour AMLCDs for head-down displays. Chapter 5 discussed issues such as standardization, obsolescence, and the use of ruggedized COTS display products, including the possibility of resizing COTS AMLCD products to aerospace display sizes.

This chapter describes the operating principles of emerging new display technologies that offer the potential for larger, higher-resolution, lighter, more reliable, and more available devices and their application to aerospace large-area Head-Down Displays, Head-Up Displays, and Helmet-Mounted Displays.

Candidate technologies

The field of display projectors has paved the way in the search for alternatives to the CRT. Since aerospace displays share a similar reliance on high-intensity lighting with these technologies, the key to a CRT replacement is most likely to be found here. As has been so painfully demonstrated, to minimize obsolescence problems it is essential to use, adapt, and apply COTS mainstream display technologies to airborne applications.

Microminiature LCDs and Digital Micromirrors Devices (DMDs) have recently revolutionized digital projector technology. Current LCD/DMD projectors produced for the consumer market can fit in a fraction of a cubic foot, cost a few thousand dollars, and can illuminate a large viewing screen in daylight. This is a very marked advance from where the technology was 10 years ago. These technological breakthroughs in the commercial projector market are at the forefront of candidates for the development of new, cost-effective, and robust

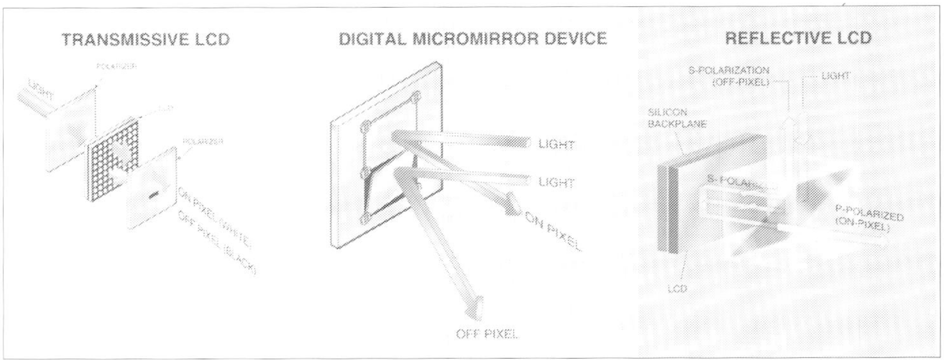

Fig. 9.1 *Microdisplay technologies*

airborne displays (**1**). There are currently three microdisplay technologies employed in display projectors, and these are shown in Fig. 9.1.

Liquid crystal displays

There are a number of possible liquid crystal configurations that may be employed for projector light valves. One of the most popular is the twisted nematic configuration. The basic principle of an LCD is that, in the absence of an electric field, light is able to pass through the LCD and therefore the LCD is 'on'. However, when an electric field is applied, the molecular orientation of the liquid crystal no longer allows for the passage of light; therefore, the LCD is 'off'. This twisted nematic configuration has a higher contrast ratio and reliability than other liquid crystal arrangements and is the predominant configuration for transmissive LCDs.

MicroLCDs can be categorized according to the positioning of their light sources:

- *Transmissive LCDs* have been the dominant display device for conference room projection applications. In a transmissive display, a light source is situated behind the liquid crystal light valve, projecting through the liquid crystal. They are a mature product, but suffer from resolution limitations in the sizes that are relevant for airborne display applications, and image contrast falls rapidly off-axis.
- *Reflective LCDs*, also known as *Liquid Crystal On Silicon (LCoS)* displays, are rapidly emerging as the technology of choice for new low to medium end-applications. In a reflective display, the light source is positioned so that light reflects from the LCD onto the projection screen. These devices utilize a backplane of conventional crystalline silicon technology that is aluminized to increase the reflectivity. An LCD material is applied and sealed with a cover glass.

The pixel configuration is somewhat different between displays, but the geometry for either may be readily achieved in practice. One important advantage of a reflective LCD is that it allows for the switching matrix to be placed behind the focal plane. In this position it does not occupy space in the image path, so the device pixels may be placed closer together, resulting in a higher image resolution.

Digital Micromirror Devices (DMDs)

The DMD is the primary competitor to the LCD for the image-modulating component of a projector, appearing more widely in medium to high end-applications. Compact projection systems are possible when a single DMD device is operated in a colour sequential mode giving improved performance when compared with a three-image LCoS solution.

The DMD had its origins in a Texas Instruments programme that was aimed towards producing a high-performance printing array. The device was originally implemented as an analogue scattering array, that is, the mirror movement would be linearly dependent on the device input signal. This proved not to be viable because many components within the device exhibited hysteresis – the tendency to stick in one position. Consequently, Texas Instruments redeveloped the device to be digital. In this implementation the mirrors have two possible positions, either left-facing or right-facing. By altering the frequency of mirror modulations, different intensity levels or greyscales can be produced.

The digital micromirror device, or Digital Light Projector (DLP) as TI prefers to name it, has certain advantages over the LCD. The switching time is of the order of 15 μs, making it roughly 1000 times faster than a liquid crystal, with the result that possible difficulties associated with image lag are markedly reduced. Further, the contrast ratio is higher and the device may be more readily ruggedized for aviation environments. On the negative side, the device must be operated with a point light source, which is more difficult to implement than the fluorescent backlights used with transmissive liquid crystal arrays.

High-intensity light sources

What makes the CRT so useful, and therefore so difficult to replace, is that it functions as both an image modulator and a light source. Though the LCD and DMD have proved much more reliable image modulators than the CRT, there is still a need to find a viable new light source. Typical projector applications employ high-intensity arc lamps, but there are many obstacles to overcome in adapting arc lamps to avionics applications. For example, the acceleration of a high-performance aircraft can distort the arc, reducing the amount of light that can be used in projection. Also, a means must be developed to dim the lamp over several orders of magnitude (at least 10 000:1) to match the display luminance to the ambient light levels found in an airborne application.

Other light sources, such as Light-Emitting Diodes (LEDs) and high-frequency fluorescent lights have also been tested, but at present lack the necessary brightness and contrast ratio because of high levels of light divergence. The laser is another possibility, but reliability is uncertain in an airbourne environment.

Organic LED

Light-Emitting Diodes, based upon semiconductors such as gallium arsenide, gallium phosphide, and, most recently, gallium nitride, have been around since the late 1950s. They are mostly used as indicator lamps, although they were used in calculators before liquid crystals, and are used in large advertising signs, where they are valued for very long life and high brightness.

Crystalline LEDs are expensive, and it is very difficult to integrate them into small high-resolution displays. However, there is a wide class of organic compounds, called conjugated organics or conjugated polymers, that have many of the characteristics of semiconductors. Just as with crystalline semiconductors, *p*-type and *n*-type organic materials can be combined to make light-emitting diodes whereby a current passing through a simple layered structure produces visible light with high efficiency. While the early diodes did not have sufficient efficiency or life to be commercially attractive, the progress in improving these factors has been astounding, probably owing in part to the infinite variety of organic compounds that can be synthesized.

Organic LED (OLED) technology is now emerging as having potential benefit for both direct-view and microdisplay applications. In both cases, OLEDs offer higher efficiency and lower weight than competing liquid crystal displays, since they do not require backlights or reflective light sources, but high luminance and long life are proving difficult to achieve.

Virtual retinal displays

The design and implementation of existing crewstations constrain the transfer information through the human–machine interface. Conventional panel-mounted displays [Head-Down Displays (HDDs)] and the limited Field of View (FoV) provided by Head-Up Displays (HUDs) require a pilot to view and comprehend a three-dimensional world on the basis of the two-dimensional display of information.

Retinal scan technology offers the unique ability to deliver a high-resolution, full colour information-rich virtual crewstation directly into the eye. However, concerns have been expressed about safety.

The Transmissive LCD

The operating principle of the Transmissive LCD (T-LCD) microdisplay is very similar to the direct-view AMLCD except that the device is of course much smaller and designed for projection applications. A typical T-LCD is shown in Fig. 9.2, and can be thought of as the electronic equivalent of a 35 mm transparency. It is illuminated (as a direct-view AMLCD) by a very intense light source. A light-blocking layer is incorporated to protect the thin film transistors (TFTs) being affected by the light source which otherwise would act as phototransistors.

The transmission of a T-LCD is about 10–20 per cent. The contrast is by no means uniform, as shown in the polar plot of Fig. 9.2, and rapidly reduces beyond about 10°.

The Reflective LCD

The functional principles of a Reflective LCD (R-LCD) projector are shown in Fig. 9.3. The architecture is fairly standard for professional projectors. The components are relatively inexpensive and multisourced. The assembly has to be accurately aligned and be rigid both thermally and mechanically for aerospace applications.

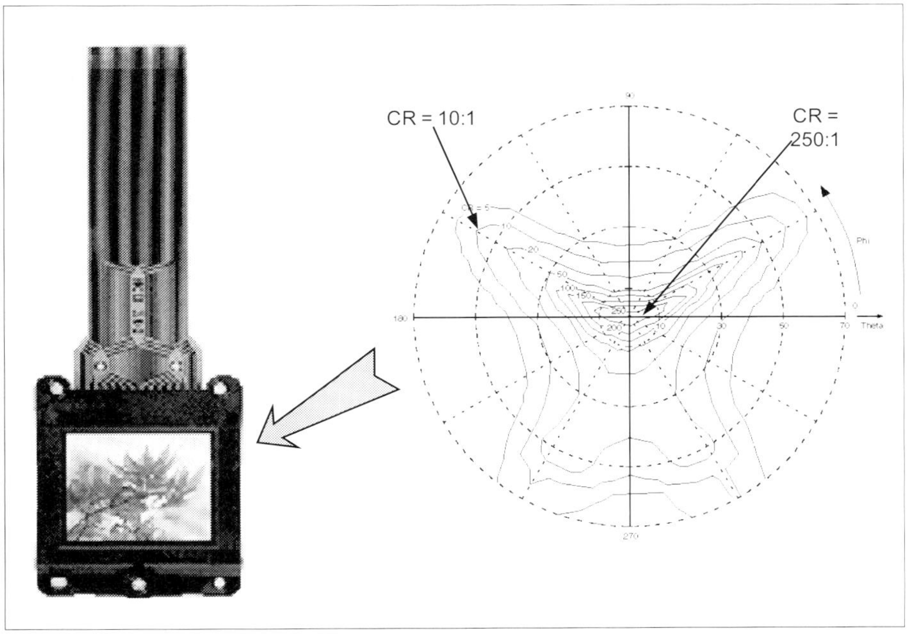

Fig. 9.2 *Typical transmissive LCD device and contrast plot*

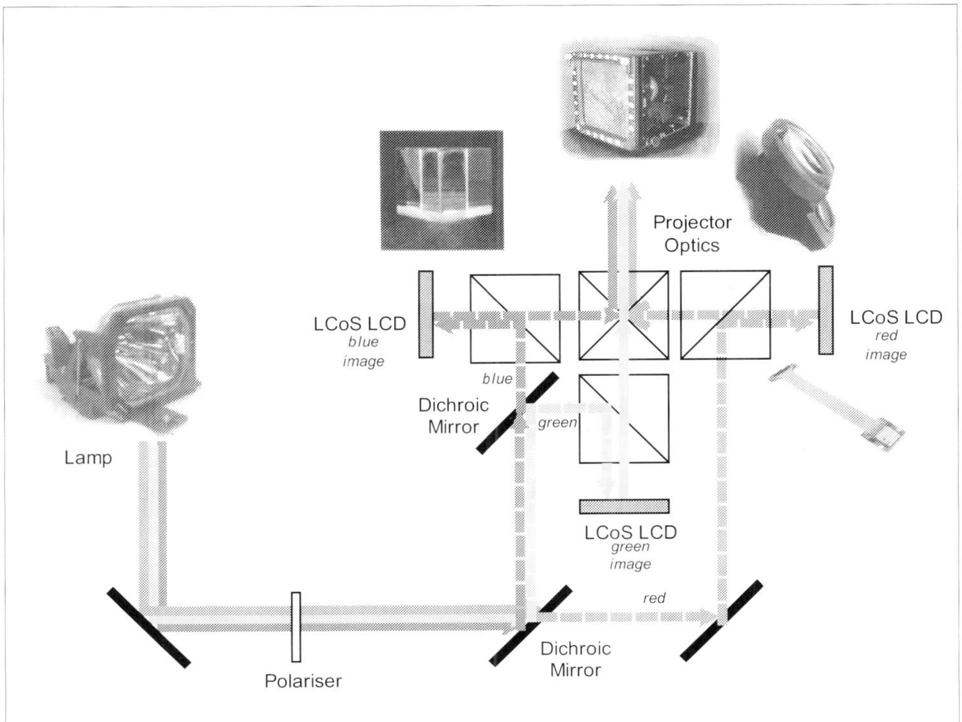

Fig. 9.3 *Reflective LCD projector (Kaiser Electronics)*

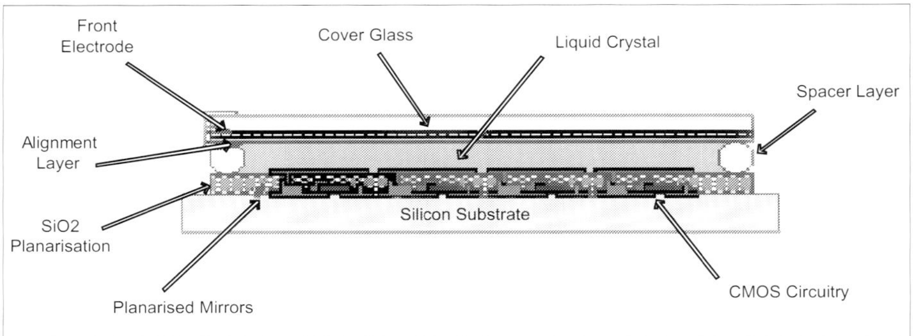

Fig. 9.4 *Reflective LCD (LCoS) device structure*

Reflective LCD, or Liquid Crystal on Silicon (LCoS), devices themselves are multisourced and are in widespread use in tabletop projectors. The imaging device, shown in Fig. 9.4, is essentially semiconductor wafer technology using LC principles to polarize reflected light through a simply attached liquid crystal cell placed over the device.

For a monochrome display (as in an HUD or an HMD), only a single device is required, illuminated by a monochromatic source. For a full colour display, typically three LCoS devices are employed, with beam-splitting optics to generate three R:G:B images which are recombined and projected onto the screen. The switching time of LCoS devices is too slow to permit the sequential colour arrangement employed with DMD devices (described later) to be used. The spectrum of a bright, white light source is split into its three primary components and illuminates three reflective micro-LCDs (or LCoS devices). The three images are recombined and through a projector lens are focused onto a rear view screen. The white light source is typically a mercury arc lamp. Dimming has to be achieved by mechanical/optical shuttering in the light path. The lamp will dissipate about 60 W, so forced air cooling is required.

The reflective LCoS LCDs are conventional CMOS processing on a silicon substrate. Examples of typical devices from a variety of manufacturers are shown in Fig. 9.5. The optical components are conventional. The device size is similar to a Pentium III. A glass cover plate retains the AMLCD material. The projector lens is conventional design but is optimized to the image throw distance (screen size). One projector fits all, with a reoptimized projection lens. The lamp is about 1 in^3, the image engine (optics plus micro-LCDs) about $4 \times 1 \times 1$ in, and the projection lens about 2 in^3. The projection angle is about 90°.

The digital micromirror device

The Digital Micromirror Device (DMD) comprises an array of silicon micromachined mirrors on a silicon wafer. The mirrors are tilted to modulate the angle of reflection of light. Because of the improved reflection efficiency of DMD compared with LCoS and the fast switching speed, it is possible to build a projector with a single DMD device obtaining full colour rendition using sequential colour processing. This arrangement eliminates the need for a beam splitter and the subsequent problems of alignment/reconvergence of the three

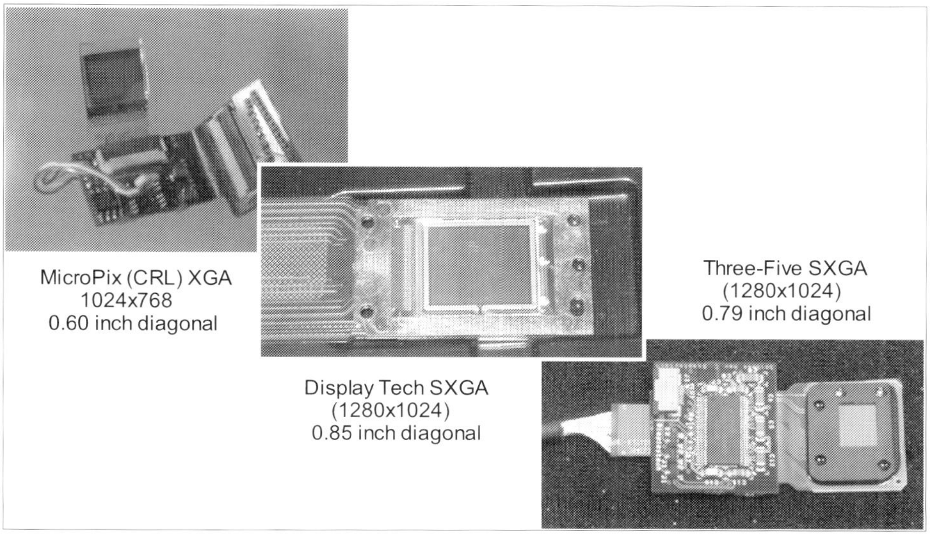

Fig. 9.5 *Examples of LCoS devices*

colour images. The result is a significantly clearer image than LCoS technology, and in smaller volume.

The development of the DMD by Texas Instruments has yielded a robust, reliable, high-performance device that can now be packaged with other components to produce a range of display solutions. This resultant Digital Light Projector (DLP) brings together the DMD with a light source, optics, colour filters, and a projection lens to produce an image on a display surface (i.e. rear projected onto a screen). The arrangement is now widely used for PC and cinema projection systems (**2**).

The DMD is an array of aluminium micromirrors as shown in Fig. 9.6. These are monolithically fabricated over an array of CMOS random-access memory cells, each of which corresponds to a micromirror. This allows each mirror to be individually addressed, causing it to rotate approximately 10°, limited by a mechanical stop. The micromirror superstructure is fabricated through a series of

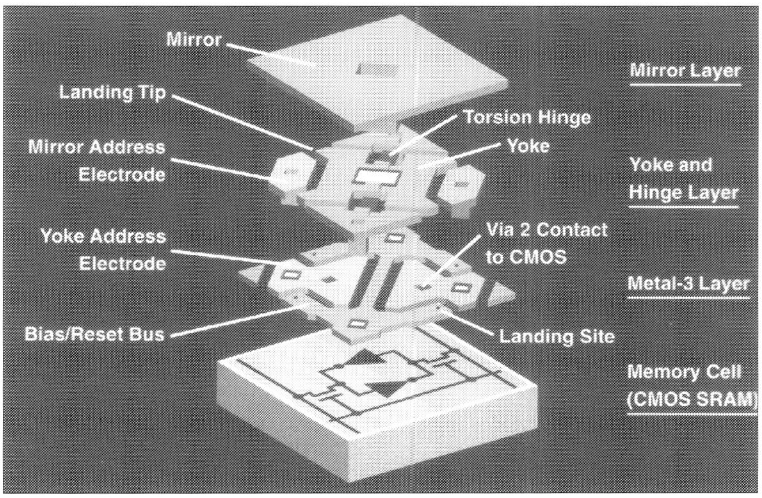

Fig. 9.6 *DMD pixel (Texas Instruments)*

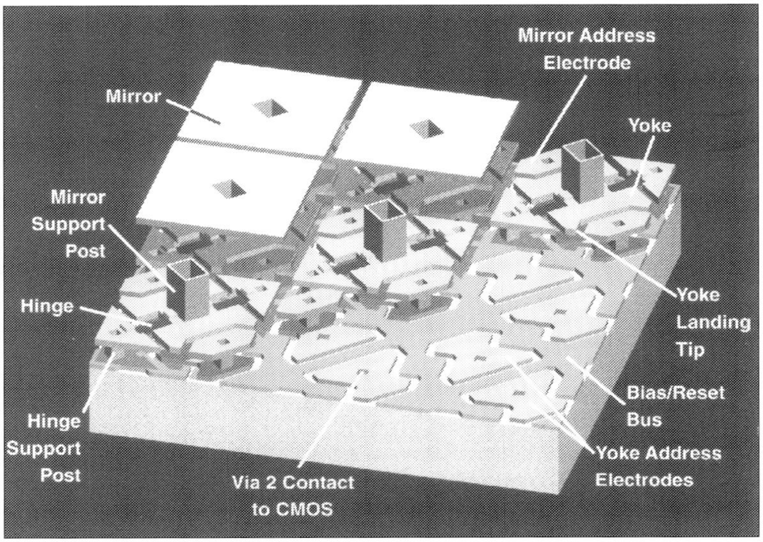

Fig. 9.7 *DMD array (Texas Instruments)*

aluminium metal depositions, oxide masks, metal etches, and organic spacers. The organic spacers are subsequently etched away to leave the micromirror structure free to move.

The mirror is tilted as a result of electrostatic attraction between the mirror structure and the underlying memory cell. The mirror and yoke are connected to a bias/reset voltage. The address electrodes are connected to the underlying CMOS memory. Movement of the mirror is accomplished by applying a bias voltage to the mirror/yoke structure according to data stored in the memory cell. When this occurs, the mirror is attracted to the side with the largest electrostatic field. To release the mirror, a short reset pulse is applied to the mirror, which excites the resonant mode of the structure, and the bias voltage is removed. The combination of these two occurrences results in the mirror leaving the landing site. The mirror lands again when the bias voltage is reapplied.

Each micromirror is 16 μm square. The array, shown in Fig. 9.7, places each micromirror on a 17 μm pitch, leaving a gap of less than 1 mm between the micromirrors. This results in a greater than 90 per cent 'fill factor' and is one significant advantage of the DMD. The configuration of the array is flexible, depending on the application. DMD array sizes have been produced in the following configurations: 640×480 (VGA resolution), 800×600 (SVGA resolution), 1024×768 (XGA resolution), and 1280×1024 (SXGA resolution), as well as demonstrations of other array sizes.

A bright light source illuminates the DMD array. Depending on the state of each micromirror, light is either reflected towards the projector objective lens or towards a light absorber on the wall of the projector as shown in Fig. 9.8.

Full colour operation is obtained by interposing a colour wheel comprising red, green, and blue colour filters between the light source and the DMD device as shown in Fig. 9.9. Complex algorithms in the associated processor/DLP chip set provide signals to each DMD pixel to produce a full colour image with at least 256 grey levels per colour by suitable mark-space modulation of each pixel in synchronism with the rotation of the colour wheel.

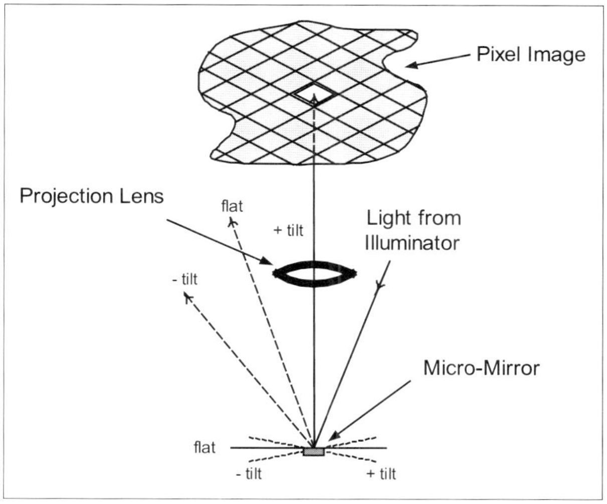

Fig. 9.8 *DMD light path (Texas Instruments)*

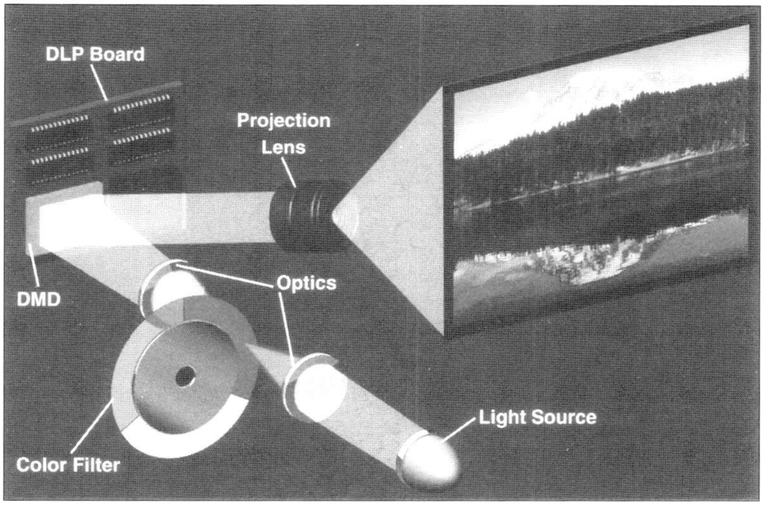

Fig. 9.9 *DMD projector (Texas Instruments)*

DLP benefits

The DMD fill factor of greater than 90 per cent, coupled with the fact that each pixel displays all three of the primary colours, gives the DMD an appearance of a finer pitch than any other device of comparable resolution. When an operator works with the display at close ranges, like a monitor, this reduces the graininess of the image being displayed. The result is an image that appears contiguous and photograph-like in quality.

The switching speed of the DMD is of the order of 15 µs. This allows the total image to be updated at speeds that eliminate the interframe image retention associated with other technologies. The fast update rate of the DMD eliminates the smearing and other motion artefacts.

Reliability tests of the DMD have demonstrated greater then 100 000 operating hours and more than 1 trillion mirror cycles. Texas Instruments has performed extensive testing of the DMD device at various operating temperatures and its performance is well documented. To guarantee an operating life of 10 000 hours, TI suggests a continuous operating temperature of less than 70 °C.

Independent studies have tested the DMD to much higher transient temperatures. Other display technologies are more sensitive to extremes. It is well known that LCD devices do not respond well to low-temperature extremes. Both LCDs and LCoS devices exhibit a phase change at high temperatures.

The robustness of the DMD against vibration and shock make this technology a good candidate for rugged environments. There has been considerable testing of the DMD under vibration environments. The small mirror mass pushes the resonate frequency of the spring/mass system above 100 kHz. This is far above normal mechanical environments. The electrostatic forces involved in the actuation of the mirrors far exceed the acceleration forces. Accelerated testing of entire DLP projection systems has shown the display can withstand random vibration environments of over 25 g r.m.s. while operating.

By far the largest criticism of single-device DLP systems has been sequential colour. This problem is well understood and has been effectively addressed. Original DLP systems used a three- or four-segmented colour wheel that was synchronized to the incoming video. These systems are responsible for the early sequential colour criticism. Since then, much research has been performed in the area of video processing and colour wheel optimization. The biggest gains in eliminating the sequential colour criticism were in the areas of colour wheel design, video processing, and device operation.

Organic Light-Emitting Diodes (OLEDS)

The operation of a LED is based upon the fact that semiconductors can be of two types, *p*-type or *n*-type, depending upon whether dopants pull electrons out of the crystal, forming ‘holes’, or add electrons. A LED is formed when *p*-type and *n*-type materials are joined. When a voltage is applied, causing electrons to flow through the structure, electrons flow into the *p*-type material, and holes flow into the *n*-type material. An electron–hole combination is unstable; there is too much potential energy to be stored. As a result, they combine and release the energy in the form of light. This can be a very efficient way to convert electricity into light.

Light-emitting diodes, based upon semiconductors such as gallium arsenide, gallium phosphide, and, most recently, gallium nitride, have been around since the late 1950s. These crystalline LEDs are expensive, and it is difficult to integrate them into small high-resolution displays.

Organic compounds, called conjugated organics or conjugated polymers have many of the characteristics of semiconductors. They have energy gaps of about the same magnitude, they are poor conductors without dopants, and they can be doped to conduct by electrons (*n*-type) or holes (*p*-type). Initially, these materials were used as photoconductors, to replace inorganic semiconductor photoconductors, such as selenium, in copiers. About 15 years ago, Ching Tang and Steve Van Slyke of Eastman Kodak discovered that, just as with crystalline semiconductors, *p*-type and *n*-type organic materials can be combined to make light-emitting

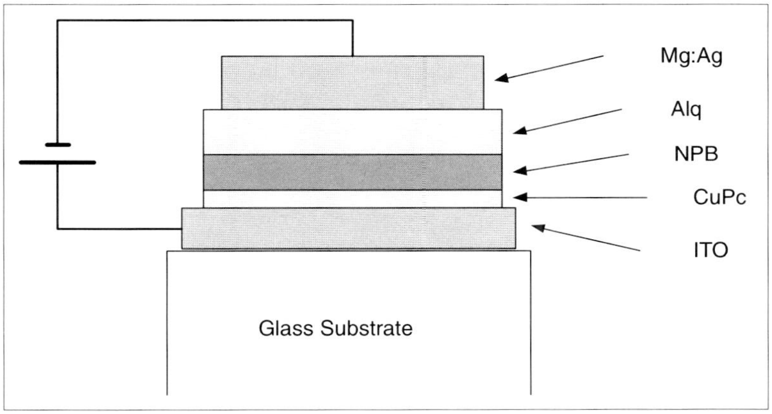

Fig. 9.10 OLED emitter (eMagin)

diodes in which a current passing through a simple layered structure produces visible light.

While the early diodes did not have sufficient efficiency or life to be commercially attractive, the progress in improving these factors has been astounding. In contrast with crystalline semiconductors, the OLED fabrication processes allow the possibility of making very high-resolution displays. Also, full colour has been quickly achieved, whereas it has taken almost 30 years to get reasonably efficient blue LEDs.

A typical OLED of the Kodak variety is formed by starting with a transparent electrode, which also happens to be a good emitter of holes, indium–tin oxide (ITO). The ITO is covered with a thin layer of copper phthalocyanine, which passivates the ITO and provides greater stability (see Fig. 9.10). Then, the *p*-type material, for example, a naphthaphenylene benzidine (NPB), is deposited, followed by the *n*-type material, for example, aluminium hydroxyquinoline (Alq). Finally, a cathode of magnesium–silver alloy is deposited. All of the films can be evaporated, making fabrication very simple. Electrons and holes recombine at the interface of the *n*-type and *p*-type materials and emit, in this case, green light. A voltage of 5–10 V is sufficient to drive enough current to cause a very bright emission, which is another virtue of the technology, since low-voltage circuits are easier and less expensive to fabricate.

More recently, Burroughes *et al.* of Cambridge University developed a similar type of light-emitting diode using polymers, specifically polyparaphenylene vinylene (PPV), as the emitter. This device is variously called a Polymer LED (PLED) or a Light-Emitting Polymer (LEP). Polymer devices employing polyfluorene-based materials have achieved efficiencies at least as good as those of molecular OLEDs, but as yet their lifetimes are inferior.

The stated advantage of the polymer devices is the ability to spin the layers on and, in some cases, to pattern the films with photolithography. Molecular films, on the other hand, are typically soluble in both water and common organic solvents, so patterning with conventional photolithographic techniques is not feasible. This creates a real challenge for achieving high-resolution simple matrix displays and displays with fine stripes of different emitters for full colour displays.

Matrix displays with high information content can be made from OLEDs in two ways. Passive matrix arrays can be formed by patterning the lower electrode,

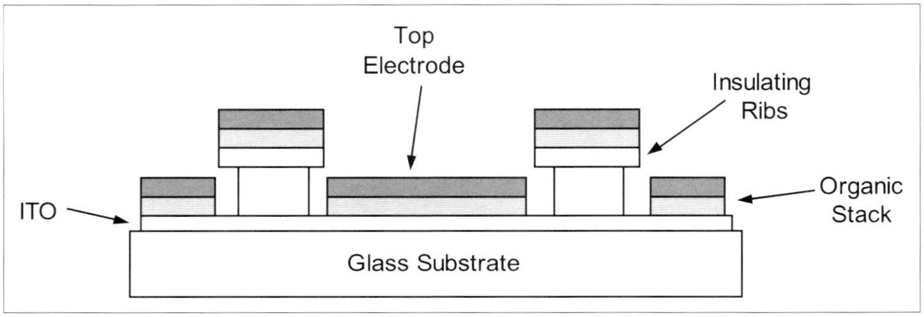

Fig. 9.11 OLED matrix structure (eMagin)

for example the ITO layer on glass, into stripes. Then a shadow mask structure is formed perpendicular to the electrode stripes, using photolithography. At this point the organic stack, followed by the top electrode, can be deposited over the whole array. The shadow mask structure ensures that the top electrode will not be continuous but rather will consist of stripes similar to those of the bottom electrode. Figure 9.11 shows schematically how such a passive matrix can be constructed without subjecting the organic layers to photolithographic processing. Pioneer used this approach in manufacturing the first commercial OLED displays.

A second approach to making high information content matrix displays is to employ an active matrix in which the row and column structure is built into the substrate using standard semiconductor techniques. In this case the finished substrate has an array of discrete electrodes, each one corresponding to a point in the matrix. The organic stack can be deposited, followed by a transparent electrode, over the whole array; no further patterning is necessary. This latter approach is eminently suited to making microdisplays, because a small silicon chip can be used as the substrate and the necessary driver circuits can be incorporated into the silicon chip along with the matrix structure. Figure 9.12 is a schematic rendering of this type of display structure. Note that in this case the OLED must be up-emitting, i.e. the top electrode, usually the cathode, must be transparent, whereas most OLEDs have a transparent bottom electrode.

Colour displays can be achieved in various ways. Alq can be doped with dyes which shift the emission colour (undoped gives green), and other emitter materials can be used for other colours, so stripes of different emitters can be deposited through shadow masks to create pixels with red, green, and blue subpixels. This approach has been used where the pixels are relatively large, but it becomes

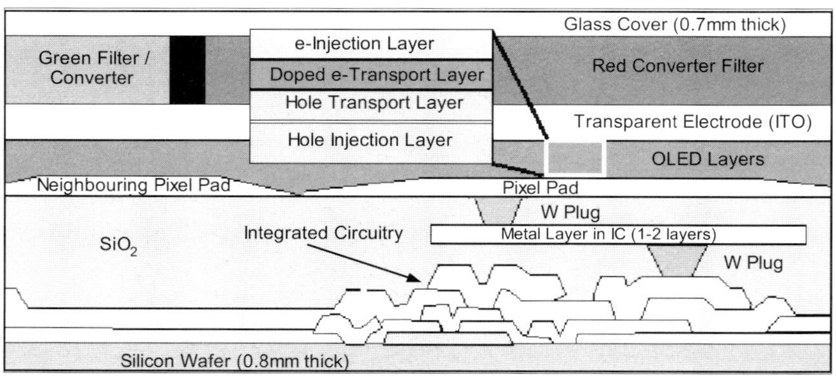

Fig. 9.12 OLED matrix structure (eMagin)

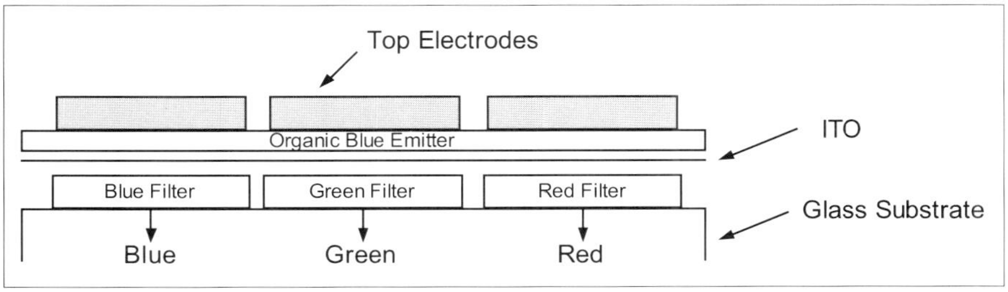

Fig. 9.13 *OLED colour changing media (eMagin)*

impractical for microdisplays, where the subpixels can only be a few microns in size.

A second approach is to dope the emitter with more than one dye, so that it emits white light. Passive colour filters, such as are used in liquid crystal displays, can then be used to create the red, green, and blue subpixels. The filters can be patterned on a separate plate, using photolithography, then overlaid onto the white emitter array. This is the easiest way to make a colour OLED display, although it does waste light in the filters, since at each subpixel only about one-third of the white light passes through the colour filter.

A better method for achieving colour is the use of fluorescent, or colour-changing, media in place of filters. In this case a blue emitter is used, and the blue light is converted to green or red by the Colour-Changing Media (CCM). This is depicted in Fig. 9.13. Less light is wasted in this case relative to the use of filters, provided the CCM materials have high conversion efficiency. As with the filters, the CCM materials can be patterned by photolithography, and some materials have efficiencies of 50–80 per cent. Colour purity with CCMs can be quite good, but if filters are used in conjunction with the CCMs the colour gamut can be excellent (**3**).

In summary, OLED technology can be used to great benefit for both direct-view and microdisplay applications. In both cases, OLEDs offer higher efficiency and lower weight than competing liquid crystal displays, since they do not require backlights or reflective light sources. These are important attributes for head-mounted and portable products. Issues of poor life and low luminance are being overcome. A typical OLED microdisplay device is shown in Fig. 9.14 (**4**).

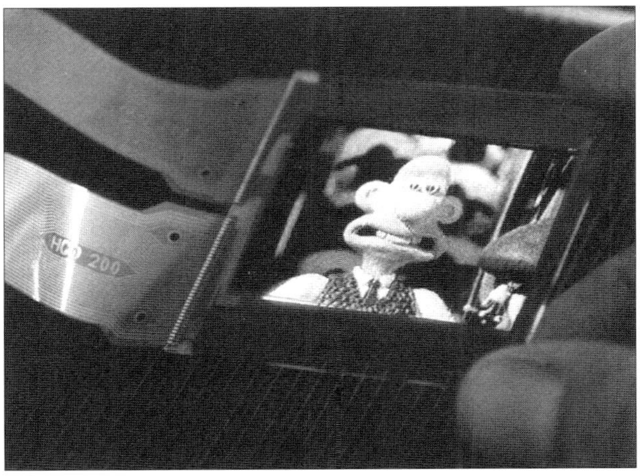

Fig. 9.14 *OLED microdisplay (Cambridge Display Technology)*

Rear projection Head-Down Displays

Rear projection is a possible alternative technology to direct-view AMLCD, able to provide large-area head-down instrument displays. Rear projection has the advantage of being independent of the size vulnerability of using COTS direct-view displays.

The technology is scalable. A single optical engine design combined with slightly modified folded projection optics can be used for a variety of display sizes from, say, 4 in to 32 in diagonal.

The technology is obsolescence proof. As the commercial market drives higher-resolution and more efficient display devices (LCoS or DMD), improved image generation components can be incorporated without major redesign. There are no single high-cost components in the product design. Furthermore, the technology allows the crewstation designer freedom to meld the image area to the shape of the instrument panel: square, rectangular, and even curved, conformal shapes are possible.

Fig. 9.15 *Projector optical elements*

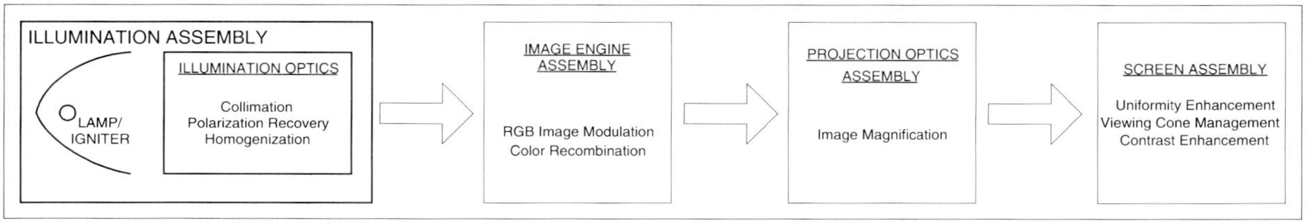

The basic optical elements of a projection system are illustrated in Fig. 9.15 and comprise:

- *Illumination assembly*. Every projection system using a non-emissive image source must have a source of illumination that contains sufficient energy in the red, green, and blue wavelengths. Traditional commercial movie projectors employ large arc lamps. Conference room projection systems most commonly employ high-pressure mercury lamps. Other lamp technologies are in use (e.g. xenon and metal halide) but do not offer efficiency equivalent to the mercury lamps. For most applications the lamp should have the characteristics of a point source; that is, the light should emit from as small an area as possible. A typical high-pressure mercury lamp has an arc length of the order of 1.3 mm. The illumination optics shapes the lamp output for optimum performance of the image engine and projection optics. Collimation and shaping are performed to create a beam of near-parallel rays that will illuminate an area just slightly larger than the display devices.
- *Image engine assembly*. This assembly represents the heart of any projection system. While numerous architectures are possible, the choice of display device technology is the single most important influence in architecture selection.

  Options are:
  - reflective LCD – Liquid Crystal on Silicon (LCoS),
  - Digital Micromirror (DMD).

- *Projection optics assembly*. This high-performance optical assembly magnifies the image and provides a focused image at the rear of the screen, free of chromatic aberration, distortion, and other optical defects.
- *Screen assembly*. The screen assembly plays a significant role in presenting the pilot with a high-quality image that is viewable in the luminous environment of the cockpit. For high-performance avionic applications, when compared with commercial applications, the screen must provide:
  - improved uniformity,
  - shaping and steering of the viewing cone,
  - enhanced contrast.

Figure 9.16 shows an open-box view of a typical rear-projection Head-Down Display. The lower section of the box contains the image engine and lamp. The projection lens is at the rear of the box. Figure 9.16 also shows the general arrangement in plan view.

A typical rear-projection HDD will comprise the functional elements shown in Table 9.1. In this case the image engine is a digital micromirror with a colour wheel. These elements would be replaced by a beam splitter and three-colour LCoS arrangement in a reflective LCD design.

The image source is either an XGA (1024×768) or preferably an SXGA (1280×1024) device. Some of the pixels would be masked in a square form factor display. The light source is positioned so that it can be replaced through an access door on the housing.

A 6 × 6 in rear projection HDD is being evaluated in the F/A-18E/F Super Hornet. Manufactured by Kaiser Electronics, the display was first flown on 22 May 2002 at the Naval Air Warfare Test Centre, China Lake, California. The display used reflective LCD LCoS technology.

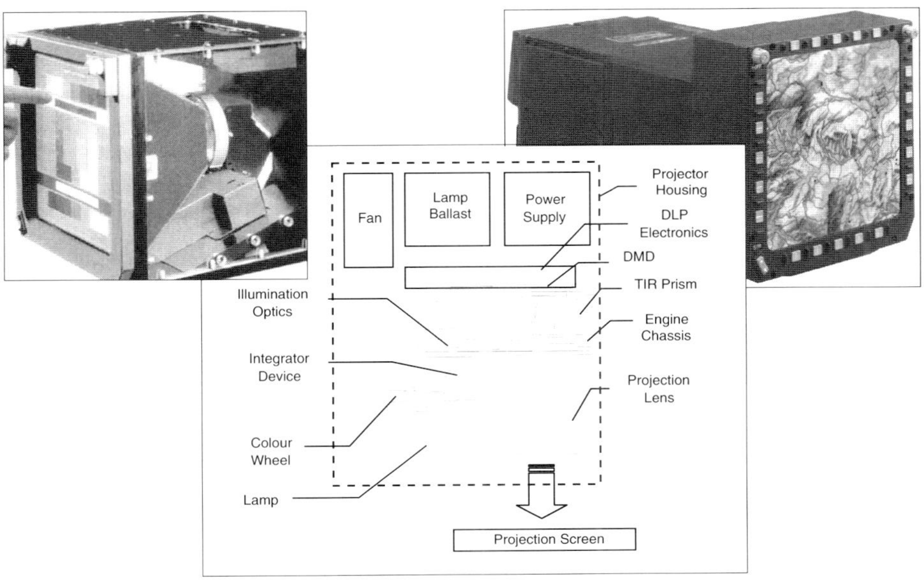

Fig. 9.16 Rear projection head-down display (Kaiser Electronics)

Table 9.1 Rear projection HDD functional elements (typical)

Item	Description
Fan	Cools lamp, ballast, optics, and DLP
Power supply	Supplies power to lamp ballast, DC voltages to electronics
Lamp ballast	Provides proper voltage and current to lamp, includes ignitor (~20 KV ignition)
Lamp	120 W high-pressure mercury arc lamp
Colour wheel	Provides sequential RGBW colour, includes motor and timing sensor
Integrator device	Integrator rod or light pipe used to harmonize light for uniform screen illuminance
Illumination optics	Optical elements used to image end of integrator onto DMD at correct magnification (~×3)
TIR prism	Glass prism, two piece with angles set for total internal reflection on illumination path and transmittance on projection path
DMD	Digital micromirror device
DLP electronics	Digital input, formats data for input to DMD
Engine chassis	Mechanical platform used to position and hold DLP, illumination optics, lamp, colour wheel, and projection lens
Projection lens	Provides high-resolution magnified image of DMD at the projection screen
Projector housing	Mechanical platform holds power supply, ballast, fans, and optical engine
Projection screen	Glass or plastic screen with internal diffuser, neutral density filter coatings or films onto which image of DMD is projected

The principal advantage of projection display technology over direct-view technologies such as AMLCD and CRT is the ability to use the same display subsystem for multiple applications in a variety of formats. This ability to scale the display can significantly reduce development and recurring unit costs and dramatically improve time to market. The use of common system elements not only enables scalability, it can also significantly reduce cost of ownership by providing low-cost replacement parts.

Figure 9.17 shows the same image engine applied to three product designs, a 6 × 6 in format display, a 10 in × 8 in rectangular portrait display, and a 'big picture' 24 in × 10 in rectangular display. The latter incorporates two image engines which are arranged to overlap in the central region to provide redundancy.

The 'big picture' display is a unique architecture in that it uses multiple-display engines in tandem. The potential redundancy of this architecture makes it attractive. Overlapping portions of the optical engines achieve the expanded width of the display. The projected image can be thought of as three separate areas: two individual areas on the left and the right, and an overlapped area in the centre. One of the engines can image the centre portion and one of the side areas, while the other engine images the remaining side. The alignment of the two engines is then less critical. This is the obvious mode of operation, but it may also be possible to have both engines image the centre section simultaneously and increase the centre brightness, though greater attention will need to be given to mechanical alignment. Rear projection is the currently favoured technology for the Joint Strike Fighter (JSF) crewstation, using a multiple projector arrangement to achieve a 20 in × 8 in 'big picture' display as shown in Fig. 9.18.

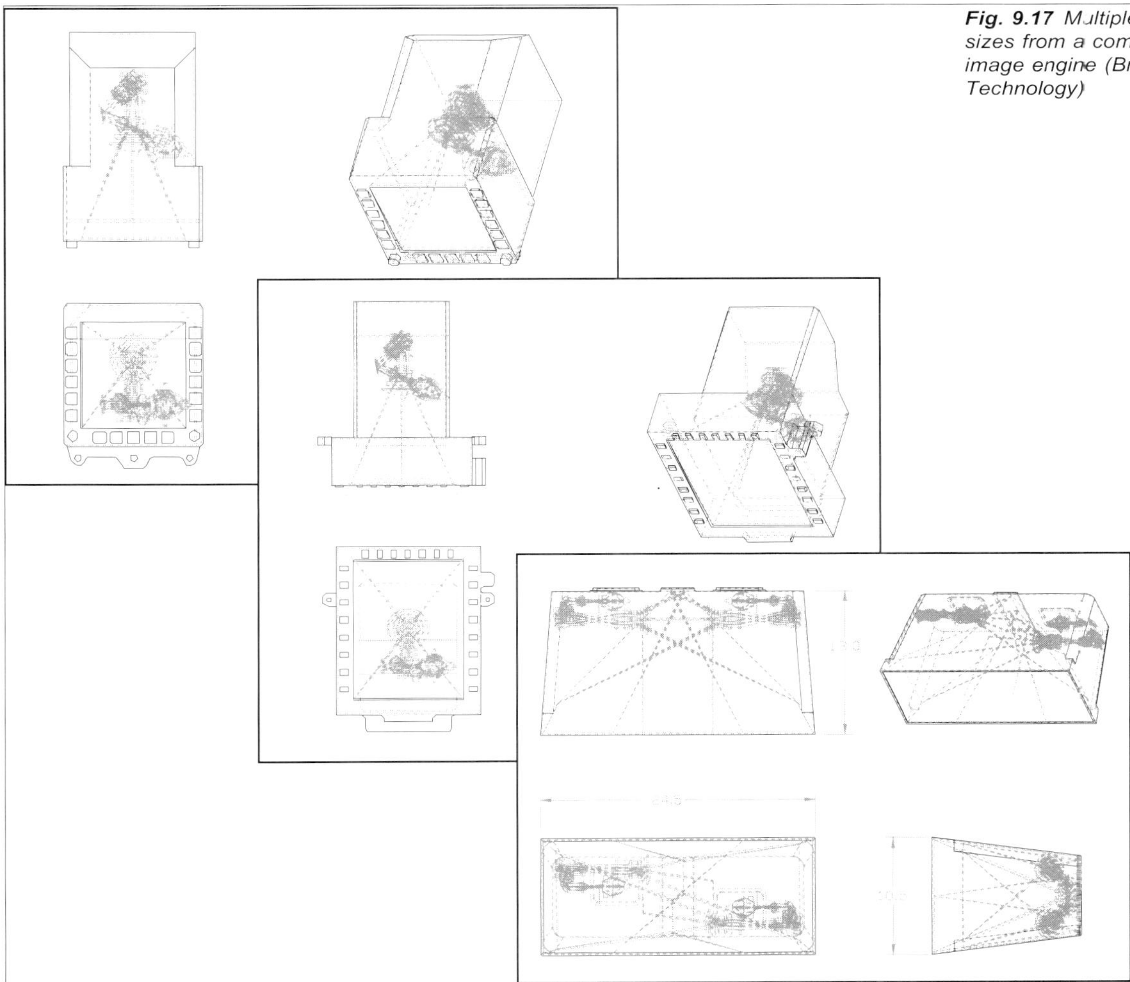

Fig. 9.17 *Multiple display sizes from a common image engine (Brilliant Technology)*

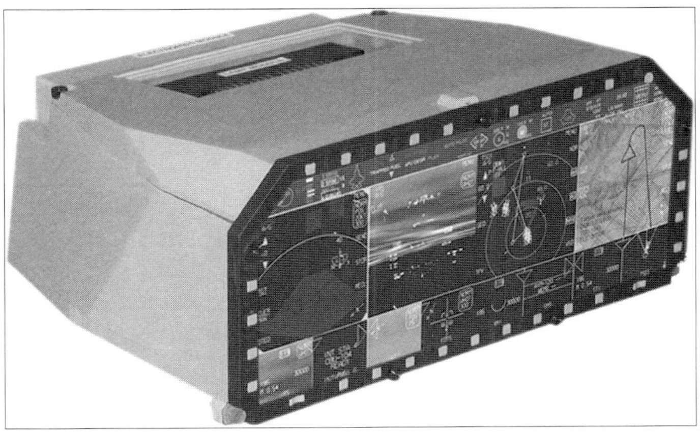

Fig. 9.18 *Multiple projector 'big picture' HDD (Kaiser Electronics)*

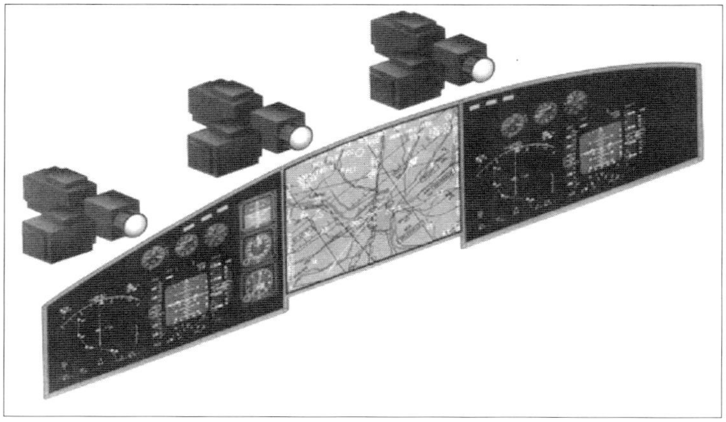

Fig. 9.19 Transport aircraft crew station using multiple rear projectors (Kaiser Electronics)

The same principle could be applied to the crewstation of a transport aircraft. A conceptual arrangement is shown in Fig. 9.19, in which three projectors span the instrument panel. Again, the potential image area from each projector could partially overlap, offering better coverage in the event of failure of one or more of the projectors than possible with a direct-view AMLCD panel. Each projector could in itself incorporate redundancy for critical components; as always the most likely part to fail is the light source.

Projection screen

Projection displays require the use of a diffusive screen to distribute the light from the image at the screen into the viewing volume. In addition to the diffuser, screen assemblies often include a Fresnel lens to reduce or eliminate the angle of incidence variation. For a uniformly illuminated screen, the diffuser and Fresnel lens (if used) determine the perceived uniformity, or illumination roll-off. In addition to the diffuser and Fresnel lens, a screen assembly also often includes an absorptive substrate on the viewer side of the diffuser. Diffusers, especially low-gain diffusers, have significant diffuse backscatter. The absorptive substrate reduces the brightness of ambient light, which is backscattered to the viewer, while having a lesser effect on the brightness of the desired transmitted image.

Viewing envelope is directly related to diffuser gain. The choice of screen gain must be balanced between achieving the desired brightness and providing adequate performance over the viewing envelope. High-gain screens will have higher brightness but smaller viewing envelopes. In addition, high-gain screens may have higher perceived roll-off owing to the variation in viewing angle across the screen.

Specular reflectance is determined by the antireflection coating of the outer surface of the screen substrate.

Screen speckle is a concern for any projection display system. Speckle refers to perceived brightness granularity. This effect is highly dependent on the diffuser technology, but, regardless of technology, speckle is worse for higher-gain screens.

Illumination

Two possible illumination technologies are currently available:

- LED illumination. One possible light source for small screen sizes is an array of solid-state LEDs. This offers several potential advantages:
  - The colour wheel is eliminated.
  - Improved colour space.
  - Greater efficiency, leading to reduced power requirements.
  - Longer life compared with arc lamps.

  The most straightforward application of LEDs would be an assembly of three arrays of individual LED packages, one for each colour, with dichroic combiners, to replace the arc lamp.
- Arc lamp. For larger display sizes the current technology of choice is a mercury arc lamp. To achieve a wide dimming range requires an optomechanical mechanism that can attenuate the system output. A number of methods are possible, including those shown in Table 9.2.

Table 9.2 Arc lamp dimming

Method	Attenuator mechanism	Switching mechanism	Issues
Switchable ND filters	Neutral-density filters, reflective type	Mechanical	Mechanical reliability, filter bleaching, dimming probably in discrete steps
Polarizing filters	Polarizer film or coating	Mechanical	System may need to be weakly polarized. Mechanical, reliability, optical complexity, filter bleaching, cost, full-on throughput. Filter dynamic range ~1000:1
Mechanical apodization filter	Variable-sized aperture in pupil or image plane	Mechanical	Complexity, reliability
Solid-state apodization filter	DLP or RLCD based pupil apodization filter	Electronic	Has advantages including programmability, contrast enhancement, etc. Full-on throughput ~70 per cent. Dimming range > 20 000:1.

The solid-state Head-Up Display

The interest in replacing the CRT in both military and civil HUDs is growing, driven by the perception that a flat panel source is lower cost, is more reliable, and allows a more compact unit, moreover with a direct digital drive. However, there are many problems to be resolved before a practical system can be fielded. The luminance and contrast requirements are very demanding; a problem compounded by the low efficiency of the display devices. The current indications are that in the near term a transmissive configuration with an RF coupled or Light-Emitting Diode (LED) backlight shows most promise, but that longer term microlasers allied with the reflective LCD/DMD configuration should be a better solution (**5**).

Performance requirements

LCDs/DMDs are effectively area array light valves and require an external source of linear polarized light. Depending upon the type, they may be backlit if a transmission LCD or frontlit if a reflecting LCD/DMD. In either case the HUD system will have to mimic the performance of the CRT solution, shown in Table 9.3, to be acceptable.

Table 9.3 Summary of desired HUD characteristics

Attribute	Value
Colour	Green at 545 nm
Luminance	55 480 cd/m^2 (1600 fL)
Ambient luminance	Dimming range 20 000:1
Contrast ratio	200:1
Operating modes	Raster with one grey shade reserved for symbology
Resolution	1000 line resolution with a 4:3 aspect ratio
Field of view	25°V × 30°H
Exit pupil	150 mm
HUD optical relay efficiency	12%

The transmissive system

This solution provides the most simple approach. Essentially, a powerful backlight is placed behind a transmissive LCD and the whole is coupled to the HUD optical relay system (see Fig. 9.20). The advantages are that the system offers a very shallow design at a low cost.

The LCD

Transmissive monochrome micro-LCDs are available from a number of sources. To match existing HUD optics design and to provide the desired resolution, the T-LCD should have the characteristics defined in Table 9.4.

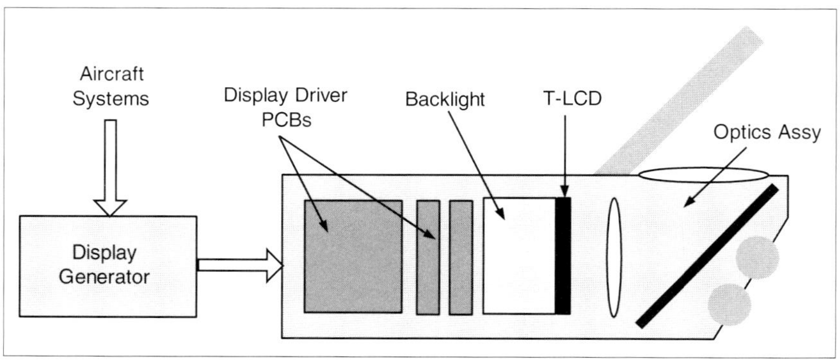

Fig. 9.20 Scheme of transmissive LCD HUD (BAE SYSTEMS)

Table 9.4 Summary of ideal T-LCD characteristics

Ideal size	76 mm diagonal
Resolution	1280×1024
Aspect ratio	4:3
Light resistance	Light blocking layer
Filters	Monochrome no filters

Table 9.5 Comparison of candidate backlights – BAE SYSTEMS

Backlight	Advantages	Disadvantages
RF coupled backlight	High efficiency	Concern over RF emissions
	Good dimming range	Single point failure
LED backlight	Cheap	Broad spectrum
	Multiple failures acceptable	Lambertian emitter
	Wide dimming range	Packaging not ideal
	Solid state	Highest luminances are red and amber
	High efficiency 70 lm/W	
Laser diffused	Saturated colour (532 nm)	Speckle
	High radiance/luminance	Size of laser and power supply
	Narrow spectrum	Low system efficiency
	Solid state	Dimming range
	May be remote via a fibre	Single point failure
		Expensive
		Eye safety limits
		Consideration of DEW
Tungsten filament lamp	Mature technology	Lambertian emitter
	Cheap	Colour filtering reduces output
	High luminance 5000 cd/cm^2 at 3200 °K	Life poor
		Best used with yellow and red
	Medium efficiency > 25 lm/W	
Arc lamps	High luminance	Broad spectrum emitter
	Small arc 1–2 mm	Lambertian emitter
	High colour temperature 5500 °K	High power needed for HUD
	Reasonable cost	Colour filtering needed
	Efficiency 70 lm/W	
Cold cathode lamps	High efficiency 50 lm/W	Inadequate luminance (50 000 cd/m^2 to date)
		Poor dimming range
Hot cathode lamps	High efficiency 90 lm/W	Poor cathode life
Electroluminescent	Simple	Inadequate luminance
	Low cost	
Field emitter		Inadequate luminance
UV pumped phosphor, e.g. Cambridge Display Technology	CRT-like characteristics	Special LCD required
		Immature technology

Fig. 9.21 *The CRL Brimar 'light engine' (BAE SYSTEMS)*

The backlight

This needs to be very intense and ideally collimated. The source colour should be a saturated green, peaking at nominally 545 nm. A narrow spectrum may be required to operate with existing HUD relay optics optimized for this. A typical frequency-selective combiner coating will have a 15–20 nm notch. Table 9.5 identifies possible light sources.

Of the backlights considered, only the RF light and the LED matrix offer a near-term solution. The microlaser has good potential and particularly has the advantage for remote location and coupling via a fibre optic. Most of the other sources are unfortunately Lambertian emitters. Coupling into an LCD panel ideally requires collimation of the light, which is not easy to do.

- ***The CRL Brimar 'light engine'.*** The RF coupled 'light engine' is shown in Fig. 9.21. A power supply generates an RF signal at about 9 MHz which is fed to a spiral planar antenna. The antenna couples into a mercury/argon filled quartz envelope and UV light is generated which stimulates a phosphor to produce light at the required wavelength. The process of photoluminescence is much more efficient than cathode luminescence (as used in a CRT). Layers of beam enhancement film steer the light into the modulator. The lamp luminance is controlled by pulse width modulation.
- ***The LED backlight.*** The LED is quite a broad emitter but this is no worse than that of a P1 phosphor used in the early HUDs. A matrix of some 50–100 LEDs is needed adequately to illuminate the T-LCD. A high degree of divergence gives poor coupling into the display, which can be overcome to some extent by using diffusers such as holographic screens or collimating optics.

The reflective system

The imaging device may be either:

- LCoS reflective LCD.
- Digital micromirror.

Reflective LCD (LCoS) and Digital Micromirror (DMD) devices have the advantage that they are more efficient than T-LCD.

A polarized light source would be ideal, as light would not be wasted at the first pass through the fold mirror. The following light sources can be considered:

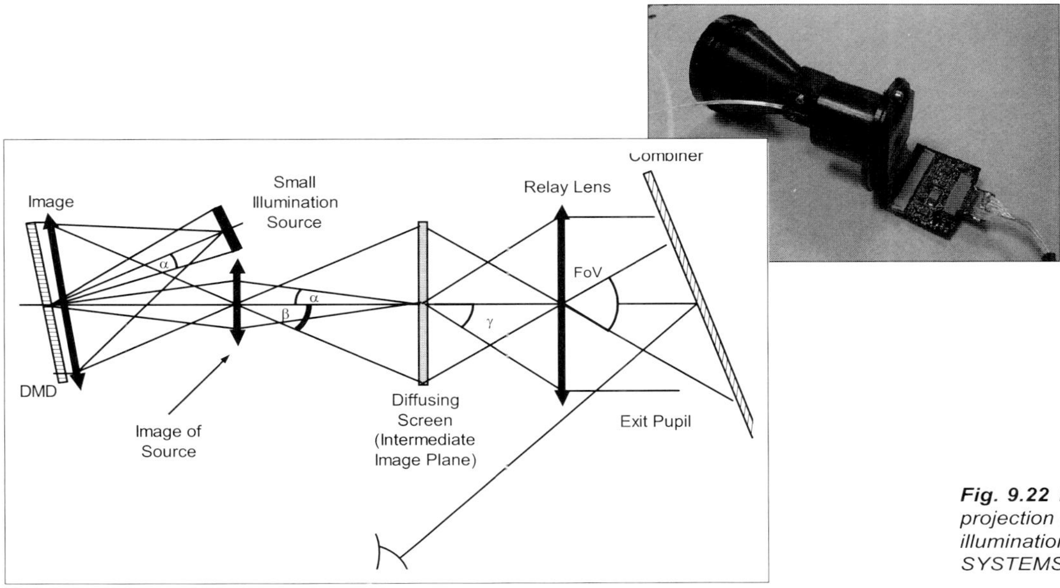

Fig. 9.22 Reflective projection engine with laser illumination (BAE SYSTEMS)

- Lasers, which can be polarized, are potential candidates. Good saturated colours. No filters needed.
- Arc lamps emit over a broad spectrum, and optical efficiency is impaired if they are filtered to produce the requisite spectrum. Additional UV and IR filters may be required to eliminate radiation harmful to the LCD. The resultant colour is not narrow band. Stability may be a problem. They emit over a wide angle.
- LEDs are an attractive alternative because they are potentially cheaper and less complex. However, their local coupling into a reflective device is difficult.

Figure 9.22 shows the BAE SYSTEMS (Patented) reflective projection engine with laser illumination.

Table 9.6 Power of illuminant source for transmissive and reflective configurations

	Illuminant source			
LCD	**Laser**	**Tungsten lamp**	**Arc lamp**	**General green Lambertian source**
Transmissive	1.3 W	350 W	125 W	13 W
Reflective	600 mW	175 W	62 W	6.5 W

Notes. The required luminance to the eye is 5480 cd/m^2.
The projected solid angle with a FoV of $30° = 2\pi(1 - \cos 30/2) = 0.214$ sr.
The exit pupil is 150 mm, giving an area of 177 cm^2.
Flux required: $5480 \times 0.214 \times 177 \times 10^{-4} = 20.75$ lm. If a dual combiner is used, then the efficiency of 12 per cent gives a flux of $20.75/0.12 = 173$ lm.
Transmissive configuration – assume the LCD is 20 per cent efficient, then the source flux will be $173/0.2 = 864$ lm.
Reflective configuration – assume the efficiency is 40 per cent, then the source flux will be $173/0.4 = 433$ lm.

Backlight requirements

The important issue is to determine the flux required from the illuminating source to achieve the level of luminance at the eye. Table 9.6 gives illuminant source power figures for transmissive and reflective configurations.

The microlaser has the best promise as an illuminator, but, although the laser head is coming down in size, there is still the need for a very regulated power supply, which is necessarily quite large. The overall volume and power consumption is substantially more than the CRT and HVPS which it seeks to replace. The laser power required is readily available and the narrow bandwidth is a positive advantage.

The light source in the Brimar RF coupled 'display engine' was invented at CRL in the United Kingdom and is being developed further under a UK Department of Trade and Industry CARAD programme. The module achieves good luminance and has promise of being a practical solution.

The simplicity of an LED backlight encourages development.

Technology demonstrator

Figure 9.23 shows a technology demonstrator HUD, which can host the transmissive or reflective 'display engine'. The concept of replacing a CRT with a flat panel display has been shown to be somewhat more difficult than the original simple idea of a lamp behind an LCD. Both the LCD and the light source have individual problems of performance. Moreover, coupling the two together efficiently and in turn coupling the 'display engine' to the HUD relay optic is difficult to achieve efficiently.

Figure 9.24 shows a comparison of the space envelopes required for the transmissive and reflective approaches as against the original CRT technology.

Fig. 9.23 *Technology demonstrator LCD HUD (BAE SYSTEMS)*

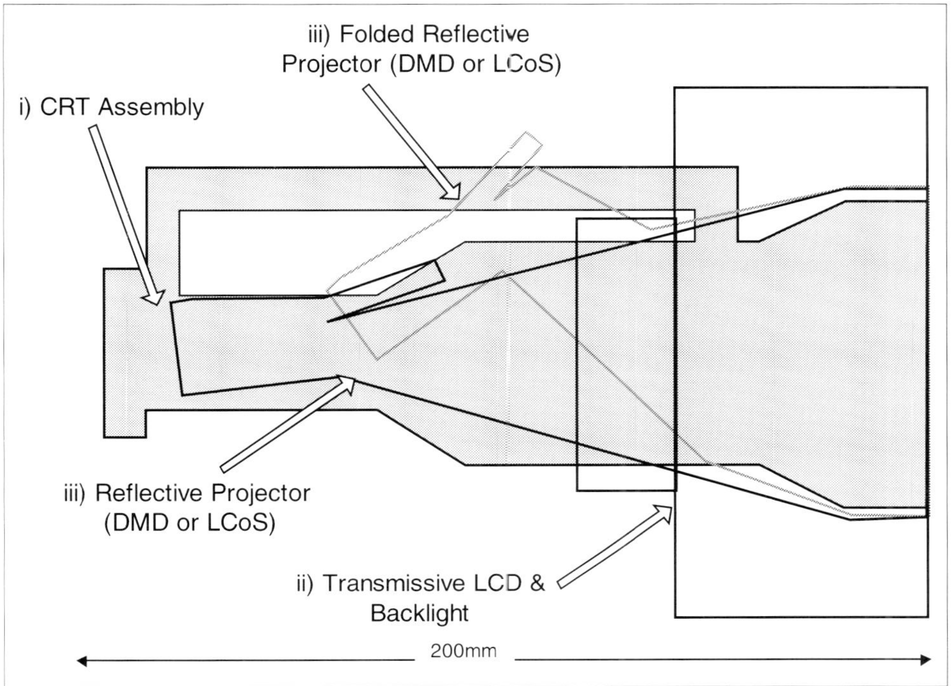

Fig. 9.24 *Comparison of the packaging of the transmissive and reflective approaches (BAE SYSTEMS)*

The solid-state Helmet-Mounted Display

Current CRT technology HMDs impose large design burdens on the engineer and can be difficult to use in the operational environment. They are bulky and heavy and require high-voltage, expensive, quick disconnect interfaces. A new, second generation of HMDs is in development using solid-state technology which will provide greater performance and lower cost of ownership owing to high reliability. Most importantly, helmet size and head-borne weight will be reduced.

Although fixed-wing and rotary-wing operational requirements are merging, the area of helmet and head-borne weight still remains unique:

- Tactical fixed-wing aircraft operate in high-*g* environments. The pilot, upon ejection from the aircraft, is confronted with extraordinary forces. The helmet must be capable of sustaining high-*g* forces and ejection wind blasts of 600 knots. This dictates a robust but lightweight helmet.
- Helicopter pilots, on the other hand, rarely encounter high-*g* loads. They always arrive at the scene of the crash with the aircraft. Consequently, they need a sufficiently robust helmet capable of protecting the crew member during a crash. Head-borne weight is not so significant, except not to inflict injury during the crash sequence.

Recent studies indicate that the maximum head-borne weight for a helicopter pilot should not exceed 5.5 lb (2.5 kg), and for a fast jet fighter pilot it should not exceed 4 lb (1.8 kg). New solid-state technology will assist in meeting these exacting requirements.

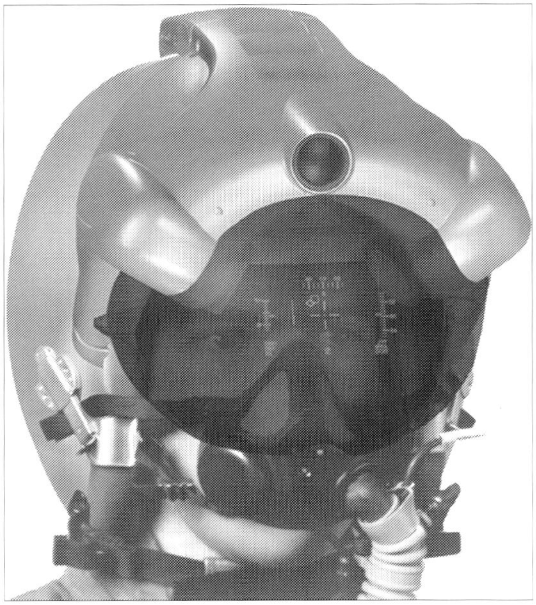

Fig. 9.25 *JSF Helmet-Mounted Display (Rockwell-Collins)*

Second-generation HMDs being developed for the Comanche Helicopter and the Joint Strike Fighter (JSF), shown in Fig. 9.25, are now using transmissive AMLCDs as the primary image source (**6**). The image sources are two 1280 × 1024 transmissive AMLCDs manufactured by Kopin. They achieve a contrast ratio of 100:1 with a response time of less than 9 ms. T-LCDs typically have a transmission of approximately 10 per cent. A small LED backlight has been developed that can produce greater than 26 000 fL light output which is sufficient to achieve good readability in a 10 000 fc (108 000 lux) day environment. The LED requires around 0.8 mA and dissipates approximately 1.5 W per T-LCD, a significant achievement. The electrical power requirements in the region of the pilot's head have decreased from approximately 13.5 kV required by the CRTs to only 5 V.

The use of T-LCDs allows the use of a smaller optical chain, shown in Fig. 9.26, resulting in less complex optics, reduced weight, and a more compact design that has a minimum impact on the helmet profile. Advances in compensation film technology solve problems of contrast and greyscale variation across the viewing plane. For helicopter applications, the design also allows the use of a standard aviator helmet, which further reduces costs while achieving a well-balanced HMD that is compliant with weight and CofG criteria.

The remote display processor incorporates a Pentium-based processor which receives incoming video from the sensor suite, digitally processes and overlays the video signals, and performs helmet tracking and input/output (I/O) control functions. The processor has a flexible architecture to accommodate future growth and display technology insertion. A fibre-optic data bus increases speed and reduces latency. A real-time image-warping engine performs dynamic scaling and warping of high-resolution video data. Dynamic image warping is required when using an AMLCD to compensate for the geometric distortion of an off-axis optical design.

Fig. 9.26 JSF HMD optical arrangement (Kaiser Electronics)

Virtual retinal displays

The Virtual Retinal Display (VRD) offers the potential for a virtual cockpit. This initiative started with military cockpit applications in mind, where significant strides are being made to link the pilot more effectively to computer-generated information by immersing him in a virtual world of 3D vision, 3D sound, and 3D touch. It is unlikely that that degree of realization will ever be appropriate for a civil transport flight deck, but the technology allows the projection of images directly into the eye of the pilot.

The technology, developed and patented by Microvision in Seattle, Washington, promises to provide the flight crew with high-resolution, full colour, and daylight-viewable images wherever the pilot is looking. This is an entirely new display technology, not a means to devise new ways to enhance or miniaturize existing display technologies. The VRD paints an image directly onto the wearer's retina using a modulated, low-power beam of laser lights. Operating at extremely low power, there is no danger to the eye.

To create the image, the VRD uses a photon source (or three for a full colour display) to generate a coherent beam of light. The use of a coherent source (such as a laser diode) allows the system to draw an extremely small diffraction-limited spot on the retina. The light beam is modulated to match the intensity of the image being rendered. The low-power beam of laser light is projected onto the retina in a raster scan pattern. To draw the raster, a horizontal scanner moves the beam to draw a row of pixels, and then the vertical scanner moves down one line for the next row.

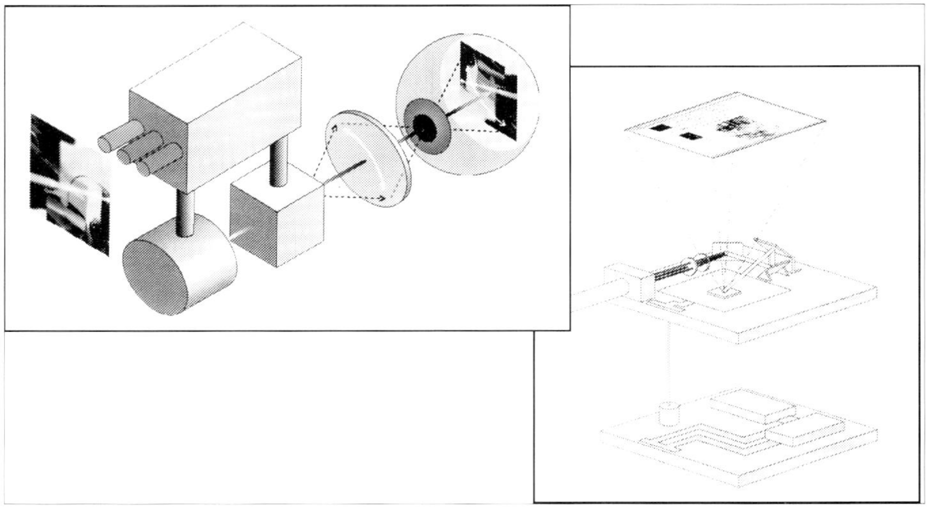

Fig. 9.27 Virtual retinal display components (Microvision Inc.)

The VRD comprises four functional elements, shown in Fig. 9.27:

- *Drive electronics.* The drive electronics receive and process signals for the image source.
- *Light source.* The VRD uses very low-power light sources to create one pixel at a time. For full colour images, three light sources (red, green, and blue) are modulated and merged to produce a pixel of the appropriate colour.
- *Scanners.* Mechanical horizontal and vertical scanners project one pixel at a time through the viewing optics of the eye to the retina. This in effect paints an image by rapidly moving the light source across and down the retina in a raster pattern.
- *Viewing optics.* Refractive, reflective, and diffractive optical elements are used to expand and transmit the rapidly scanning beam through the lens of the eye onto the retina to create the virtual image within the viewer's eye.

The VRD overcomes two very significant problems experienced with conventional head-mounted display technologies – weight and power.

Potentially the VRD does not require an intermediate image on a screen. The only required components are the photon sources, the scanner, and the optical projection system. Small, light, and low-power photon sources are used (such as laser diodes). Scanning is accomplished with a small micromachined electro-mechanical scanner. The projection optics are incorporated into the front reflecting surface of a pair of glasses.

The VRD is finding applications in healthcare, automobiles, and aerospace. A monochrome (red) system (known as Nomad™ and shown in Fig. 9.28) is being evaluated by a number of interested parties, including civil aircraft operators. It has a field of view of $23° \times 17°$ and a resolution of 600×600 pixels (SVGA). The head-mounted unit weighs 500 g (18 oz).

VRD, along with other candidate technologies, is being evaluated by the US Army as part of the Virtual Cockpit Optimization Programme (VCOP). VCOP will initially target military rotorcraft, but the techniques developed could ultimately have application in both military and commercial jet aircraft (**7**).

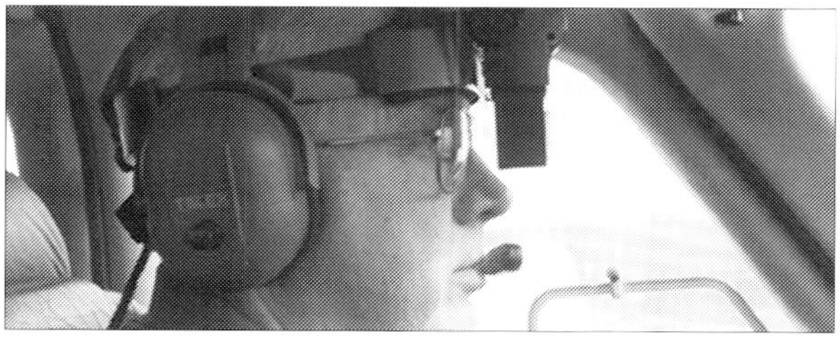

Fig. 9.28 VRD Head-Mounted Display (Microvision Inc.)

The virtual cockpit system will consist of a powerful system processor and graphics generator that collects and processes flight data, mission cues, and sensor imagery from the aircraft systems and then merges and presents the information to the pilot in a graphical head-up mode. Integrated with head-worn systems, the information from aircraft systems is projected over a pilot's visual field.

The requirement analysis addresses the following elements:

- Aircraft selection and mission scenario definition for flight testing demonstration.
- Virtual cockpit layout and control and display requirements.
- Pilot/Vehicle Interface (PVI) definition, which will include head tracking, 3D audio, digital map virtual display with pathway in the sky symbology, tactical situation displays, and other virtual displays. At a later stage, the PVI will incorporate voice recognition and speech synthesis.

The concept of the virtual cockpit is to present flight data, sensor imagery, and other mission cues to a pilot in a highly graphical and intuitive manner, making the aircraft easier and safer to fly while improving mission performance. Flight reference data, sensor imagery, and other symbology will be displayed, referenced to the outside world, including 'threat zones', and superimpose a safe 'pathway in the sky'.

The Army's vision of the virtual cockpit also includes a 'what you see depends on where you look' concept. As the pilot looks up and out of the cockpit, various types of targeting, navigational, or terrain overlays would appear. When the pilot looks in a downward direction, he may see 'virtual' instruments projected onto the eye which replace many of the existing dials and multifunction displays that are in cockpits today.

A prototype Helmet-Mounted VRD display (see Fig. 9.29) has been developed for military rotorcraft applications.

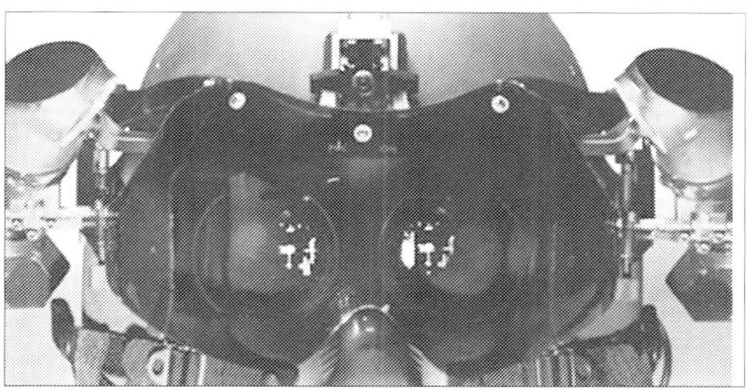

Fig. 9.29 Prototype VRD helmet for rotorcraft applications (Microvision Inc.)

The cardinal point specifications are as follows:

Field of view	30° vertical × 52° horizontal binocular with a 30° overlap
Colour	Monochrome green (532 nm)
Brightness	Approx. 2000 fL
Resolution	1355×960 pixels (SXGA)
Weight (with HGU56/P helmet)	2.4 kg (including cable)
Safety	No high voltages on head, meets ANSI standards for eye safety
Operational features	Compatible with eyeglasses, oxygen mask

References

(1) **Tisdale, G.** and **Billings, M.** (2001) Tomorrows HUD, *Flight Deck International*, February.

(2) **Hornbeck, L. J.** *Digital Light Processing and MEMS: Timely Convergence for a Bright Future*, Texas Instruments.

(3) **Howard, W. E.** eMagin Corporation.

(4) **Collins, L.** Roll-Up Displays: Fact or fiction, *IEE Review*, February.

(5) **Bartlett, C. T.** The *Flat Panel Head Up Display*, BAE SYSTEMS Avionics Limited, Rochester, Kent.

(6) **Casey, C. J.** *Emergence of Solid State Helmet Mounted Displays in Military Applications*, Kaiser Electronics.

(7) **Whiston, S.** (1999) Future flight, *Flight Deck International*, September.

CHAPTER 10

Advanced Display Formats

Introduction

This chapter discusses the evolution of display systems and information formats to meet the operational challenges of the future flight deck. Earlier chapters have charted the evolution of the flight deck from its earliest beginnings through electromechanical instruments to the all-glass flight deck of today (see Fig. 10.1).

Side-by-side PFD and ND displays and formats are now the accepted norm for the civil cockpit. Most recent cockpit designs have six display surfaces at or above 6×6 in usable screen area. There is a trend to use larger and fewer rectangular AMLCD displays based on COTS laptop PC products ruggedized to meet the air transport environment.

The aviation industry has an enviable and well-earned safety record, but this safety record can and should be improved even further. Given that flight crew error is cited as a primary factor in such a large percentage of accidents involving transport category airplanes, addressing flight crew error becomes a primary target for improving safety. Recent accidents and incidents highlight difficulties in the interaction between flight crews and advanced flight deck automation. Of the more than 200 heavy air transport accidents involving hull loss or fatalities in the past 10 years, more than 50 per cent were associated with either Controlled Flight Into Terrain (CFIT) or during the approach and landing phases of flight.

The incidence of CFIT accidents has not significantly improved in spite of the introduction of the 'glass flight deck'. CFIT accidents generally result from a sequence of events that link together as a chain to produce the end, often fatal, result. To prevent a CFIT accident, it is necessary only to break the chain at any one link. It has been said that, to combat CFIT, we must be more effective at managing the last moments or even seconds of an imminent event.

To accomplish this, three principles must be recognized:

- There will forever be sequential combinations of events conspiring to produce the chains that lead to accidents.

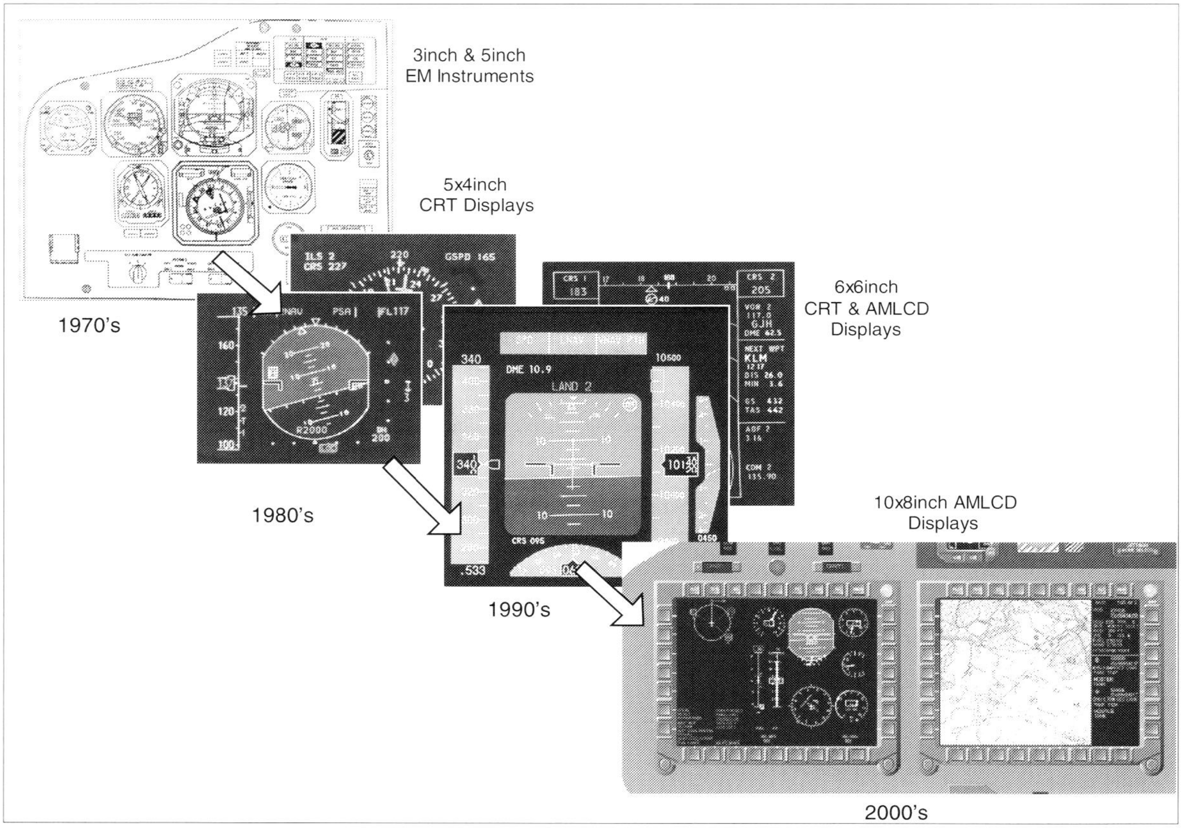

Fig. 10.1 Civil flight deck display evolution

- The primary reference of human beings in the world, the sense of sight, has evolved to be instantaneous, intuitive, and 3D.
- CFIT accidents have historically occurred in all phases of flight and cannot be predicted.

The skies are becoming ever more crowded. There are now over 200 000 aircraft on the civil register worldwide (including privately owned aircraft). More than 4 million fare-paying passengers travel by air every day. In the United States there are over 55 000 flights on a typical day, with more than 6000 aircraft in the air at any one time. In Europe the traffic density is similar. There are over 20 000 flights in a typical day, with more than 2800 aircraft in the air at any one time. The expectation is that air traffic in Europe will double by 2015. Near-miss incidents are rising. In-air collisions are fortunately rare, but are high-profile news with devastating consequences to the individuals involved and to the industry as a whole. If nothing is done to improve the rate of CFIT accidents, then by 2015 there will be one CFIT accident every 2 weeks.

In the context of civil air transport, this chapter reviews the Future Air Navigation System (FANS) initiative being undertaken better to manage the airspace and the presentation of information to support precision approaches using procedures based on Required Navigation Performance (RNP). The chapter then discusses current enhancements to improve crew situational awareness and thereby reduce CFIT accidents and mid-air collisions using Terrain Avoidance (TAWS) and Traffic Collision Avoidance (TCAS) systems.

The last 30 years have seen a revolutionary change in display technology. However, information display change has been evolutionary. Modern display formats still have many vestiges of their predecessors. This is understandable in a safety-conscious industry in which each new step is carefully thought through, tried, and tested. However, with modern computing power and graphics technology the capability is readily and affordably available to make significant advances in information presentation to enhance crew situational awareness, reduce workload, and hence reduce the risk of accidents. As the volume of air traffic increases and as battlefield digitization requires military aircraft to cooperate in a system of systems, it will be essential to make this revolutionary step.

3D and 4D Head-Down Display formats are being developed to meet these challenges, and the Head-Up Display is being introduced in civil transport applications. Alternative means are being introduced to facilitate interaction between the crew and the avionics, including voice commands and cursor control.

In the context of the military crewstation, the potential impact on the display suite of battlefield digitization, smart weapons, and force mix concepts using manned and unmanned air vehicles is bringing an information explosion in the crewstation. The next generation of fast jet fighter aircraft epitomized in the JSF will employ large-area Head-Down Displays and a Helmet-Mounted Display but no Head-Up Display.

In June 1996 the FAA undertook a review of the interfaces between flight crew and modern flight deck systems (**1**). The FAA Human Factors Team recognized that it was necessary to look beyond the label of flight crew error to understand why the errors occurred and consider issues more broadly, including the functionality of the underlying systems.

The FAA Team identified issues that show vulnerabilities in flight crew management of automation and situation awareness:

- Pilot understanding of the automation capabilities, limitations, modes, and operating principles and techniques. 'Why did it do that?', 'What is it doing now?', and 'What will it do next?'.
- Differing pilot decisions about the appropriate automation level to use or whether to turn the automation on or off when unusual or abnormal situations are encountered.
- Flight path awareness, including insufficient terrain awareness (sometimes involving loss of control or controlled flight into terrain) and energy awareness (especially the low-energy state).

The HF team reported that these vulnerabilities appear to exist to varying degrees across the whole fleet of transport category airplanes, regardless of the airplane type, the manufacturer, or the operator. The team recommended that systems design take full cognizance of the following principles:

- Minimize human error. It is impossible to prevent all human error without removing the human flexibility and adaptability that contributes significantly to safety. It is the negative consequences of error that need to be eliminated,

not necessarily the errors themselves. However, it is still desirable to minimize errors that are design or system induced.
- Increase error tolerance. The systems should be designed to aid the flight crew to detect errors when they occur. Also, the systems should be designed such that errors that do occur have bounds on the undesirable consequences that result.
- Avoid excess complexity as perceived by the user. The systems should be designed to support the flight crew, and should not be perceived as unnecessarily complex.
- Increase system observability, especially by improving system feedback.
- Evaluate new technology or operational changes introduced into the aviation system, especially the flight deck, for their effect on human performance.

Many specific recommendations have also been made in a study sponsored by the Royal Aeronautical Society on the future flight deck (**2**). Flight instruments are required to substitute for the visual cues of the outside world and to inform the pilot what the aircraft is doing. The displays to achieve these objectives must be simple, easy to operate, and capable of interpretation with the minimum of thought, especially in rapidly changing situations. They should appear natural and intuitive so that in high-stress situations the pilot can rely on basic interpretative skills, releasing mental capacity to deal with operational problems.

This report highlighted the following concerns with current information formats:

- The Airspeed Indicator (ASI), Altimeter and Vertical Speed Indicator (VSI) have suffered from being adapted to strip displays. In this form they lack the pattern information that is given by a rotating pointer on a fixed circular scale, the angular position of which can convey so much information at a glance.
- There appear to be two conventions for the ASI scale, with high numbers at the bottom or at the top. With the former design solution, pitching the aircraft will cause the speed and altitude scales to move in opposite directions, which may be confusing during take-off and initial climb.
- The VSI has sometimes been given a reduced role in the latest flight systems, and it tends to be lost in the clutter of the PFD. It should be clear and easy to read, uncluttered, and of adequate size.
- It has already been noted that one of the benefits of modern technology has been the introduction of the pictorial Navigation Display (ND). Supplementary heading information in the form of a compass rose or arc should be available below the 'attitude ball' on the PFD. While it remains necessary to use ground-based aids such as VOR for non-precision approaches, or to resolve map shift, a full compass rose display should be available to each pilot on which VOR and/or ADF needles can be fully displayed.
- The vertical profile is commonly displayed by writing the height constraints by the appropriate waypoint on the ND. An alternative method, which provides a separate vertical profile display in the lower part of the ND, might be more natural and easier to use.

Future air navigation system (FANS)

The following is an overview of the FANS initiative. Though FANS is not a display system, the need for intuitive interaction with FANS data will drive the HMI of next-generation flight deck design. For more detailed information, the reader is directed to reference (3) and to the publications of Air Transport Regulatory Authorities.

Air transportation is a global operation. In the United States there are 15 000 trained air traffic controllers who operate in 25 Air Traffic Control (ATC) centres, coordinated by a single National Control Centre. In Europe there are currently 65 air traffic control centres. No single point of coordination has yet been established, though there is considerable effort being expended to rationalize European airspace.

A worldwide initiative was launched in 1997 to implement a phased approach to a radical overhaul of the Air Traffic Management (ATM) arrangements. The initiative is known as the Future Air Traffic Management (FANS) environment. Much has already been achieved, but there remains much to do.

The objectives of FANS can be summarized as:

- Increase system capacity.
- Increase levels of safety.
- Create a worldwide continuum of airspace where boundaries are transparent to the user.
- Improve navigation and landing capabilities to support advanced approach and departure procedures.
- Facilitate the dynamic accommodation of user 3D and 4D flight trajectories.
- Increase the involvement of the user in ATM decision making.
- Improve information to all users.
- Accommodate a full range of aircraft types and capabilities.

The major milestones in the introduction of FANS are indicated in Table 10.1. The FANS objectives will be met by coordinated improvements in three principle areas, fundamental to the enhanced and safe use of the airspace:

- *Communications*. The use of digital data links will increase the flow and accuracy of information between the Air Traffic Control (ATC) centres, the Airline Operations Centre (AOC) and the aircraft itself, augmenting the current voice communications.
- *Navigation*. The use of GPS (or EGNOS) in communication with existing navigation aids will lead to a radical improvement in 3D position accuracy and allow closer yet safer spacing between aircraft while reducing the risk of Controlled Flight into Terrain (CFIT) accidents.
- *Surveillance*. The use of digital data links and ATC mode S to broadcast and receive 4D aircraft position and intended flight plan to the ground and all aircraft in the vicinity will enhance the ability to recognize and resolve potential airspace conflicts.

The acronym CNS/ATM is synonymous with FANS – Communication, Navigation, and Surveillance underpinning the future Air Traffic Management system. The ultimate objective of FANS is to provide a free flight capability in which airline/

Table 10.1 Fans implementation milestones

	Milestone
1997	The introduction of new approach procedures based on the improved performance now available from on-board navigation systems using the satellite-based Global Positioning System (GPS). These new approach procedures will be based on Required Navigation Performance (RNP)
1998	The introduction of Reduced Vertical Separation Minima (RVSM) from 2000 ft to 1000 ft initially between flight levels FL330 (33 000 ft) and FL370, then subsequently between FL290 and FL410
1998	RNP procedures extended to include en route navigation in the United States and Europe (RNP-5 in Europe), reducing lateral separation minima and permitting a corresponding increase in traffic density
2000	Traffic Collision Avoidance Systems (TCASs) mandated for all new aircraft operating in Europe
2002+	Cat I GPS precision approach procedures in the United States using a Wide-Area Augmentation System (WAAS) to enhance the accuracy of GPS data
2003+	Upgrade Air Traffic Transponders (IFF) to incorporate mode S which extends the capability of the aircraft on-board transponder to transmit aircraft current position (based on GPS) and flight plan intent in addition to ident and altitude information
2005+	Differential GPS Cat II and Cat III precision approaches introduced in the United States using the Local Area Augmentation System (LAAS) enhancement
2005+	Automatic Dependant Surveillance (ADS) introduced to broadcast aircraft 4D position and intended flight plan to ground ATC and all aircraft in the vicinity
2008+	European differential Geostationary Navigation Overlay System (EGNOS) introduced. This programme encompasses the deployment of a European satellite system – Galileo

aircraft are able to determine and fly the optimum route for their operation from departure to destination against a preplanned time of arrival. Free flight has the potential to reduce flight time with consequent savings in operating cost by planning and maintaining the aircraft trajectory in 4D with the minimum of fuel burn. It will be desirable to be able to make dynamic adjustments/replan of the flight trajectory should the need arise, e.g. to avoid weather or congestion. Upon arrival in the terminal area, the objective is to avoid (or at least minimize) the need for holding patterns by phasing the approach with that of other aircraft and by being able to make more direct approaches than traditional ILS procedures.

Inevitably, FANS will mean a fundamental change in role and responsibilities for the air traffic controllers and for the aircrew. The aircrew will become managers of systems and negotiators of airspace in their local vicinity. The air traffic controllers will become overseers of safety.

Communications

The main element of communications improvement encompasses digital data links between the aircraft itself, air traffic control, and the airline operational centre. Several communications media, standards, and capability are proposed. The principle mechanizations are:

- VHF Data Link (VDL) – line-of-sight communications used for domestic ATC flight service and surveillance.
- SATCOM and HF data link – for oceanic and transcontinental areas where line-of-sight communications are poor.

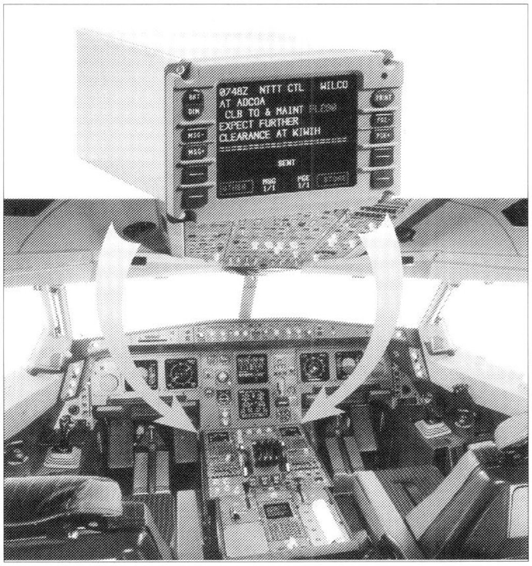

Fig. 10.2 Airbus installation of the DCDU (Smiths Aerospace)

- ATC mode S – an extension to the Air Traffic Surveillance (IFF) transponder (see later).

The operating principle is that the Airline Operations Centre (AOC) files and negotiates the aircraft 4D flight plan with the Air Traffic Control Centre (ATC) before take-off. The flight plan is then uploaded into the aircraft Flight Management Computer System (FMS) and displayed to the crew texturally on the FMS Control and Display Unit (CDU) and pictorially, as a route plan, on the Navigation Display (ND). The flight plan includes:

- Aircraft current 3D position and 4D intent in the form of future waypoints.
- Predicted arrival times at significant waypoints.
- Selected procedures en route and in the terminal area.
- Destination airport and diversionary alternates.

When FANS is fully implemented, the flight plan will be broadcast to the ground ATC and to other aircraft in the vicinity.

The aircraft flight crew, ATC, and AOC may each request changes to the flight plan. Before data link, means to request/acknowledge changes to the flight plan used audio communications, sometimes with inevitable ambiguity. With a digital data link, these vital communications can be unambiguously displayed on the flight deck, and recorded. The implementation means is aircraft and operator dependent. For instance, Boeing and their suppliers are adding additional information to the FMS CDU. Airbus are adding an additional textural display onto the flight deck, the Data Control and Display Unit (DCDU), which is located in the centre console, forward of the FMS CDUs as shown in Fig. 10.2.

Navigation

A number of navigation improvements are being introduced that will significantly improve knowledge of aircraft 4D present position and revolutionize traditional route structures and procedures. This revolution is made possible by the introduction of satellite navigation systems, generally known as Global Navigation Satellite Systems (GNSSs).

Global Position Systems (GPS) is a US military satellite radio navigation, position, and time transfer system that provides highly accurate position and velocity information and precise time on a continuous global basis to an unlimited number of properly equipped users. The system is unaffected by weather. The system currently includes a constellation of 25 satellites, five of which are always viewable from anywhere on earth. GPS operation is based on the concept of ranging and navigation from the group of visible satellites. The GPS receiver verifies the integrity of the signals through the process called Receiver Autonomous Integrity Monitoring (RAIM). The geometry of GPS satellites favours lateral fixes. Determination of vertical position is less accurate and in general requires an alternate means to refine the height estimate (usually barometric data).

GPS offers two levels of services:

- *Standard Position Service (SPS)*, which provides all users a horizontal positioning accuracy of 100 m or less to a 95 per cent confidence level (300 m to a 99.99 per cent confidence level).
- *Precise Positioning Service (PPS)*, which provides authorized users with a positional accuracy better than 10 m. The PPS service is at the discretion of the US DoD using a feature called selective availability. Currently this feature is turned off, making PPS available to all.

To become independent of the selective availability function and to improve the precision of height estimation, there are plans in the United States to introduce two safety-critical systems to augment basic GPS. The principle is to measure the performance of the GPS system at precisely surveyed sites and then transmit the dynamic error to user aircraft, which can then make appropriate corrections to their position estimations. These arrangements are often known as differential GPS. The two differential GPS systems are:

- *Wide-Area Augmentation System (WAAS)*. Precisely surveyed ground sites uplink correction messages to geostationary SATCOM communication relays to aircraft flying within the coverage area. WAAS is expected to improve GPS accuracy to around ±7 m and able to be certificated for Cat I approaches.
- *Local Area Augmentation Systems (LAAS)*. LAAS will work on the same principles as WAAS except that the ground station will be collocated with the destination airport. LAAS data will be transmitted by VHF line-of-sight data link to user aircraft. LAAS is expected to improve GPS accuracy to around ±1 m and able to be certificated for Cat II and Cat III approaches (see Table 10.2).

Europe plans to introduce a similar and interoperable capability using its Galileo constellation of geostationary satellites. The greatly improved navigational accuracy afforded by satellite navigation systems is facilitating enhanced route structures and procedures and reduced aircraft separation minima.

Table 10.2 Approach minima

Approach category	Decision height (ft)	Runway visual range (ft)
Cat I	200	1800
Cat II	100	1200
Cat IIIa	100	700
Cat IIIb	50	150

Required Navigational Performance (RNP)

New route structures are being developed on the basis of the concept of Required Navigational Performance (RNP). This concept separates aircraft on the basis of their navigation performance, not on the previous basis of separating air segments. These route structures define the RNP for various phases of the flight (see Fig. 10.3).

To use these routes an aircraft must account for its current Actual Navigation Performance (ANP) which the crew must verify is adequate for the desired route segment. RNP and ANP are usually reported to the crew on the FMS CDU together with a warning alert should the currently achieved ANP exceed the RNP. Ground stations also monitor the ANP. The RNP route structure facilitates a significant reduction in lateral separation minima when compared with previous minima constructed for traditional VOR/DME radio navigation accuracies.

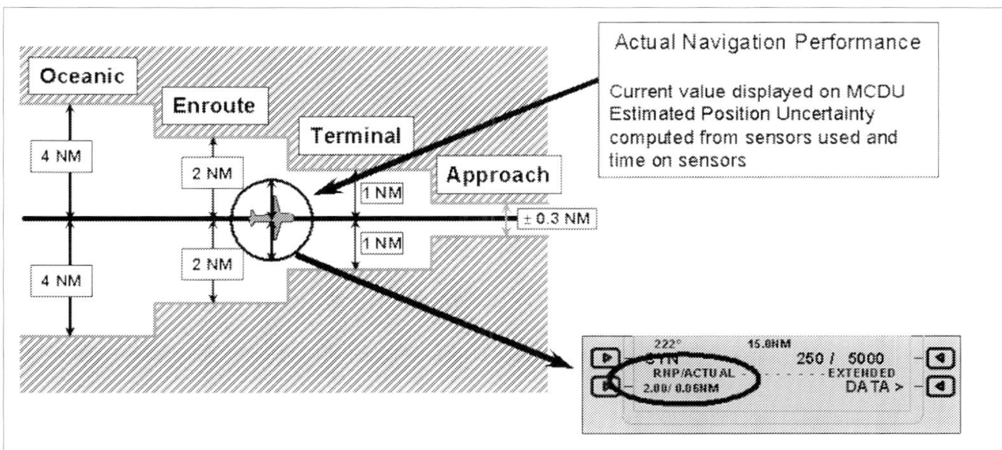

Fig. 10.3 *Required navigation performance (RNP) (Smiths Aerospace)*

Reduced Vertical Separation Minima (RVSM)

Improvements in altitude estimation have allowed the separation minima in the vertical axis to be reduced from 2000 feet to 1000 feet between FL290 (29 000 ft) and FL410 (41 000 ft). This has introduced six new flight levels for oceanic and en route domestic segments, facilitating a 30 per cent increase in traffic.

Surveillance

Surveillance enhancements include ATC transponder mode S and Automatic Dependent Surveillance modes A (ADS-A) and B (ADS-B) for use with Traffic

Collision Avoidance (TCAS) systems. The ATC transponder allows ground surveillance radars to interrogate aircraft and formulate data correlated with the radar track of each aircraft. Coincident with the Primary Surveillance Radar (PSR), a Secondary Surveillance Radar (SSR) transmits a series of interrogation pulses (on 1030 MHz) which are received by the on-board transponder. The transponder replies on a second frequency (1090 MHz) to give information relating to the aircraft.

ATC mode S extends the capability of the current ATC transponder to provide air-to-air as well as air-to-ground data link communication and hence the ability for each aircraft autonomously to interrogate and determine the precise whereabouts of other aircraft in the vicinity. Mode S also enhances the data exchange. Basic ATC transponder information (level 1) is limited to ident (call sign) and altitude. Mode S adds levels 2 to 4 with enhanced data messages to include 4D position and intended flight plan.

Automatic dependent surveillance provides similar capability, except the communications link is SATCOM or HF data link (for oceanic crossings) and VHF datalink (for transcontinental routes).

Flight management systems

An integrated navigation system is a prerequisite of FANS-based procedures. High-performance, high-integrity Flight Management Systems (FMSs) are now available that use a multisensor approach to blend the performance of all available navigation sources and provide a suitable interface with the flight crew.

A typical FMS installation is shown in Fig. 10.4, and will encompass dual FMS computers and dual Control and Display Units (CDUs) interfaced to a suite of navigation sensors. Inputs, usually dual for reasons of integrity and availability, are typically:

- Dual IN/IRS.
- Dual radio navigation aids (VOR/DME/TACAN/ILS).
- Dual satellite navigation system (usually GPS).
- Dual air data.
- Fuel system and time.

The crew interact with the flight management system through the control and display unit. The CDU, shown in Fig. 10.5, consists of an alphanumeric keypad and a display surrounded by soft-function keys. The CDU is usually mounted in the interseat console, one for each crew member. The display, originally a monochrome CRT, is now more likely to utilize a colour AMLCD.

The display format comprises a series of textural lines which display route and other information. The display is arranged as a hierarchical set of pages accessed through the multifunction keys arranged down the sides of the display area. The crew enter command and data through the alphanumeric keypad and can perform such functions as:

- Compile a route.
- Interrogate a route.
- Alter a route.

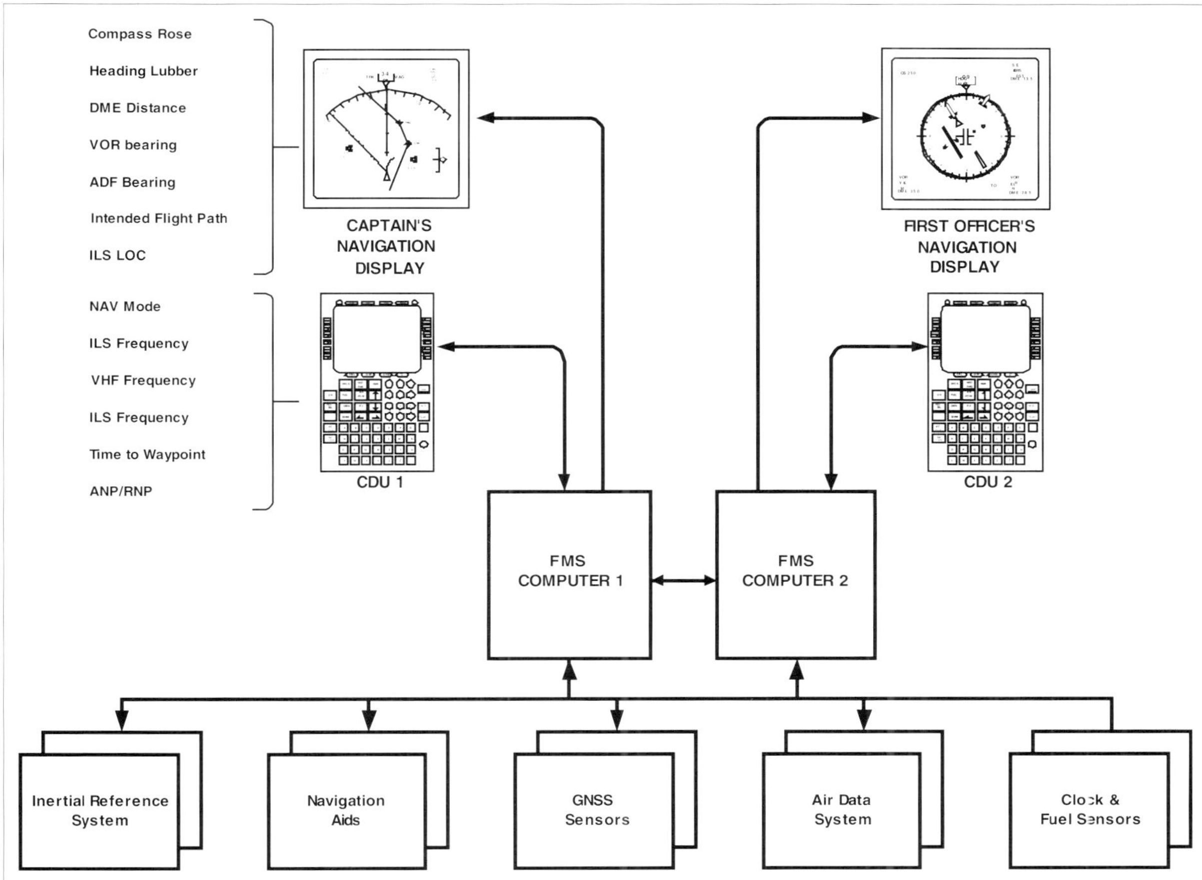

Fig. 10.4 *Flight management system*

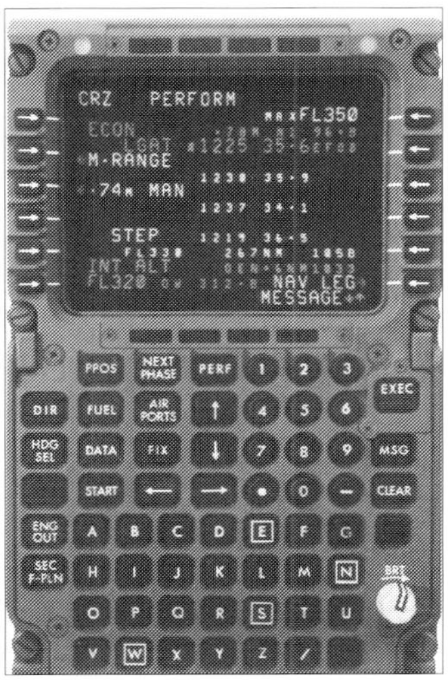

Fig. 10.5 *FMS control and display unit (Smiths Aerospace)*

Fig. 10.6 FMS trajectory computation (Smiths Aerospace)

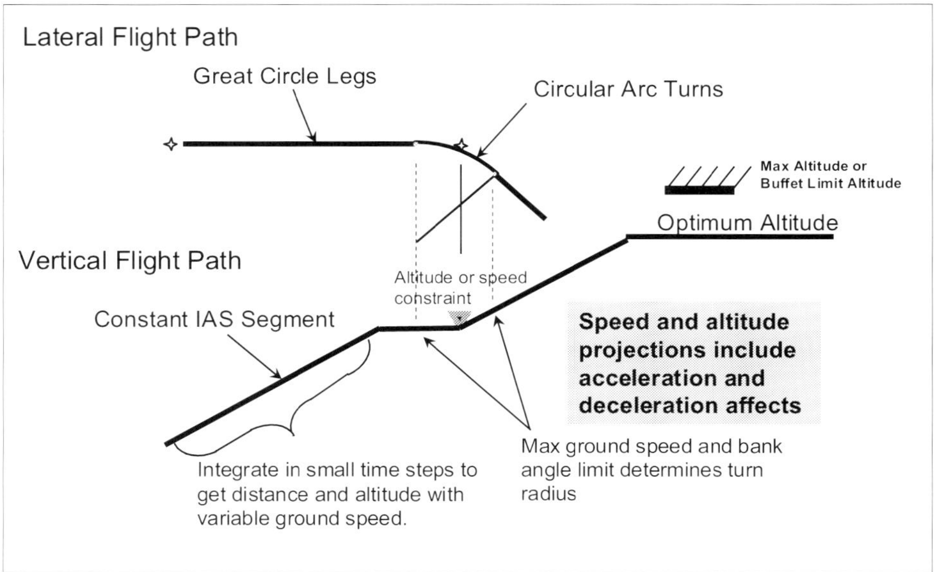

The FMS computes the best estimate of the aircraft present position from all available sensors. Based on the route structure, the FMS then performs the navigation calculations, computes the aircraft 3D/4D flight trajectory (see Fig. 10.6), and communicates the desired flight path to the crew as lateral and vertical guidance cues on the PFD display, and if coupled to the autopilot will automatically control the aircraft flight trajectory.

Within the FMS is a navigation database which comprises all the required information relevant to the flight sectors likely to be flown including departure, en route, terminal arrival, approach, and descent procedures. From all these data is assembled the current flight plan, and this is pictorially presented to the flight crew in plan form on the ND map display format (see Fig. 10.7).

Lateral and vertical guidance extends throughout the approach and descent phases, including monitoring of Actual Navigation Performance (ANP) against Required Navigation Performance (RNP) for each phase, as shown in Fig. 10.8. An alert is issued if the aircraft ANP exceeds the RNP for the flight segment currently being flown.

Modern flight management systems also interface with the aircraft on-board fuel gauging and fuel management systems and can indicate earliest and latest time of arrival based on a model of aircraft performance, which includes minimum and maximum speed profiles. Given a required time of arrival from ATC, the FMS can optimize the flight trajectory for minimal fuel burn.

High-precision approaches

The application of FANS technologies is allowing major advances to be made in the design and execution of high-precision approaches. One such example is the new approach designed and certified for runway 26 (RW26) at Juneau, Alaska. Before the implementation of the RNP approach, access to Juneau RW26 was impossible in all but the best Visual Flying Rule (VFR) conditions. The airport at

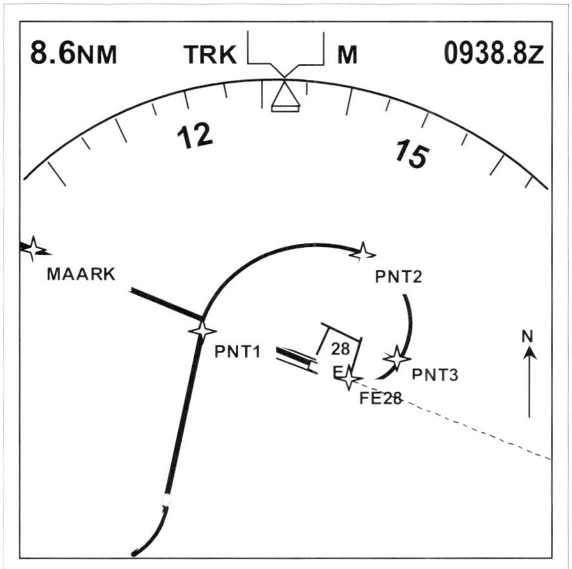

Fig. 10.7 FMS guidance and map formats (Smiths Aerospace)

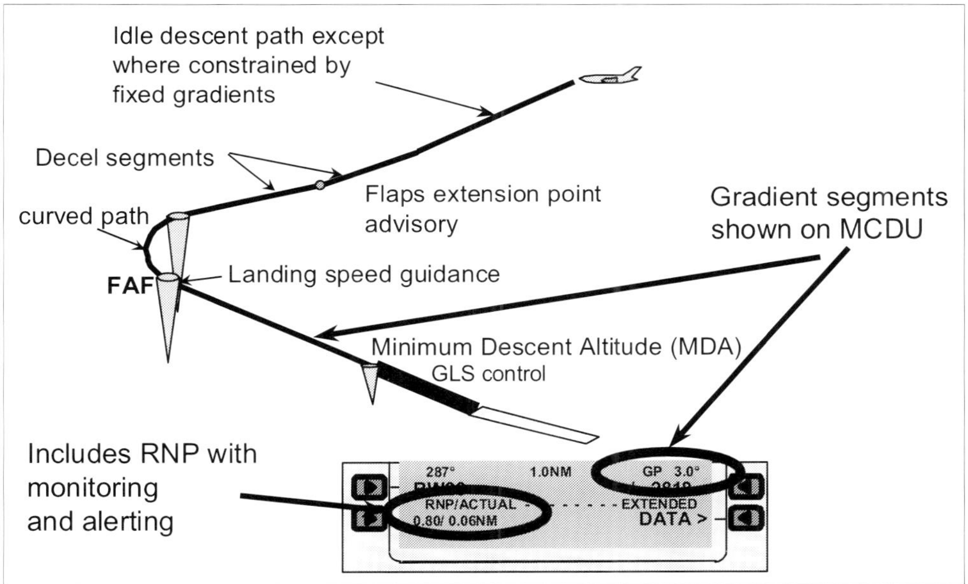

Fig. 10.8 RNP navigation and alert (Smiths Aerospace)

Juneau is surrounded by high mountains, and the approach to RW26 requires flying northwest up the Gastineau Channel to make a left-hand turn onto the runway as shown in Fig. 10.9.

The terrain geography prohibits the installation of an ILS beam for RW26. Thus, non-visual Instrument Flying Rule (IFR) conditions, the conventional approach, has required overflying the airport, then making a procedure turn to land in the opposite direction on RW8. The newly designed RNP approach shown in Fig. 10.10 enables the aircraft to execute a RW26 VMC-like approach in IMC conditions.

Fig. 10.9 Juneau airport

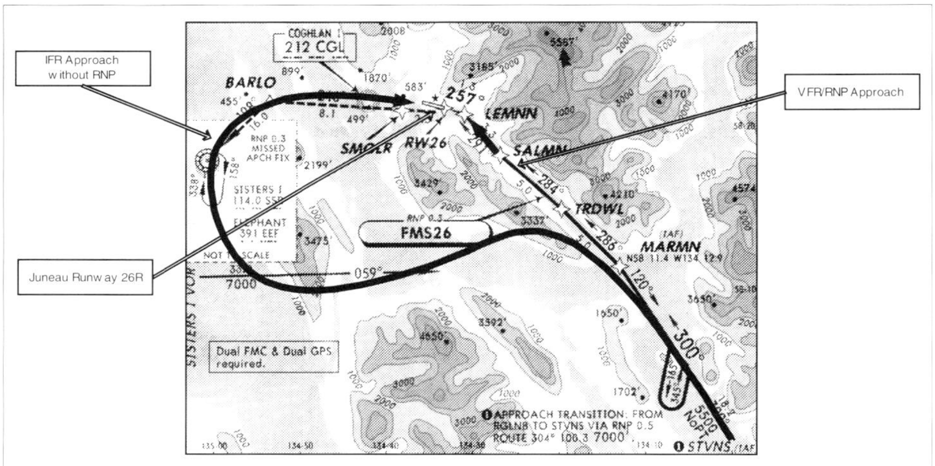

Fig. 10.10 VFR/RNP and IFR approaches to Juneau (Smiths Aerospace)

The procedure has been designed to accommodate RNP-0.15, RNP-0.2, and RNP-0.3 approaches with decision heights at 337, 437, and 1810 ft respectively. This approach was designed by the aircraft operator (Alaska Airlines), the aircraft designer (Boeing), and the FMS supplier (Smiths Aerospace).

The high-precision approach profiles shown in Fig. 10.11 provide lateral and vertical guidance down the flight path, commencing from the outer marker MARMN some 15 miles out and transitioning the middle and inner markers at predetermined heights and speeds until finally making a left-hand turn to RW26.

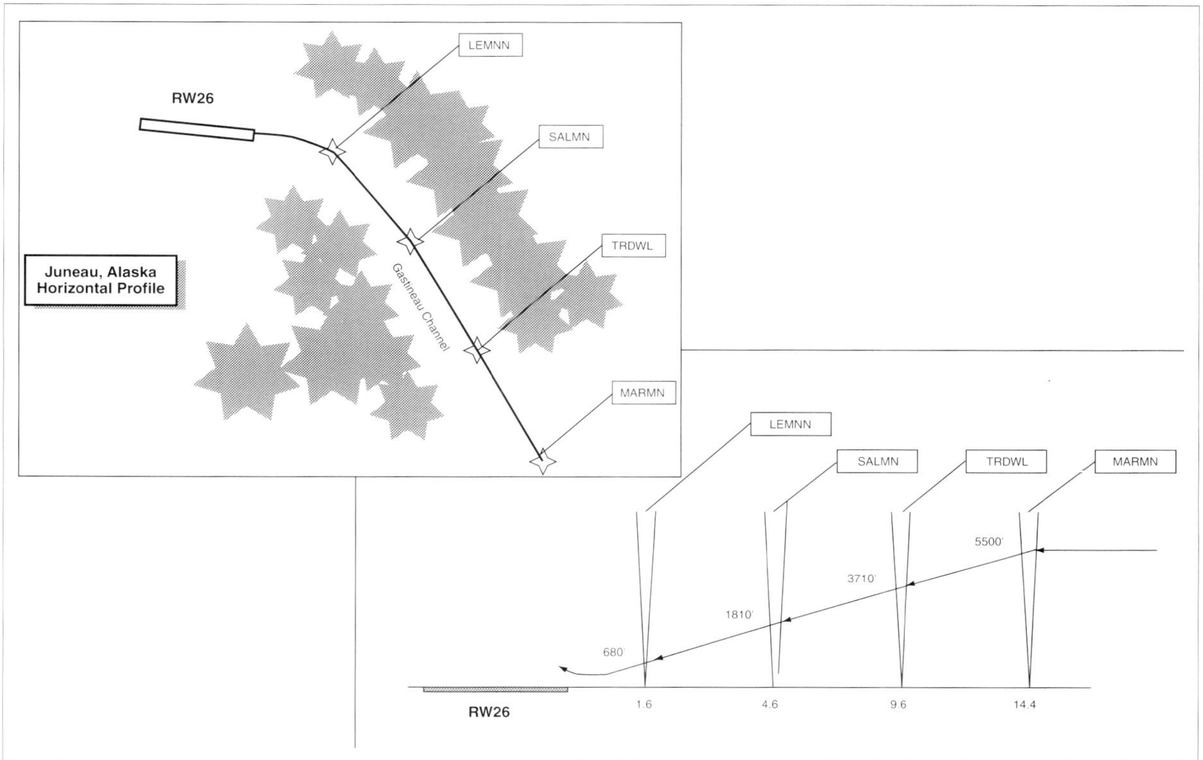

Fig. 10.11 Juneau Approach profiles (Moir and Seabridge)

Displays to the crew are the conventional PFD and ND map formats augmented with aircraft flight path trend information and with terrain information from the Terrain Avoidance Warning System (TAWS). The PFD provides lateral and vertical guidance conventionally via the flight director deviation bars. The pilot may elect to fly the aircraft manually or retain the autopilot coupled to the FMS and monitor the descent.

The map mode is displayed on the ND which shows the aircraft present position, track-up, with respect to the desired flight path, markers, and waypoints. The pilot is immediately able to relate to the display format since it provides a precise representation of the lateral approach (LNAV) profile in a style that corresponds exactly to the procedure chart he will have clipped to his control yoke. Aircraft flight path vector rate of change is displayed as a trend vector emanating from the aircraft symbol. This line will bend in a turn to indicate likely future flight path trajectory.

Terrain Avoidance Warning System (TAWS)

Terrain Avoidance Warning Systems (TAWSs) or Ground Proximity Warning Systems (GPWSs) provide visual and audible indications to the crew to warn of the risk of unintentional flight into the ground (terrain). Though not itself a display system, intuitive representation of TAWS data on the Navigation Display is an essential element to reduce CFIT accidents.

The TAWS incorporates algorithms that use radar altimeter information plus a variety of aircraft sensors to describe a safe flight envelope. Warnings are generated when key aircraft dynamic parameters deviate from values defined by the algorithms and the following scenarios unfold:

- Flight below the specified descent angle during an instrumented approach.
- Excessive bank angle at low altitude.
- Excessive descent rate.
- Insufficient terrain clearance.
- Inadvertent descent after take-off.
- Excessive closure rate to terrain (the aircraft is descending too quickly or approaching higher terrain).

Recent enhancements (also known as Enhanced Ground Proximity Warning Systems (EGPWS)) add a worldwide terrain database which is compared with the aircraft's current position and altitude. Within the terrain database the earth's surface is divided into a grid matrix with a minimum safe altitude assigned to each square in the grid. With knowledge of the aircraft present position (accurately determined by virtue of the RMP approach), and with knowledge of the aircraft current manoeuvre, the TAWS computer is able to predict potential interaction of the aircraft with the landmass. This interaction is presented on the ND map display in exact register with the flight plan as areas of colour dots with varying density, as shown in Fig. 10.12 and Table 10.3. This presentation provides the crew with confidence that the vertical flight profile is correct.

In addition, the TAWS system provides audible cues to the flight crew of approaching potentially hazardous conditions. A 'Terrain, Terrain' audible warning is given when the flight path manoeuvre results in the imminent danger of incursion into the green or medium yellow areas shown in Fig. 10.12. A 'Pull-Up, Pull-Up' audible warning is given for danger of imminent incursion into the heavy yellow or red areas.

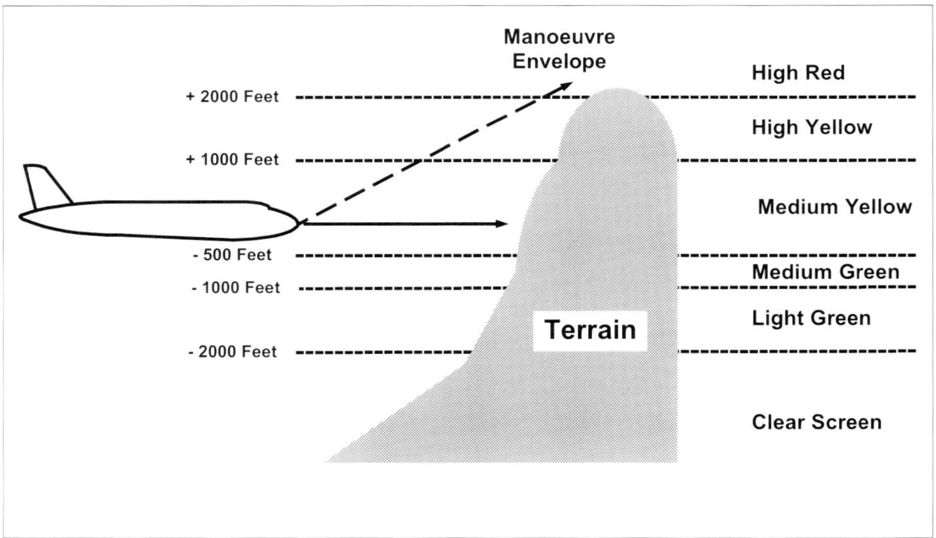

Fig. 10.12 *TAWS encoding (Moir and Seabridge)*

Table 10.3 TAWS encoding

Terrain relative to aircraft manoeuvre	Encoding
Below 2000 ft	No dots
Between 2000 and 1000 ft below	Sparsely spaced green dots pattern
Between 1000 and 500 ft below	Medium spaced green dot pattern
Between 500 ft below and 1000 ft above	Medium spaced yellow dot pattern
Between 1000 ft and 2000 ft above	Densely spaced yellow dot pattern
2000 ft above	Densely spaced red dot pattern

Traffic Collision Avoidance System (TCAS)

The Traffic Collision Avoidance System (TCAS) was developed in prototype form during the 1960s and 1970s to provide a surveillance and collision avoidance system to help aircraft avert collisions. However, it has taken many years to develop the communications system and the deconfliction algorithms to the current point where the system can be used with confidence. Though not a display system *per se*, TCAS is included here since visual alerting is an essential aspect of surveillance and deconfliction.

TCAS operates without support from ATC. The TCAS equipment on-board each aircraft will detect the presence of nearby aircraft equipped with transponders that respond to ATC modes A and C (ident plus altitude) or S (ident, altitude plus flight plan data) interrogations. Once detected, TCAS determines range and bearing, then tracks and continuously evaluates the threat potential of these aircraft to own aircraft.

TCAS surveys traffic approximately 15–40 nm forward, 5–15 nm aft, and 10–20 nm either side, as well as altitudes within 8700 ft of own-aircraft altitude. It can typically track up to 30 aircraft, and usually displays up to 25 of the highest priority threats. TCAS is unable to detect aircraft that are not carrying an appropriately operating transponder or have unserviceable equipment.

The system comprises two elements, a surveillance system and a collision avoidance system. The collision avoidance system predicts the time to, and separation at, the intruder's closest point of approach (CPA). These calculations are undertaken using range, closure rate altitude, and vertical speed.

TCAS provides an audible Traffic Advisory (TA) to alert the crew if safe boundaries are violated and displays warning symbology on the Navigation Display (ND). The TA is advised 20–40 s before point of closest approach. The crew must visually identify the intruding aircraft and are permitted to alter their vertical flight path by up to 300 ft.

TCAS I does not offer any deconfliction advice but does provide the crew with vital data in order that they may determine the best course of action. TCAS II with mode S offers a more comprehensive capability. TCAS II determines the relative motion of both aircraft, determines the appropriate course of action, and issues a Resolution Advisory (RA) via mode S to the crew of both aircraft to execute the necessary manoeuvres to maintain a safe vertical separation between them. The RA is advised within 15–35 s and is based on a 5 s crew reaction time. Deconfliction symbology is presented on the ND display.

Table 10.4 TCAS symbology

Threat		Traffic symbol
Beyond proximity advisory range (PA)	◇	Open white diamond
Within ±1200 ft and 6 nm	◇	Solid white diamond
Within 20–48 s of CPA	○	Solid yellow circle
Within 35 s of CPA	□	Solid red square

When TCAS is tracking an intruder, a symbol will appear on the ND display at the appropriate range and bearing (pictorially). The symbology (based on the intruder threat) is coded in accordance with Table 10.4.

If available, the traffic symbol will also have an altitude tag which displays:

- Relative altitude (±).
- Trend arrow if vertical rate is greater than 500 ft/m.

TCAS II resolution Advisories are:

- 'CLIMB, CLIMB, CLIMB'.
- 'DESCEND, DESCEND, DESCEND'.
- 'REDUCE CLIMB, REDUCE CLIMB'.
- 'CLEAR OF CONFLICT' when potential collision situation is resolved.

TCAS II only permits vertical deconfliction manoeuvres. TCAS III will shortly be introduced which will include lateral deconfliction manoeuvring also.

TCAS II will soon be made mandatory for civil airliners with a weight exceeding 15 000 kg or 30 or more seats operating in Europe. This will be extended to aircraft exceeding 5 700 kg or more than 10 seats, probably by 2005.

The vertical situation display

Currently, flight crews must assimilate vertical situation information from various sources to create a mental picture of the vertical profile. These sources include barometric and radio altitude readouts, the vertical speed indicator, ground proximity warning systems, terrain depiction systems, navigation information from the FMS, and navigation charts.

To help prevent CFIT and approach and landing accidents, Boeing has created a clear graphical picture of the aircraft vertical flight path that enhances the flight crew overall situation awareness (**4**). This Vertical Situation Display (VSD) works in conjunction with the TAWS terrain-mapping feature to provide flight crews with an intuitive presentation of the vertical situation relative to the surrounding terrain and the final approach descent path. The VSD depicts a profile, or side view, of terrain and flight path data.

The VSD is designed to maximize safety while minimizing required changes to aircraft hardware and airline flight operations and training. It can be implemented within the constraints of available space on existing aircraft displays.

A VSD graphically represents a view of the vertical profile of the aircraft. The Boeing VSD depicts a swath that follows the current track of the aircraft and therefore is referred to as a track-type VSD. When selected by the flight crew, it appears at the bottom of the navigation display (Fig. 10.13).

Fig. 10.13 VSD location on the flight deck – The Boeing Company (Reprinted from AERO Magazine by permission of The Boeing Company.)

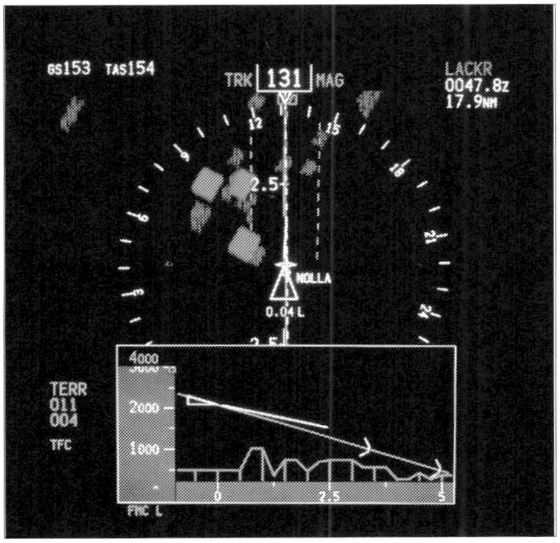

Fig. 10.14 Example of a side-looking VSD format – The Boeing Company (Reprinted from AERO Magazine by permission of The Boeing Company.)

Figure 10.14 shows an example of the Boeing side-looking VSD. The basic features of this VSD include altitude reference and horizontal distance scales, an aircraft symbol, a vertical flight path vector, terrain depiction, navigation aids, glideslope depiction, and various information selected by the flight crews – selected altitude, minimum decision altitude, and selected vertical speed predictor.

The computation of the track-type VSD includes an algorithm that expands the swath in the direction of the turn to give flight crew additional terrain look-ahead in a turn. The swath width actually is dynamic and varies as a function of navigation accuracy requirements and whether or not the aircraft is turning. Consequently, the swath is depicted on the map with two dashed lines. The information contained between these two lines is the information depicted on the VSD. This results in a display that is more intuitive to flight crews. Wherever appropriate, symbols from other formats are incorporated into the VSD using the same representation shape as the corresponding information on the primary display format.

The VSD can be used to assess path stability and speed stability. To facilitate speed stability, a new symbol is introduced on the VSD. The range-to-target speed symbol is a green dot that shows where excess speed will be dissipated along the

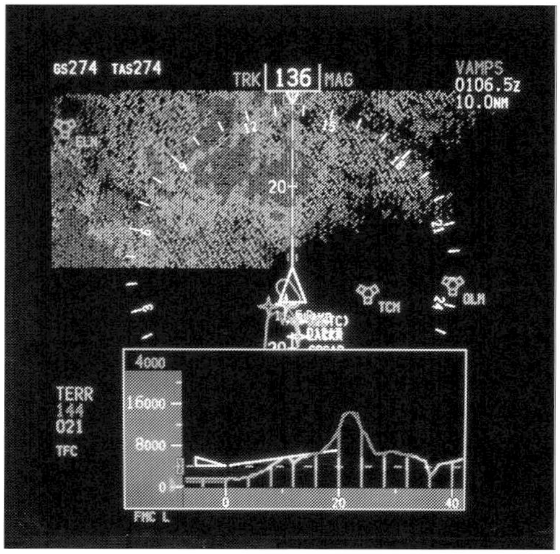

Fig. 10.15 Aircraft heading into terrain (TAWS caution) – Boeing (Reproduced from AERO Magazine by permission of The Boeing Company.)

vertical flight path vector. If excess speed is not an issue, then the symbol will not appear on the display.

The main benefit of the VSD is improved safety. The VSD gives flight crews an intuitive view of the vertical situation just as the current map display provides an intuitive depiction of the lateral situation. In conjunction with the other safety features of the flight deck, this increased vertical situation awareness helps prevent CFIT and approach and landing accidents and incidents.

The VSD depicts terrain information from the TAWS or other on-board sources from another perspective. The TAWS generates a lateral view of the surrounding terrain and provides terrain proximity alerting. The VSD depicts the vertical dimension of the terrain (Fig. 10.15) which will allow crews to recognize possible terrain conflicts more readily before a TAWS alert is generated.

The VSD also depicts the final approach segment of the intended path of the

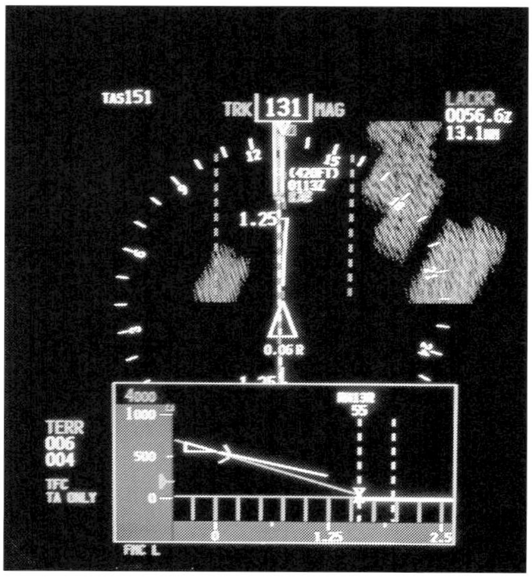

Fig. 10.16 Final approach – Boeing (Reproduced from AERO Magazine by permission of The Boeing Company.)

aircraft to the runway (Fig. 10.16), thereby assisting flight crews as they establish the glidepath. Terrain alerting from the TAWS is disabled gradually during this phase of flight to eliminate nuisance alerts, but the VSD is available throughout.

The VSD also complements the increased use of constant-angle, area navigation and Required Navigation Performance (RNP) approaches by providing immediate validation of the selected approach path and allowing full-time monitoring of the aircraft position relative to the selected glidepath. As low-altitude, in-cloud manoeuvring becomes commonplace and RNP criteria allow better utilization of restricted airspace, the VSD will serve as an invaluable confirmation of aircraft performance.

The VSD will provide additional operational benefits. Earlier recognition of terrain clearance problems facilitates more timely go-arounds and earlier CFIT avoidance. In addition, with improvements in vertical awareness, flight crews have an improved ability to monitor the vertical path. Earlier recognition of unstabilized approaches helps reduce the number of go-arounds and missed approaches. Because many unstabilized approach problems are manifested during the landing phase of flight, earlier recognition should also reduce the number of hard landings, runway overruns, brake fires, and tyre failures.

Enhanced 3D and 4D situational awareness displays

The precision approach to Juneau, Alaska, demonstrates the gains to be made using FANS technology. The display presentation to the crew uses extensions to the well-tried and established PFD and ND formats which provide essentially two orthogonal 2D windows on the world. The VSD adds the very valuable dimension of an intuitive vertical descent profile.

However, present-day formats are still awash with numbers, letters, and symbols. These parameter descriptions are derived from tools used by scientists and engineers to abstract principles of reality for mathematical analysis. They suit that purpose well, but are an underuse of the CRT/AMLCD media to portray real-time spatial awareness. The pilot still has to perform mental gymnastics with 2D information presented in two orthogonal axes on his PFD and ND displays to formulate a 3D/4D picture of the real world.

The objective is to provide pilots with consistent, intuitive, 3D real-time and predictive pictorial information, which requires no time for mental integration in order to intervene most effectively during the last moments of a critical circumstance.

Early research into intuitive 3D and 4D graphical information displays coined the term ‘highway in the sky’. ‘Highway in the sky’ itself first flew experimentally on a Calspan Laboratory C-130 aircraft in 1983. Many research programmes have further developed the ‘highway in the sky’ concept.

With precise knowledge of aircraft present position, a high-fidelity terrain database, and modern computer graphics processing, it is now possible to generate an accurate 3D representation of the terrain viewed from the flight deck. Moreover, if an element of aircraft predicted manoeuvre is incorporated into the display, then it can be argued that the display is 4D since it includes an element of time. Display formats have been developed and flown experimentally that predict the next 60 s or

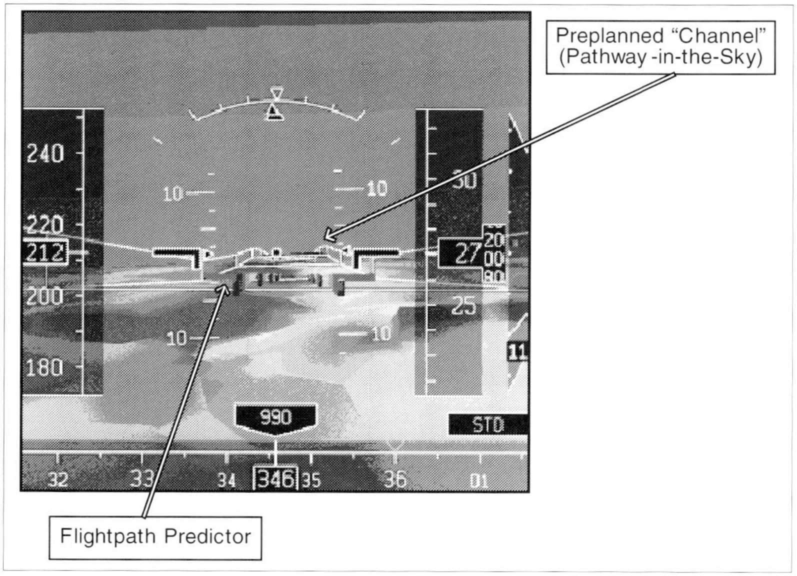

Fig. 10.17 *4D primary flight display (VDO Luftfahrtgerate)*

so of flight set in the context of the real-world terrain to achieve constant, complete, naturally intuitive, and predictive spatial awareness. Simply put, the 'highway' tells you where you should go and the terrain tells you where you should not.

In 1995 the University of Munich, Germany, carried out a significant programme of work together with VDO Luftfahrtgerate Werk of Frankfurt. An aircraft was equipped with 4D displays and evaluated taxiway, take-off, approach, landing, and terrain avoidance manoeuvres with much success (**5**).

4D experimental primary flight display

The experimental primary flight display shown in Fig. 10.17 replaces the simple sky/ground shading of the traditional PFD format with a 3D perspective of the outside world terrain. It promotes better situational awareness for the pilot and an early prediction of danger, especially the proximity of terrain. The 3D presentation includes a synthetic picture of the terrain, with features such as obstacles, rivers, and towns displayed as clearly as the horizon and runway. The display changes in real time according to the aircraft position and direction.

Surrounding the display are the traditional PFD tape-scale indicator symbols for speed, height, heading, etc., to ease transition to the new format. Special symbols for flight control are integrated into the display. A flight path predictor symbol (which can be likened to a series of aircraft symbols in future perspective) indicates the anticipated motion of the aircraft on the basis of its current manoeuvre.

The desired preplanned flight path (the 'highway in the sky') has been experimentally flown in a number of styles. The following are examples of possible implementations for various phases of the flight.

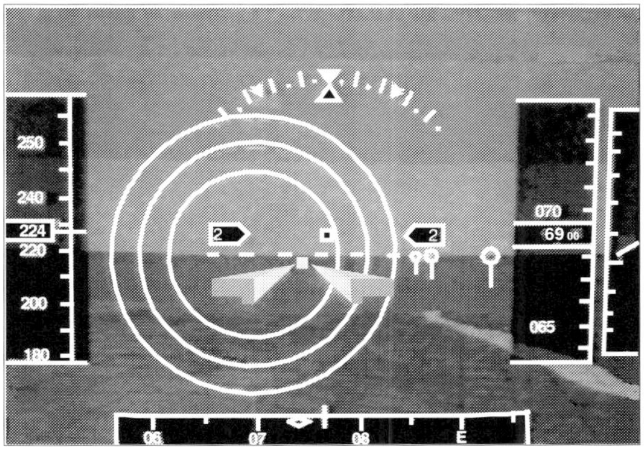

Fig. 10.18 Perspective PFD – en route

En route

In the format shown in Fig. 10.18, waypoints are shown as simple hoops or gates through which the aircraft is planned to pass. The hoops are placed on vertical poles at the correct height and location and in perspective against the synthetic terrain. As the aircraft flies towards the waypoint, the hoop will enlarge until the aircraft flies through the hoop. The hoop will appear above or below the aircraft symbol if the aircraft has to ascend or descend to the waypoint. In the terminal area the hoops will become more densely packed and may be augmented by intermediate rings for better indication of the desired flight trajectory. The ring diameter would be proportional to the required RNP for the procedure of that flight segment.

Descent

In this phase the potential for CFIT becomes more significant and, as shown in Fig. 10.19, the terrain is colour coded in accordance with TAWS conventions as appropriate to the aircraft manoeuvre to indicate any terrain features that could endanger the current flight path.

Fig. 10.19 Perspective PFD – descent

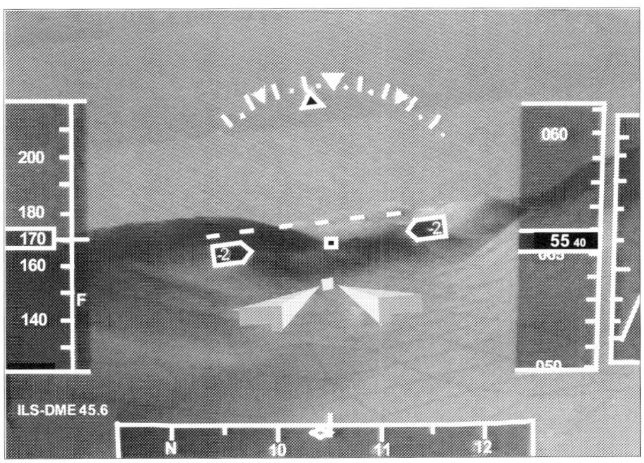

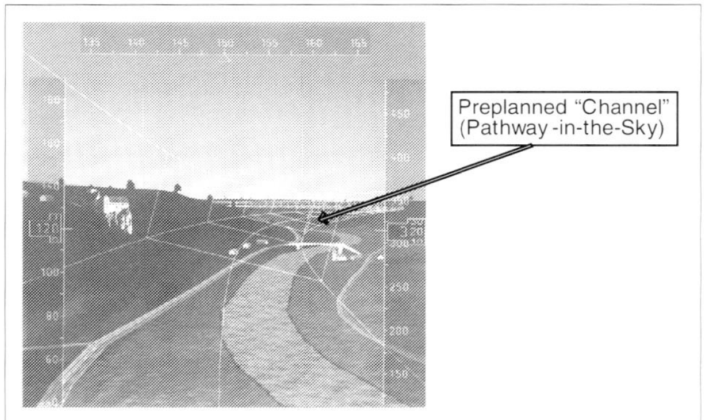

Fig. 10.20 *Perspective PFD – approach (VDO-Luftfahrtgerate)*

Approach

During the approach phase the desired flight path becomes a wire-frame model or channel, as shown in Fig. 10.20, more precisely to indicate the 'pathway in the sky'. If an ILS approach, the wire frame would indicate glideslope and localizer centre-line and allowable deviation. If an RNP approach, the wire frame would indicate RNP accuracy.

Landing

The channel/predictor combination provides an early warning of any deviation from the intended flight path and supports the pilot during manual approaches, facilitating precise, rational, and intuitive handling of the aircraft (see Fig. 10.21). Terrain warning is indicated by TAWS encoded colour changes of any features with heights above the predicted flight path.

4D experimental navigation display

The experimental 4D navigation display shown in Fig. 10.22 provides orientation of the aircraft with respect to its surroundings from a bird's eye perspective. It

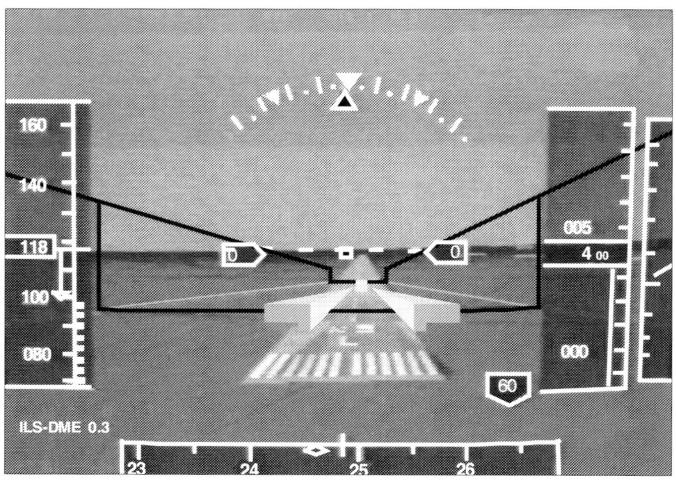

Fig. 10.21 *Perspective PFD – landing*

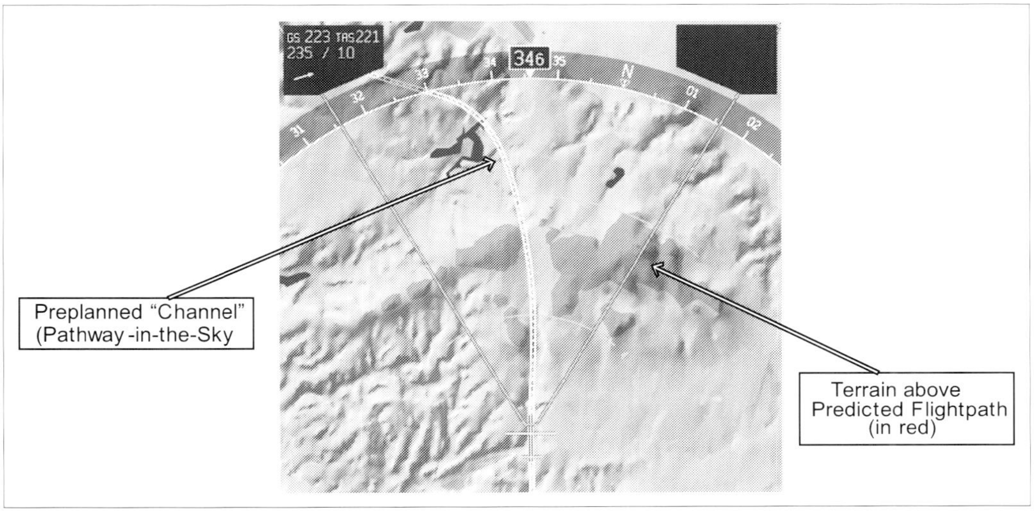

Fig. 10.22 4D navigation display (VDO Luftfahrtgerate)

incorporates a plan view of the terrain database to a higher fidelity than that of current TAWS but with the same principles and colour coding.

All flight safety related information is displayed and coded in a manner consistent with the primary flight display. The preplanned flight path is shown overlaying the terrain data. All other features of the map display are similar or identical to current-generation map displays.

An alternative ND map presentation is being developed by DASA, Germany, shown in Fig. 10.23, which depicts an aircraft flight plan superimposed on a 3D perspective view of the underlying terrain from a point above and behind the aircraft. The display also includes a vertical profile of the intended flight plan.

The display format is intended to be used in conjunction with a planning aid to facilitate simple and intuitive in-flight replanning. Possible mechanizations include touch screen or cursor control.

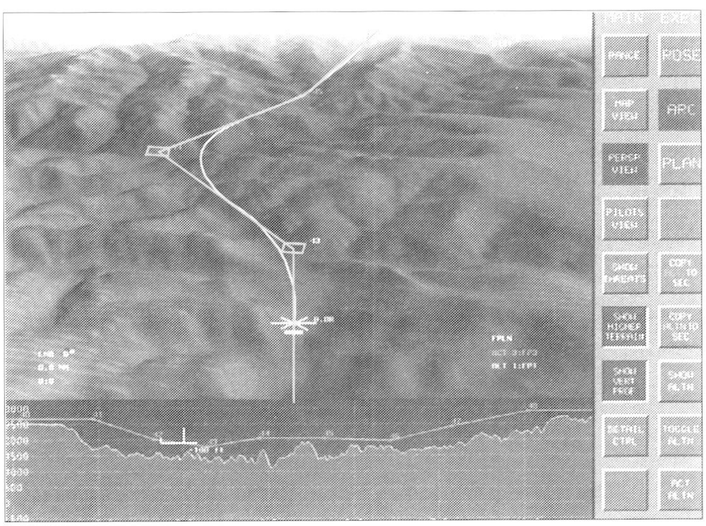

Fig. 10.23 4D perspective navigation display (DASA)

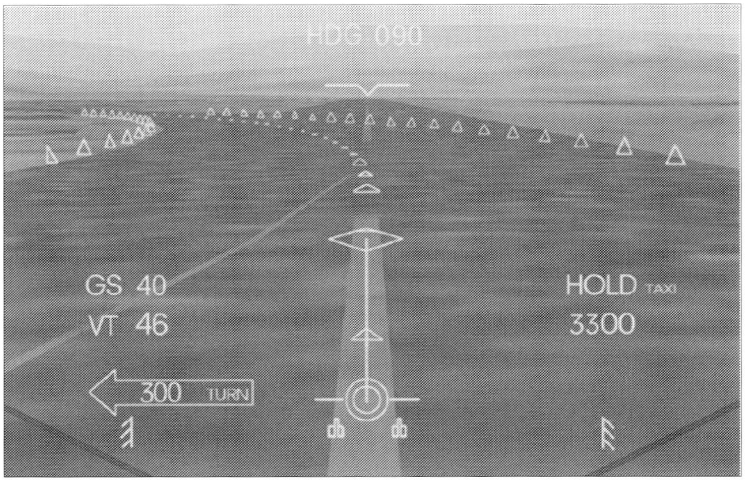

Fig. 10.24 Airfield navigation display (Rockwell-Collins)

The objective of this format is to eliminate navigation errors linked to data manipulation using alphanumeric keypads by enabling the pilot simply to touch or point to and drag the desired waypoint on the screen to make a route change. The perspective view of the terrain is a constant and intuitive reminder to check terrain clearance before executing a route change.

Airfield navigation

It is important also to address the problem of airfield navigation. The job is not done until aircraft can reliably navigate gate to gate. There have been serious accidents caused by failure correctly to navigate and obey instructions when crossing active runways on the airfield itself.

Modern navigation systems provide accurate position information to within a few metres. A terrain database including the airport would facilitate airport navigation using 2D and 3D formats in a similar manner to that described above. An example format is shown in Fig. 10.24.

Review of the other work in the field includes:

- A 'pathway' system flown by NASA on its Boeing 737 aircraft.
- Graphic perspective displays flown by the University of Technology in Delft, Netherlands, on its Cessna Citation II jet.

Large-area displays for transport aircraft

Technology is now able to provide large-area displays able to present the information content of the VSD and potential new 3D and 4D formats discussed above, overlaid with TCAS and TAWS data more flexibly and in a less cluttered manner than previously possible.

Future flight decks are likely to use larger-format AMLCDs as technology advances (see Fig. 10.25) or some alternate emerging technology such as rear projection (see Fig. 10.26). Larger display sizes offer the potential to combine the PFD and ND onto a single composite surface. However, the display system

Fig. 10.25 Simulator flight deck with four 13×11 in AMLCD displays

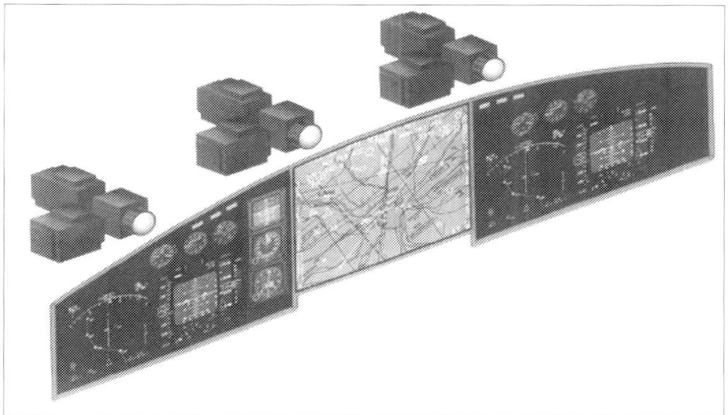

Fig. 10.26 Conceptual flight deck with projection displays (Kaiser Electronics)

architecture must be cognitive of the redundancy, availability, and integrity requirements demanded by the regulatory authorities.

The stereoscopic Head-Up Display

Head-Up Displays (HUDs) provide the pilot with a means to view real-world cues simultaneously with on-board flight information. The major advantages of Head-Up Displays for the pilot are:

- Pilot workload is reduced when the overall piloting tasks require Head-Up, outside-the-cockpit flight references.
- The expanded scale of the Head-Up Display data and their overlay on the external visual scene allow the pilot to fly more precisely.
- A conformal display allows the pilot to assess the aircraft performance directly.
- Essential flight information presented on the Head-Up Display reduces eyes-in-the-cockpit time during critical flight manoeuvres.

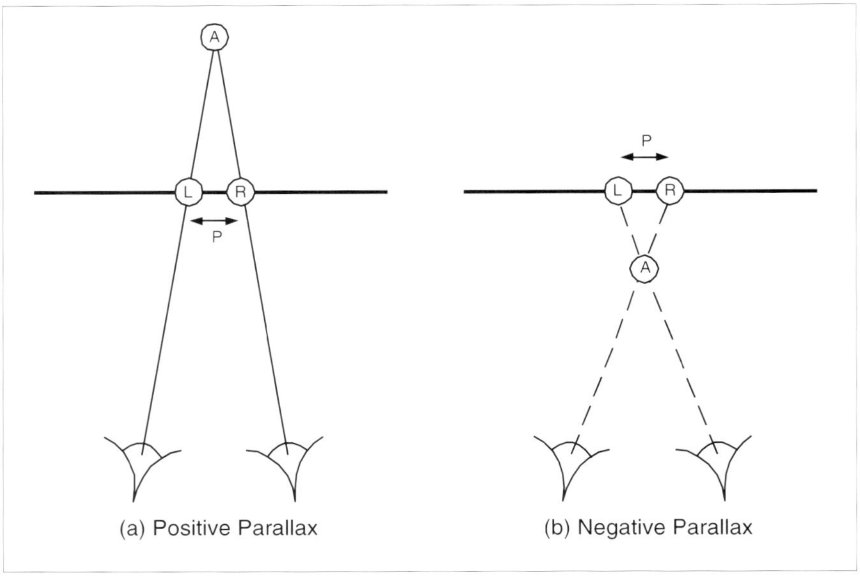

Fig. 10.27 Parallax in stereoscopic displays

Head-Up Displays provide the pilot of an aircraft with a means to view real-world cues simultaneously with on-board flight information. While colour and intensity coding is state of the art in modern glass cockpits, head-up displays are still monochrome. Furthermore, the symbols have to be based solely on lines to maintain see-through capability.

These limitations, together with collimation of the symbology to a certain distance, make all indications appear like one object in a single depth plane. The introduction of stereoscopic information presentation in the Head-Up Display promises an improved human–machine interface by providing intuitive separation and combination of indications.

A stereoscopic Head-Up Display has been developed and integrated into a fixed-base flight simulator. First tests indicate improved information perception in Head-Up Displays by adding stereoscopy (**6**).

The perception of stereoscopy is caused by retinal disparity. Retinal disparity occurs because each of our eyes sees the world from a different point of view. The eyes are, on average for adults, 2.5 in or 64 mm apart. The disparity is processed by the eye–brain system into a single image of the visual world. The ability of the mind to combine two different, although similar, images into one image is called fusion, and with normal binocular vision results in perception of depth called stereopsis. The effect of disparity is used in stereoscopic displays.

Figure 10.27 shows two projections on the plane of the screen for displaying the object A at two different positions in depth. Object A is visible on the display screen for the left eye at position L and correspondingly for the right eye at position R. The distance P between the two projection points L and R on the screen plane is called the parallax. The uncrossed or positive value of the parallax (a) will produce images appearing behind the screen. With the intersect of the eyes' axes, the parallax points are said to be crossed or negative and objects with negative parallax appear to be closer than the plane of the screen (b).

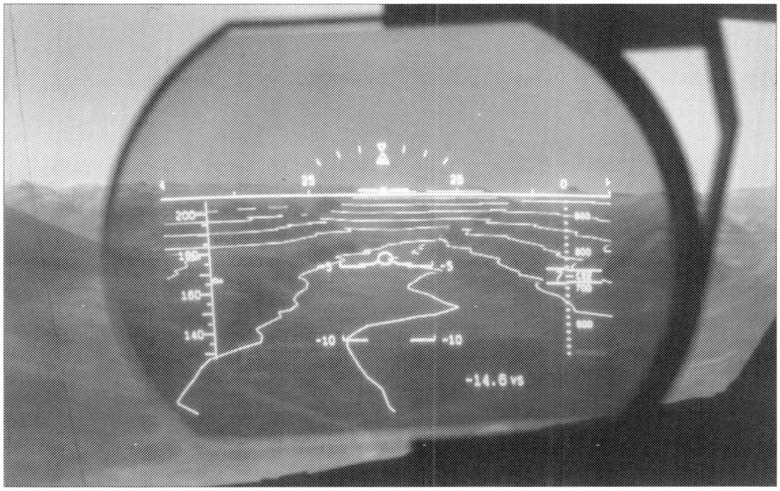

Fig. 10.28 *Possible stereoscopic HUD symbology (Darmstadt University)*

In the simulator used for this evaluation, stereopsis was achieved using a state-of-the-art civil HUD made by Thales and introducing the left and right eye images on a time sequential basis. The pilot (evaluator) wears a pair of shutter glasses, which are synchronized to the display so that the appropriate image is seen by each eye.

To evaluate the potential advantages of a stereoscopic Head-Up Display, stereoscopic symbology has been defined on the basis of current aviation Head-Up Display symbology. Typical symbology is shown in Fig. 10.28.

The 2D symbology comprises the following elements: speed, altitude, aircraft reference, horizon line with heading indication, pitch ladder, bank scale, vertical speed, radio altitude, wind indication, and flight path vector with guidance cue. The representation of 3D elements is possible for the runway outline, ILS or tunnel-in-the-sky symbology, taxi guidance, and terrain visualization.

The use of stereoscopy offers the potential advantages.

- Improved perception of information in the outside view owing to the symbology displayed in different planes of depth.
- The spatial separation of information in depth will avoid the erroneous merging of symbols. This will lead to a decluttered display.
- Warnings can be coded by positioning in depth (as colour is not recommended for Head-Up Displays).
- The preplanned flight path, taxi guidance information, obstacle, or terrain can be visualized at its spatial position.
- New 3D symbology can be introduced for a more natural and intuitive perception.

Voice control

Speech recognition technology is becoming accepted in the office environment. Recognition accuracy is acceptable for the technology to be seen as a positive contribution to improving efficiency.

In the military arena, fast jet fighters are applying speech recognition technology also. The Eurofighter Typhoon has a speech recognition system with a vocabulary of 100 words. It can be used (after training) by male and female pilots of any ethnic origin from the four nations participating in the programme. The recognition accuracy under all operating conditions (including the extreme stress of battle) is better than 98 per cent. Needless to say, this technology is significantly more sophisticated than the speech recognizers available for office laptops.

In Eurofighter Typhoon, the direct voice control system is used for routine command functions associated with the on-board systems such as radios, communications, navigation aids, etc., and allows the pilot to keep his hands on throttle and stick and his eyes out, without having to search for and select panel-mounted push-button controls for these functions. All commands are replayed visually on the HUD for confirmation before execution.

In the more benign environment of the civil transport aircraft, a derivative of this technology would be capable of controlling many non-flight safety critical functions, freeing up crewstation space for larger information displays and reducing pilot workload.

Touch screen

Touch screen technology is now also accepted for many information system applications. As discussed earlier, touch screen technology has been used experimentally by DASA, Germany, to evaluate a more robust means to enter flight plan data.

The overhead panels are a significant area of the flight deck and are yet to be modernized. Mostly, they still comprise discrete indicators, knobs, and switches to control some of the critical aircraft systems (electrics, fuel, fire extinguishers, etc). Candidate technologies to modernize and streamline this area of the cockpit include touch screen displays. Images on the displays would represent synoptic diagrams of the aircraft systems with indications of the current state of controls. Intuitive on-screen touch selection could then be used to adjust and change control state.

Cursor control

Cursor control is a more likely technology for the primary flight displays. Touch screen requires support for the hand before making a selection and tactile feedback once a selection has been made. Yet at the same time the surfaces of the primary flight displays must be kept clean and free from any scratch marks which would otherwise detract from visibility in direct sunlight. Cursor control avoids these issues.

Cursor control is likely to be implemented as either a tracker ball or as a force feedback joystick. In either case, the control algorithms must provide a robust controllable mechanism accurately and quickly to move the cursor to the required screen position to select the desired function.

Military fast jet 'big picture' crewstation

The digital battlespace and the operational environment demand the ability to present significantly more information to the flight crew about the movement of own forces, potential threats, and targets. The modern theatre of war can be extremely complex. The individual aircraft is no longer operating alone, or with his wingmen prosecuting the same mission, but is a member of a system of systems, each prosecuting a subagenda of the main mission objective. The battlespace may include at the same time manned and unmanned air vehicles. Their individual roles will be varied: they may be reconnaissance, surveillance, air superiority, defence suppression, ground support, etc. The subagenda may rapidly change. What was the main objective of the local mission may be overtaken by other events. What was not a threat may suddenly become a threat. Accuracy of up-to-date information and clarity of presentation is vital to avoid the risk of collateral damage and friendly-fire incidents.

Information about the battlespace is collected from a variety of intelligence sources (AWACS aircraft, reconnaissance UAVs, local ground forces, etc.), assimilated, processed, and then retransmitted on a need-to-know basis to forces in the battlespace by datalink. The local tactical situation, shown in Fig. 10.29, is concerned with the local mission objectives and protection of the vehicle itself. These aspects are best dealt with head-out using a Helmet-Mounted Display which can provide all-around visual support with symbology to aid threat identification, target designation, and weapon delivery plus sufficient primary flight data to fly the aircraft safely.

The global situation, shown in Fig. 10.30, is concerned with an awareness of the whole battlespace in so far as it is relevant to the mission and the continued safety

Fig. 10.29 *Tactical situation*

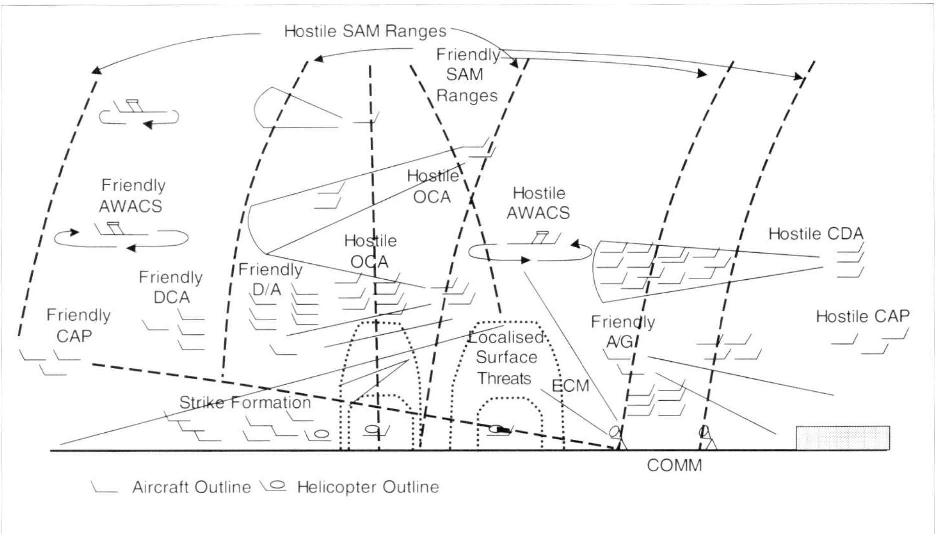

Fig. 10.30 *Global situation*

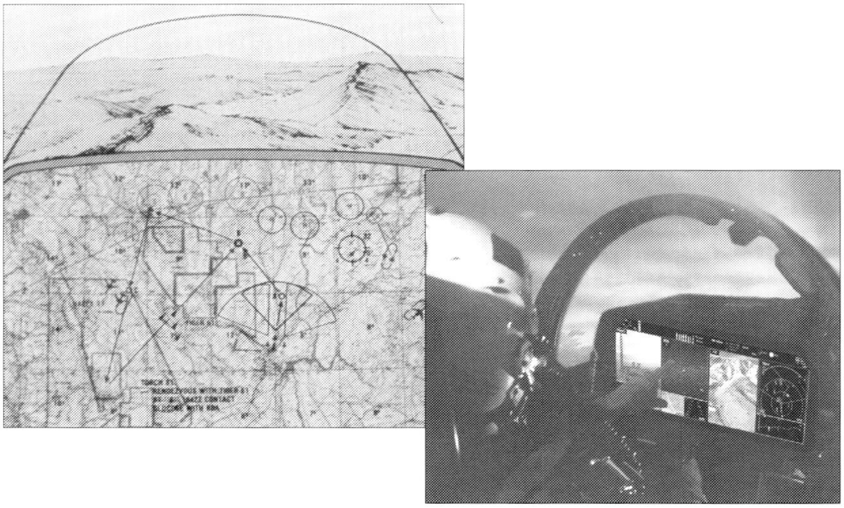

Fig. 10.31 *'Big picture' display (The Boeing Company and Kaiser Electronics)*

of the aircraft and crew. These aspects are best dealt with by a large-area head-down display. Coined the 'big picture' by Gene Adam at McDonnell Douglas, the display ideally fills the whole instrument panel and provides a contiguous view of the global situation in a manner that is totally intuitive and in a frame of reference to which the pilot can easily relate.

Examples of a 'big picture' format are shown in Fig. 10.31. The main body of the image is a synthetic 3D topographical view in the aircraft frame of reference. The image shows relevant ground features overlaid with relevant, prioritized tactical information such as local surface threats (SAM sites, radar installations, etc.) and their sphere of influence. Own planned flight path and tracks of friendly and hostile forces are shown, again only as relevant and prioritized. As the mission progresses, other information can be called into view (either automatically or on

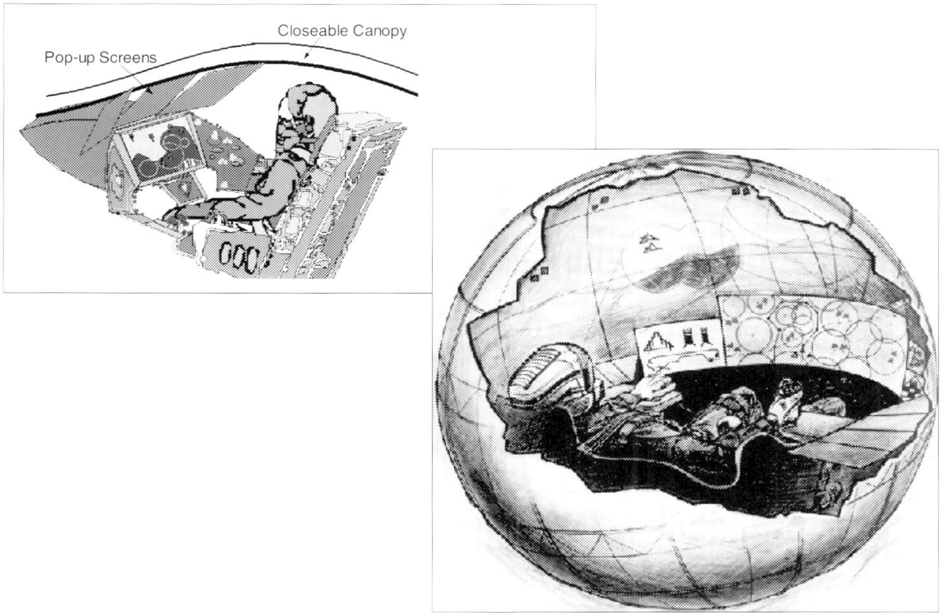

Fig. 10.32 *Closeable/ immersive cockpits (US DoD)*

demand) to show, for instance, engines/system status, weapon selections and targeting information, and sensor/weapons video.

The joint strike fighter will implement a 'big picture' Head-Down Display.

Future fighter aircraft may be potentially at risk from a number of threats in addition to conventional weapons, which may from time to time require the cockpit to be partially or completely enclosed. It has been suggested that this might be achieved by special materials within the canopy structure, and that having done so the outside worldview be provided synthetically on pop-up screens. In extremis, a totally immersive cockpit has been suggested in which the cockpit is permanently closed. These arrangements are shown in Fig. 10.32.

Unmanned air vehicles (UAVs)

At present, the military are using equipment conceived during the cold war for a scenario that no longer exists. The cost of advanced man-rated systems is high and, consequently, equipment is sometimes kept in service longer than necessary. The accompanying cost of aircrew training is very high. Moreover, since military options are often constrained by the equipment available, there can be a lack of flexibility and there may be a danger of disproportionate response. Many of the new threats cannot be addressed with conventional defence systems, and new capability is required. These new solutions must be affordable as well as effective (**7**).

Humans place constraints on air vehicles. Without a human crew there is no limit on manoeuvrability or endurance. There are no limits on shape and on the amount of intelligence carried. Dull missions present no problems, nor do dirty missions. A UAV can go into a chemically, a biologically hostile, or even a nuclear hostile environment without concern. Dangerous missions do not pose a problem since there is no life at risk.

In 1994, the global positioning system was established. The availability of accurate and continuous positional information was a major step forward and it opened a whole new range of possibilities.

Currently, there are three broad categories of UAV, though the boundaries are often blurred:

- ***Cruise missiles.*** This group encompasses a range of fire-and-forget missiles which follow a planned route and locate their targets through precise navigational means. They may be surface launched or air launched.
- ***Reconnaissance vehicles.*** This group encompasses a wide range of vehicles, from small, short-range, short-endurance vehicles to High-Altitude, Long-Endurance (HALE) vehicles. Surveillance and reconnaissance are the primary roles. These vehicles can be used for a range of missions including image gathering, intelligence gathering, data relay, weather reports, target designation, and countermeasure deployment. Some may be used as information relay vehicles to provide over-the-horizon communications from the forward battle area to the ground station in the absence of secure SATCOM links.
- ***Unmanned Combat Air Vehicles (UCAVs).*** UCAVs are intended to carry weapons, attack targets, or be weapons themselves. UCAVs can be launched from, and be recovered to, air, sea, and land. They can be remotely piloted, i.e. there is a pilot on the ground or in the air, or they can be completely autonomous. Autonomous vehicles receive a command and the vehicle works out how to comply. It then performs the task with no human intervention. UCAVs can operate independently, or in swarms, and they can be teamed with manned aircraft.

Recent events have advanced the UAV agenda. The application of existing technology is going to bring greatly increased capability, together with 2–3 orders of magnitude in reduced acquisition cost and cost of ownership. There is potential for the reduction of total defence cost while delivering more capability.

Perhaps the crewstation of the future will not be unlike the totally immersed cockpit, but in a benign environment on the ground to control more than one UAV/UCAV operating in concert (as a swarm). Perhaps the airborne crewstation will be adapted to include the control (if not the actual piloting) of more than one UAV/UCAV operating in advance of the manned vehicle to neutralize threats.

References

(1) FAA Human Factors Team report, (1996) *The interfaces between Flight Crew and Modern Flight Deck Systems*, FAA, June.

(2) *The Future Flight Deck*, Royal Aeronautical Society.

(3) **Moir, I.** and **Seabridge, A.** (2003) *Civil Avionic Systems*, Professional Engineering Publishing.

(4) **Carbaugh, D. Chen, S. Jacobsen, A. Myers, R.** and **Weiderman, J.** (2003) Vertical situation display for improved flight safety and reduced operational cost, Boeing commercial airplanes, *The Aerospace Professional*, February.

(5) **Brooks, P. E.** (1998) The promise of four-dimensional flight displays, *Flight Deck International*.

(6) **Kaiser, J. Mayer, U.** and **Kubbat, W.** *Development of a Stereoscopic Head Up Display*, Darmstadt University of Technology, Darmstadt, Germany.

(7) **Poll, D.I.A.** (2003) Is the future unmanned, *The Aerospace Professional*, April.

CHAPTER 11

Display System Architectures

Introduction

This chapter describes the display suite architectures to be found in civil transport aircraft, military fighter aircraft, and rotorcraft. The electronic components that comprise a display suite can be architected in a number of different ways to optimize performance. These are usually described as 'dumb', 'semi-smart', and 'smart' or 'integrated' architectures, named from the functionality contained within the display head itself.

The display suite can be integrated with the rest of the avionics systems in a variety of ways often commensurate with the type of data bus architecture adopted. Because of their relatively low data rates, data buses like ARINC 429 and MIL-STD-1553B are best matched to traditional avionic architectures in which functions are aligned to boxes or LRUs. This architecture has come to be known as a federated avionics. Multisource multisync data bus architectures such as ARINC 629 and AFDX Ethernet are associated with modular avionics. In this chapter we shall review applications of both types.

This chapter also describes practical realizations of Electronic Flight Instrument Systems (EFIS) and Engine Instrumentation and Crew Alerting Systems (EICAS/ECAM) as applied to Boeing and Airbus aircraft and analyses system availability, integrity, and implications for dispatch. Finally, this chapter describes display suite architectures in fast jet fighter aircraft and rotorcraft.

This chapter concludes with a review of the civil regulatory procedures and processes as they relate to display systems.

Display suite components

The constituent elements of any electronic flight instrument system, shown in Fig. 11.1, can be broadly be described as:

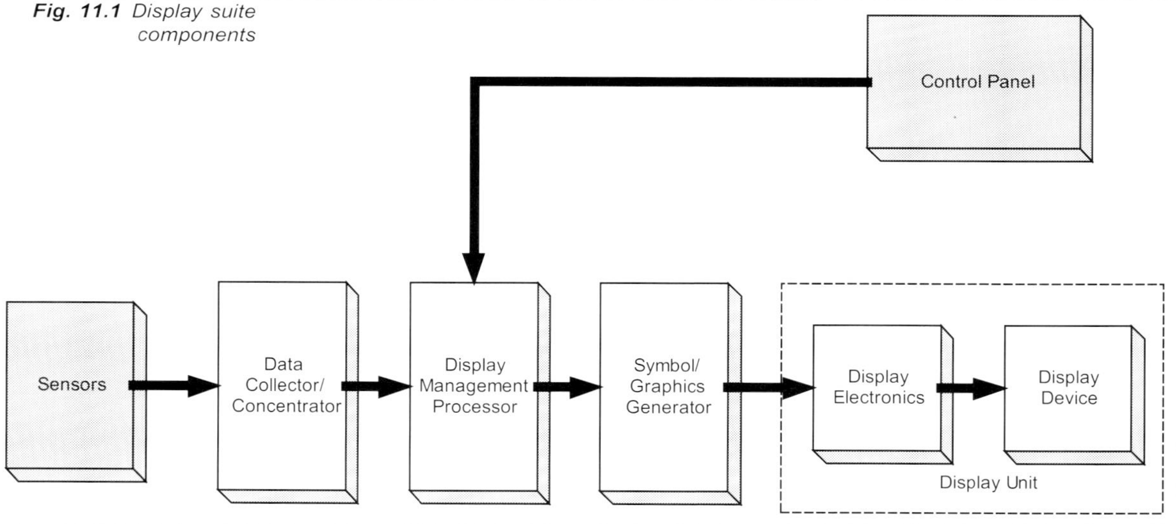

Fig. 11.1 Display suite components

- *Data collector/concentrator* which:
  - acquires the data to be displayed from the other on-board systems;
  - selects the most appropriate data sources;
  - performs data integrity checks, which might include not only checks of the source itself, but also cross-monitoring with duplicated sensors on the other side of the aircraft and correlation with other flight parameters for consistency.
- *Display management processor* which:
  - determines the display mode, submode and elements to be displayed on the basis of crew commands and selections, and modes and status of the on-board systems;
  - translates the above information into data and commands recognizable by the display graphics generator.
- *Graphics generator* which:
  - constructs the symbology and graphics for each element of the display format selected; the symbology will comprise:
    alphanumeric characters in a range of font styles and sizes,
    special symbols, pointers, and icons,
    lines of various widths and styles,
    circles, ellipses, and arcs of various widths and styles
  - performs area shade infill;
  - selects symbology and shade infill colours;
  - selects overlay priorities.
- *Display unit* which includes:
  - the display device itself (typically a CRT or an AMLCD);
  - display device electronic drive circuits to portray images from signals issued by the graphics generator;
  - for a CRT-based display unit the drive electronics will include:
    X and *Y* deflection amplifiers,
    R:G:B cathode drive amplifiers,

high and intermediate power sources for the CRT electrodes (final anode, A1, focus, grid, etc.);

- for an AMLCD-based display unit the drive electronics will include:
  AMLCD row and column drivers,
  AMLCD greyscale drive, characterization, and temperature compensation,
  backlight and dimmer control,
  heaters.

These display elements can be and have been combined in a variety of ways to form a display suite. The common arrangements are discussed below.

Dumb display architecture

The dumb display architecture shown in Fig. 11.2 was the first to be adopted. It logically followed the experience and architectures that had evolved in military CRT display systems and reflects the physical nature and complexity of driving a shadow mask CRT and the state of the art of electronics at the time of the first 'glass cockpits'.

In this architecture the *display unit* contains all, but only, the necessary electronics to support and drive the CRT itself. A shadow mask CRT is a large and heavy device and for it to operate and provide a sufficiently bright display in an airborne environment requires the image to be drawn in stroke mode. This requires complex and power-hungry electronics.

The *symbol generator* contains all the display processing electronics.

In the situation where the display device is a CRT, then the interface between the symbol generator and the display unit usually consists of specialized analogue X and Y waveforms to describe the CRT beam deflections together with digitally coded signals to describe symbol and shade infill colour.

Of course, it is not necessary to restrict this architecture to CRTs. It could and has been used with AMLCD-based display units, especially where space constraints limit the available volume behind the instrument panel. An AMLCD display complete with backlight and minimal drive electronics can make a flat panel display close to reality, having a depth of only a few centimetres. The interface to an AMLCD display unit is most likely to be a dedicated high-bandwidth digital video link encoded with red, green, and blue (R:G:B) information.

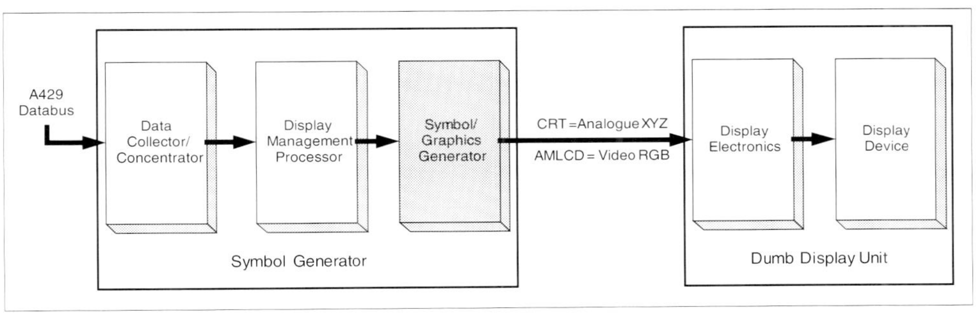

Fig. 11.2 Dumb display architecture

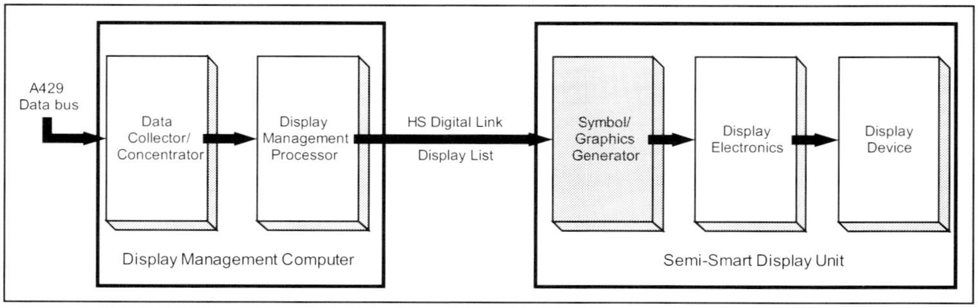

Fig. 11.3 Semi-smart display architecture

Semi-smart display architecture

The semi-smart display architecture shown in Fig. 11.3 overcomes the need for a specialized analogue/digital interface to the display unit tailored to the display device type. It was adopted first by Airbus, initially with a CRT-based display unit. In the semi-smart architecture the graphics generator is placed within the *display unit* (in addition to the display drive electronics).

The *display management computer* contains the remaining electronics. The display management computer assembles and compiles the image into a 'display list' of instructions to be executed by the symbol generator in the display unit.

The interface between the two units is digital, of intermediate bandwidth, and can use a conventional but dedicated digital link. This interface can be made display technology independent and equally applicable to CRT or AMLCD or any other display medium. Indeed, it would be theoretically possible to upgrade an installation from CRT to AMLCD if, for instance, the CRT were to become obsolete, or (more likely) from AMLCD to AMLCD with different pixel configuration as AMLCD technology advances.

Fully smart (integrated) display architecture

The third architecture is the fully integrated smart display shown in Fig. 11.4, in which all the electronics are contained within the display unit itself. This architecture is now made practical by the physical reduction in size, weight, and power of AMLCD technology when compared with its CRT predecessor. Furthermore, significant advances have also taken place in computing and graphics technology to simplify the packaging and cooling arrangements. At the same time the reliability of components has improved, so, while it could be argued that the increase in complexity of the smart display unit when compared with a dumb display unit would reduce its reliability, in practice the reliability of today's integrated display unit outstrips the reliability of the early dumb display unit and therefore does not compromise system reliability or integrity.

With federated avionic systems using point-to-point data link structures such as ARINC 429, the number of signals and hence the wire bundle and connector arrangements behind the instrument panel may compromise the benefits of an integrated smart display architecture. However, the introduction of high-bandwidth data bus structures such as ARINC 629 and AFDX Ethernet make this architecture more attractive. Boeing adopted this architecture for the Boeing 777 aircraft using AMLCD technology.

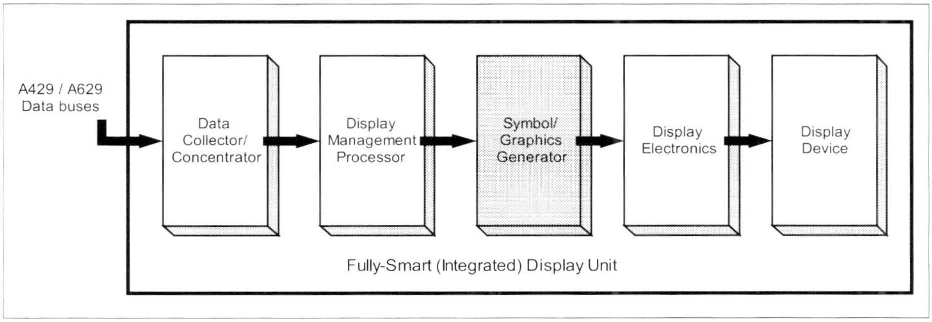

Fig. 11.4 *Fully smart integrated display architecture*

Display systems

Flight deck instruments/displays can be categorized into three types:

- *Primary flight instruments* which provide information on the ability for the aircraft to sustain safe and controlled flight.
- *Navigation instruments* which provide information about the aircraft position with respect to its surroundings in order for the aircraft to be safely directed from its point of origin to its destination.
- *Engine instruments and systems* which provide information about the aircraft systems to assure the continued safe function of the aircraft for the duration of the intended flight.

Electronic flight instrument systems (EFIS)

Systems providing EFIS Primary Flight Display (PFD) and Navigation Display (ND) functions typically have three symbol generators operating with four display units as shown in Fig. 11.5. In normal operation the third symbol generator is a 'hot spare':

- Each symbol generator is able simultaneously to produce both the PFD and ND formats.
- Each display unit sources its display format from a normal or an alternate SG source.
- Each symbol generator sources its inputs from both left-hand and right-hand sensors.
- Each display unit is able to select its display input from a normal or an alternate source.

In the event that one symbol generator fails (say SG 1), as shown in Fig. 11.6, then the affected display units (in this case PFD 1 and ND 1) select their inputs from SG 3. The formats presented on PFD 1 and ND 1 remain independent of those on the other side of the flight deck.

If one DU fails (say PFD 1), as shown in Fig. 11.7, then SG 1 will reconfigure its output so that the image on ND 1 will be a composite PFD/ND display with all requisite primary flight data including a compass.

Independent and segregated standby ADI, altimeter, and airspeed instruments are installed in the flight deck to aid resolving discrepancies and to provide critical flight information in the event of complete failure of the primary avionic systems.

Fig. 11.5 Typical EFIS architecture

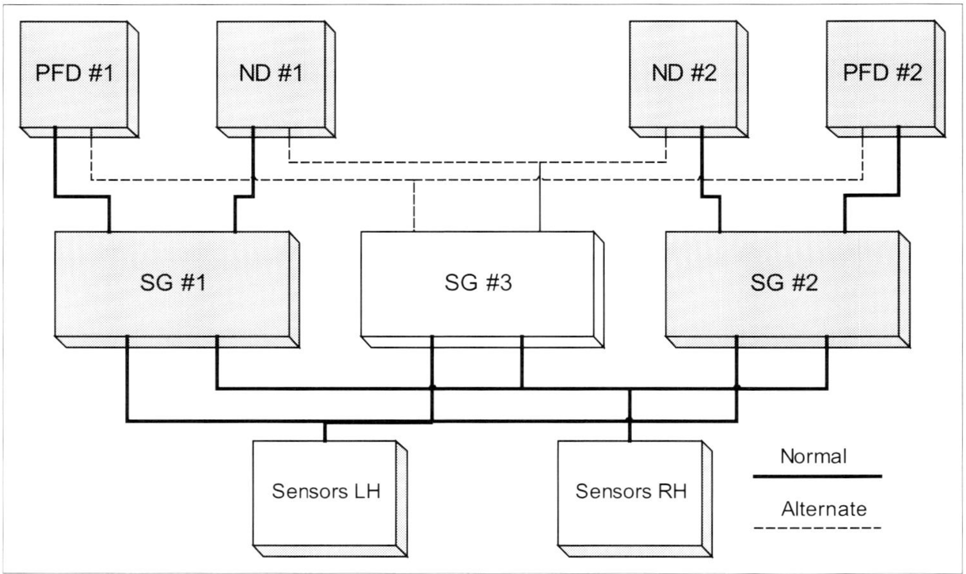

Fig. 11.6 Reconfiguration after SG failure

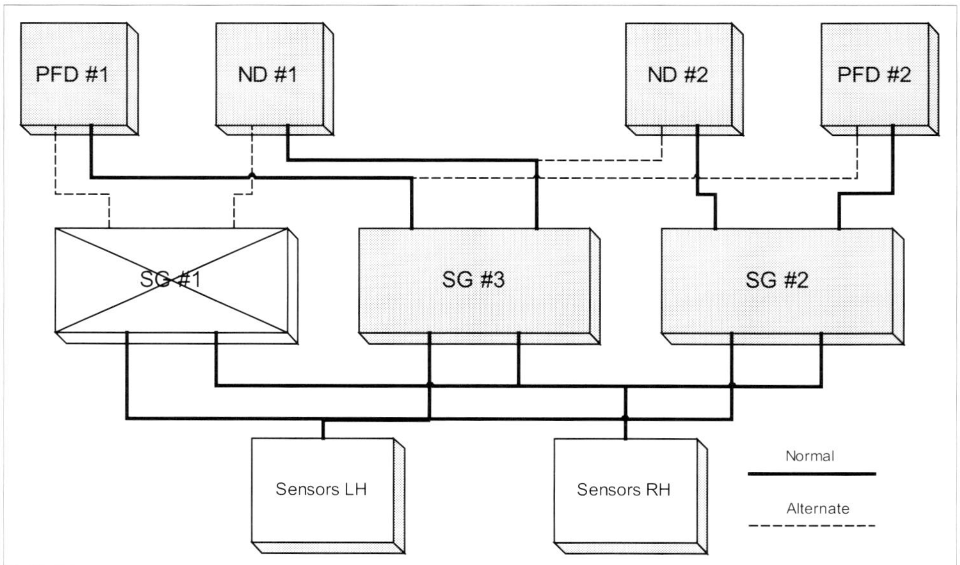

If the failure rates of the SG and DU are λ_{SG} and λ_{DU}, then the probability of complete loss of attitude, heading, airspeed, or altitude from the primary flight displays on account of failure of the EFIS system itself is

$$(\lambda_{DU})^4 + (\lambda_{SG})^3 = \text{better than 1 in } 10^{-9} \text{per hour (i.e. extremely improbable)}$$
$$\text{for } \lambda_{DU} = \lambda_{SG} = 4 \times 10^{-4} \text{ (2500 h MTBF)}$$

Strategies to assure that hazardously misleading information is not inadvertently presented to the crew include:

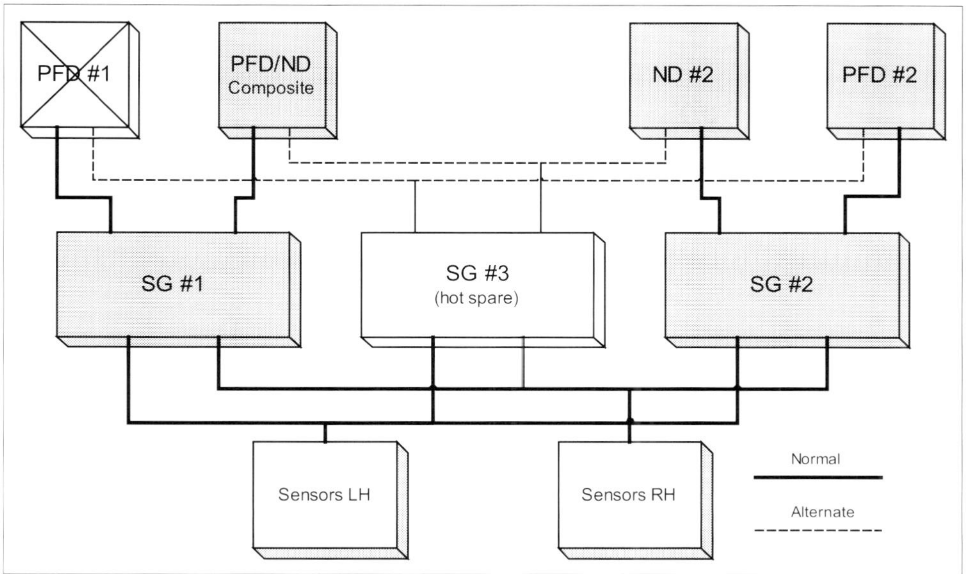

Fig. 11.7 Reconfiguration after DU failure

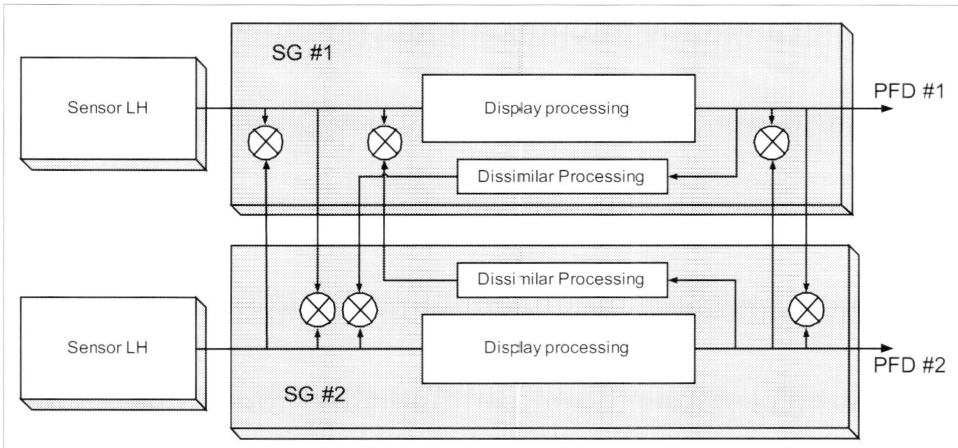

Fig. 11.8 Cross-monitoring

- Cross-monitoring of input sensor data from duplex, independent, and segregated sensors.
- Correlation of sensor data from several input sources for reasonableness against a model of aircraft performance.
- Cross-monitoring of output data to the display.

These strategies are shown architecturally in Fig. 11.8.

The integrity of the display system itself is continuously monitored by built-in-test (BIT) functions that typically check anticipated output against known stimuli. Common mode failures of the display system itself can be addressed by a number of strategies including:

- Design quality assurance (using the design process guidelines described earlier).
- Architectural features such as multiversion dissimilar processing.

Cross-monitoring of output data is usually done at a point in the graphics generator where positional information is still available digitally. Typically, this allows the read-back of start points and end-points of critical lines, arcs, and characters, and read-back of character descriptors. It is usually accepted that upstream failures in the graphics channel and the display itself from this point, even if undetected by BIT, will be fail obvious and not introduce subtle undetectable errors on critical information alone. Typically:

- Alphanumerics will be grossly distorted or illegible.
- Lines will be distorted and not join correctly with other lines to make recognizable shapes.
- Circles will be distorted.
- Shade infill will be incorrect.
- Colours will be wrong.

The frozen display is a serious potential failure mode that must be carefully analysed and addressed. The designer must include within the system secure methods to protect, warn, and remove any possibility of an undetected frozen display.

The JAR misleading data criteria can be shown by analysis to be achieved provided that the system has been designed in accordance with the JAR guidelines and at the appropriate criticality level.

Architecture examples – transport aircraft

Classic EFIS and EICAS/ECAM systems

Display systems that integrate the EFIS and the EICAS [the Electronic Centralized Aircraft Monitor (ECAM) in Airbus terminology] Airbus A330/340 as shown in Fig. 11.9. The A330/340 employs semi-smart Display Units (DUs) as described earlier. The Display Management Computers (DMCs) collect data from the aircraft systems and compile a 'display list' which is interpreted by the symbol generator within the display unit to generate the display image.

The avionics architecture is a traditional federated system. Separate Line Replaceable Units (LRUs) are provided for each major identifiable function (e.g. air data, inertial navigator, flight control, radios). In the Airbus A330/340, these systems communicate via dedicated single source – multisink ARINC 429 serial data buses. Each data link from the DMC to the DU is a dedicated high-speed ARINC 429 data bus.

Three DMCs drive the six DUs. Each DMC is able to source four display formats: the PFD, ND, and the two power/systems displays. A three-way switch selects the DU display source:

- In normal operation:
  - DMC1 sources PFD1 and ND1,
  - DMC2 sources the power/systems displays,
  - DMC3 sources PFD2 and ND2.
- In the event of failure of DMC1:
  - DMC2 sources PFD1 and ND1 and the power/systems displays,
  - DMC3 sources PFD2 and ND2.

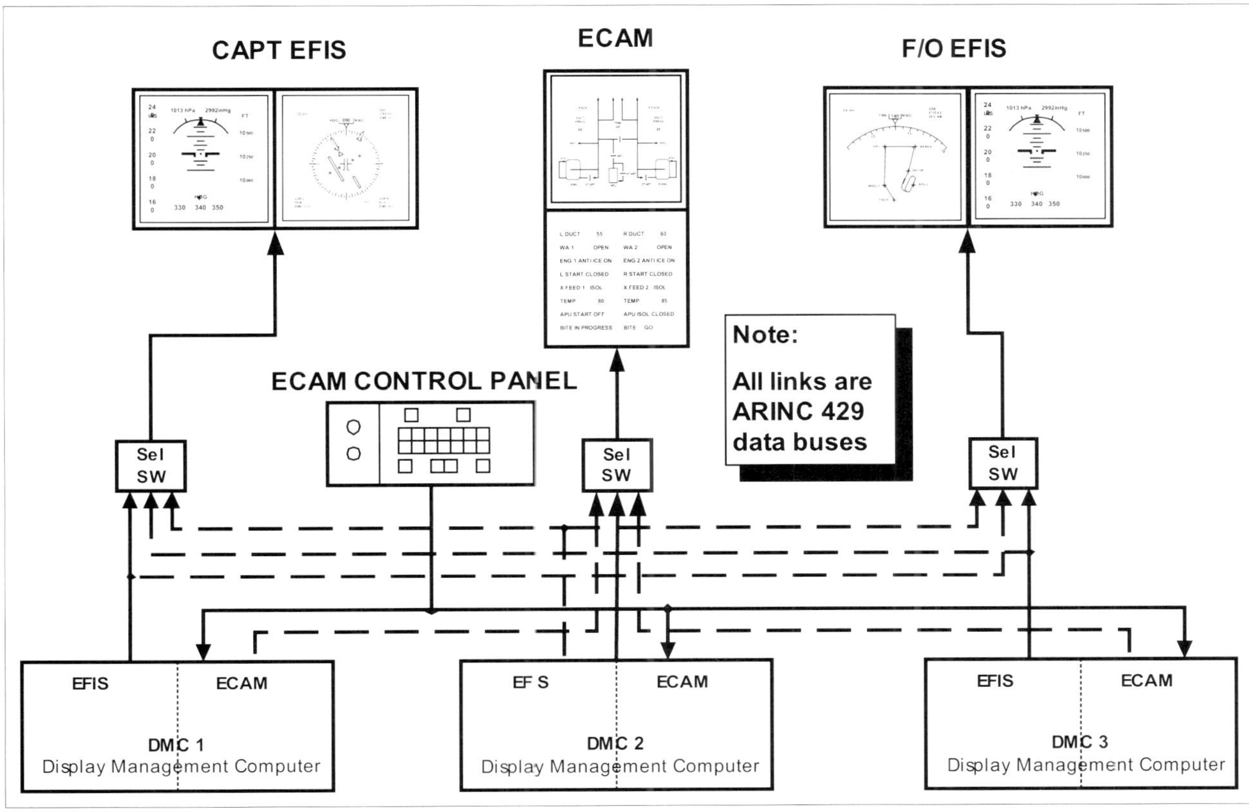

Fig. 11.9 A330/340 EFIS and ECAM architecture

- In the event of failure of DMC2:
  - DMC1 sources PFD1 and ND1, and the power/systems displays,
  - DMC3 sources PFD2 and ND2.
- In the event of failure of DMC3:
  - DMC1 sources PFD1 and ND1, and the power/systems displays,
  - DMC2 sources PFD2 and ND2.
- In the event of failure of DMC2 and 3
  - DMC1 sources PFD1 and ND1 copied to PFD2 and ND2 and the power/systems displays.
- In the event of failure of one of the pair of PFD/ND DUs:
  - the DMC constructs a combined PFD/ND format on the remaining PFD/ND DU.
- Similarly, in the event of failure of one of the pair of power/systems DUs:
  - the DMC constructs a combined power/systems format on the remaining power/systems DU.

It is important to note that, for any single DMC failure, PFD1/ND1 remain independent of PFD2/ND2. By inspection it is clear that the architecture more than satisfies the JAR criteria for parameter availability. Indeed, it can be argued that the aircraft can dispatch with only two functional DMCs.

Modular avionics display systems

The architecture of the Boeing 777 Aircraft Information Management System (AIMS), as shown in Fig. 11.10, illustrates the integration of smart display units with modular avionics. The six fully smart integrated Display Units (DUs) contain all the electronics to manage, process, and generate display formats from data transferred to them from the AIMS cabinet via high-speed ARINC 629 data buses. The data bus is connected to all DUs; two data buses provide redundancy.

The AIMS cabinet provides a centralized and redundant computing resource on which the avionic functions of the classic federated architecture reside as software applications. A comprehensive software operating system assures a safe and secure partitioned operating environment.

The AIMS cabinet manages failure of internal functions by reconfiguring applications to run on the remaining functional computing resources. DU failure is managed as described earlier by presenting combined formats on remaining DUs.

Fig. 11.10 *Boeing 777 AIMS architecture*

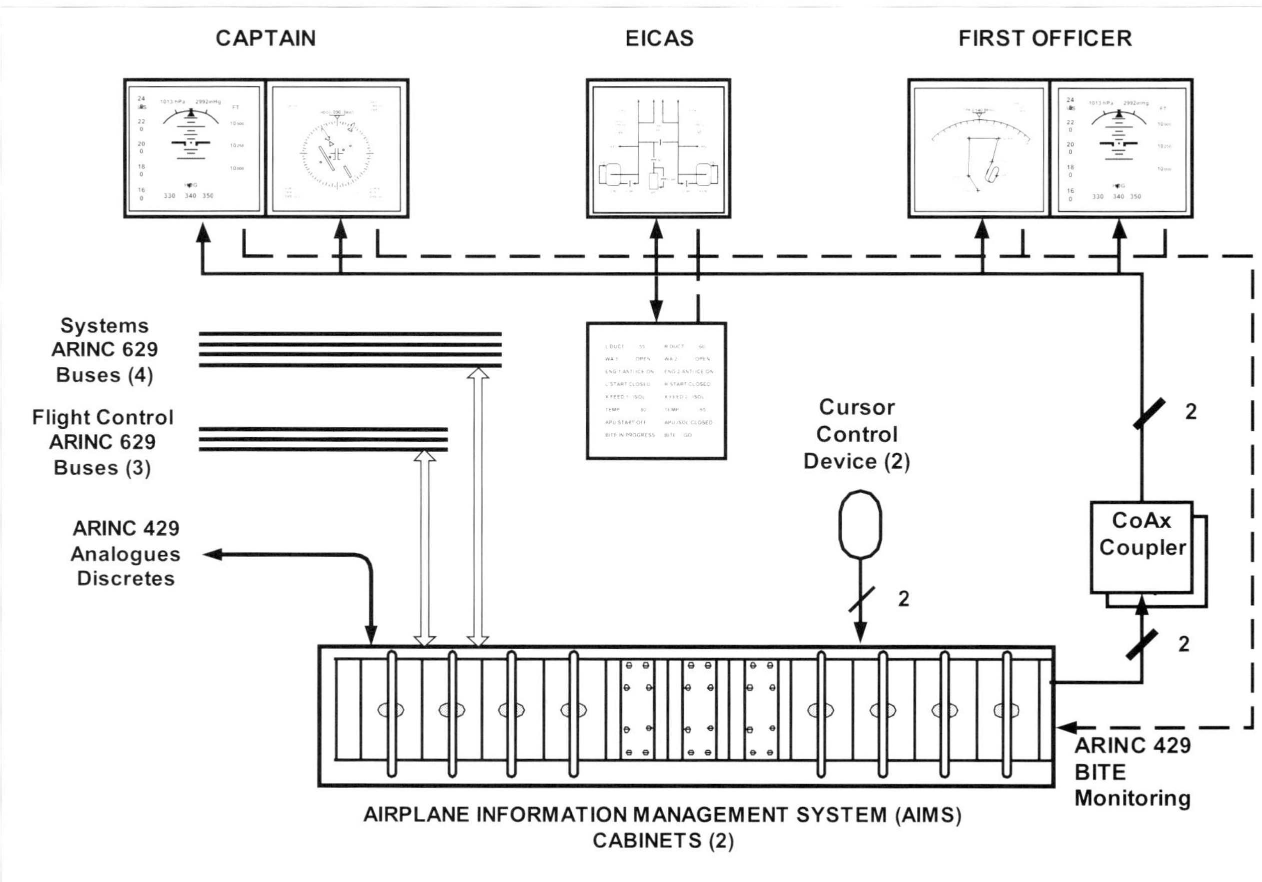

Architecture examples – military display systems

Classic federated mission systems

Figure 11.11 shows the mission systems architecture for the BAE SYSTEMS Hawk as configured for the Royal Australian Airforce Lead-In Fighter (LIF) jet trainer. The aircraft has two mission computers which acquire data and manage the aircraft sensors and weapons via two MIL-STD-1553B data buses, the avionics bus and the weapons bus.

The front seat (the trainee pilot) is equipped with a day/night Head-Up Display and three 5×5 in AMLCD multifunction Head-Down Displays. The rear seat (the instructor) is equipped with three 5×5 in AMLCD multifunction Head-Down Displays and a Data Entry Panel similar to the Up-Front Control Panel mounted on the HUD in the front seat.

The Head-Up Display operates in stroke mode in daytime and in stroke plus raster mode at night, overlaying flight/attack symbology on top of the FLIR sensor video. The HUD is a dumb device. The mission computer generates stroke symbology and combines this with sensor video to drive the HUD CRT with low-level analogue signals.

The Head-Down Displays are dumb devices. The mission computers generate raster graphics symbology which is combined with sensor video to drive the HDD AMLCDs with standard composite video signals.

Figure 11.12 illustrates the traditional federated mission system architecture of military fast jet aircraft designed in the 1980s and 1990s such as the Eurofighter Typhoon. Avionic functions are assigned to individual Line Replaceable Units (LRUs). LRUs communicate with each other via data buses grouped into

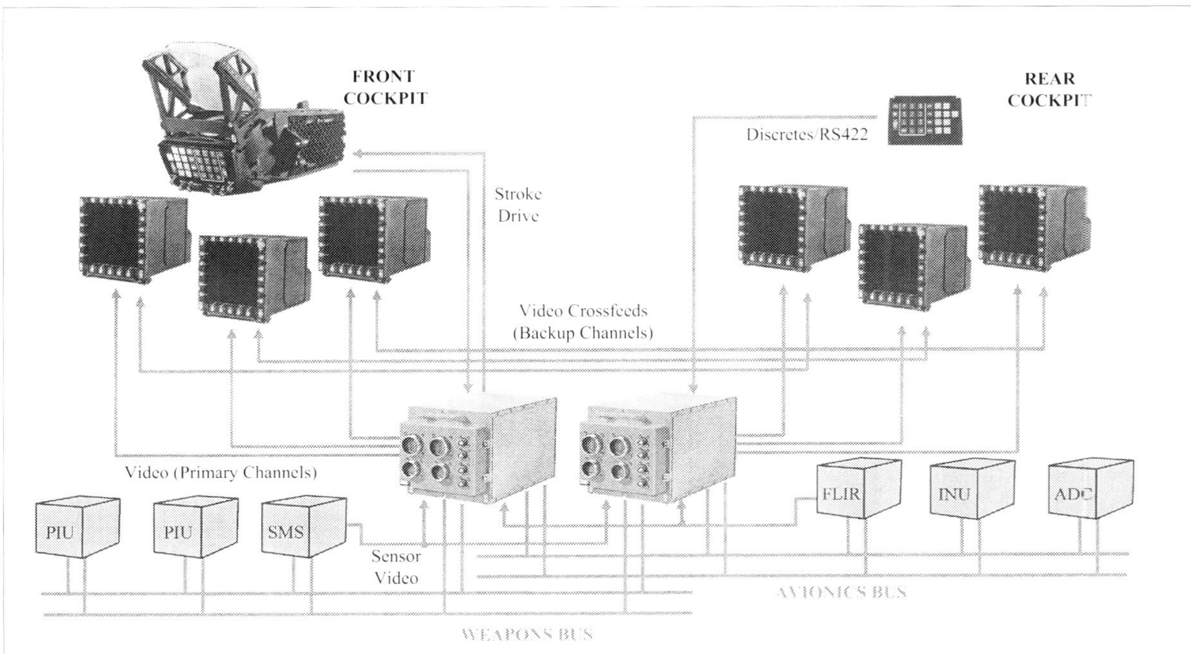

Fig. 11.11 Hawk lead-in fighter (Smiths Aerospace)

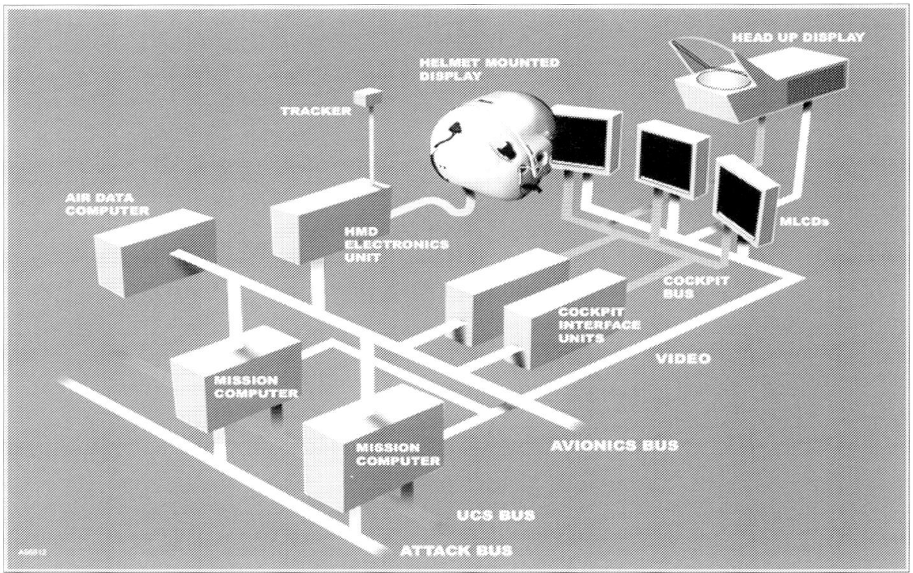

Fig. 11.12 Traditional federated displays architecture (BAE SYSTEMS)

domains. LRUs requiring data from more than one domain provide bridges between domains.

In Fig. 11.12 the cockpit displays communicate via the local cockpit bus, which includes interface units to manage the HMI interaction with the cockpit controls, switches, indicators, and communications and, in the case of Eurofighter Typhoon, voice recognition and control. Display data are computed and collated within the mission computers and translated into graphics images in a format most appropriate to the display media.

In the case of the Eurofighter Typhoon, the Head-Down Displays were initially shadow mask CRT technology but were superseded by AMLCD towards the latter end of the development programme. The HDDs are dumb; the symbology is generated in the mission computers. The interface to the Head-Down Displays is display technology independent. Eurofighter Typhoon uses conventional analogue R:G:B video; later aircraft designs sometimes encode the video digitally using a high-speed serial bus.

The Head-Up Display uses CRT technology and operates in stroke and hybrid (stroke+raster) mode to provide day/night operation. The HUD is dumb, and the stroke symbology, combined with sensor video for night-time use, is generated in the mission computer and transmitted to the HUD in specialized low-level analogue form.

The helmet-mounted display uses CRT technology and, like the HUD, operates in stroke and hybrid mode to provide a day/night capability. The 'on-helmet' electronics is minimized to achieve a low-weight helmet, and so the specialized CRT support electronics, symbol generation, and video mixing are housed in a specialized HMD electronic unit, mounted in the cockpit area. This LRU also contains the HMD tracker electronics.

Modular avionics mission systems

Figure 11.13 shows the possible architecture of a next-generation fast jet fighter such as the Joint Strike Fighter (JSF). The Head-Down Display is a single 'big picture' display, although for redundancy reasons it will contain more than one image source. The display technology will be either direct-view or projection (see Chapter 10). Display images, data and commands are transferred to the HDD via the data network, independent of the choice of HDD imaging technology. The HDD will probably combine symbolic images with sensor video, and manage image presentation and reconfiguration in the event of failure of one of its imaging sources/surfaces.

The Head-Up Display is optional. If fitted at all, its functionality and field of view will be limited. It will act as a back-up device in the event that the HMD is unavailable. It is unlikely to present sensor video. Display images, data and commands will be transferred to it via the data network, independent of the choice of HUD imaging technology.

The helmet-mounted display is day/night capable. To minimize weight, only those elements essential to the correct functioning of the helmet will be mounted on it. The other functions will be mounted off-helmet in a local electronics unit.

The mission computing, including display preprocessing and image processing, will be executed in one or more modular avionic racks. The rack provides all the

Fig. 11.13 Modular avionic architecture with 'big picture' display (BAE SYSTEMS)

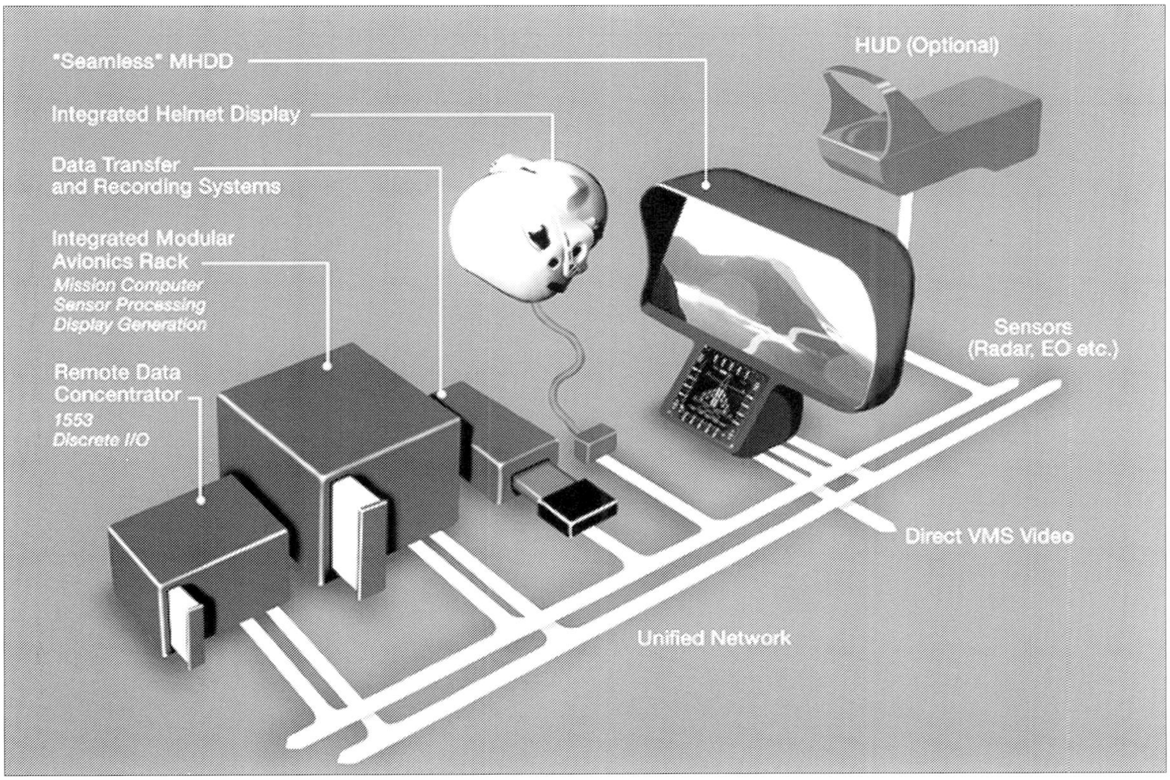

support services to a set of standardized computing modules, which include:

- Data processing.
- Signal processing.
- Graphics processing.
- Mass memory.
- Encryption.

Communication between the system elements is via a very high speed, secure, optical data transmission network.

Rotorcraft

Figure 11.14 shows the display systems architecture for the Super Lynx. The Super Lynx is a small but very capable tactical light helicopter. Space is at a premium. Therefore, a fully smart integrated display suite architecture was chosen for this vehicle. There are no additional computing boxes to be installed in the already cramped equipment bay. The displays contain integral data processing, graphics processing, symbol generation, and the display medium itself, namely AMLCD.

The engine and systems displays are 5ATI format and interface directly to the sensors and transducers of the systems for which they display information and status. Once digitized, these engine and systems data are retransmitted to the other displays in the integrated display suite.

The primary flight and tactical displays are 6 × 6 in square format. Aircraft flight information is received from the aircraft sensors and systems via ARINC

Fig. 11.14 *Rotorcraft integrated 'smart' display architecture (Smiths Aerospace)*

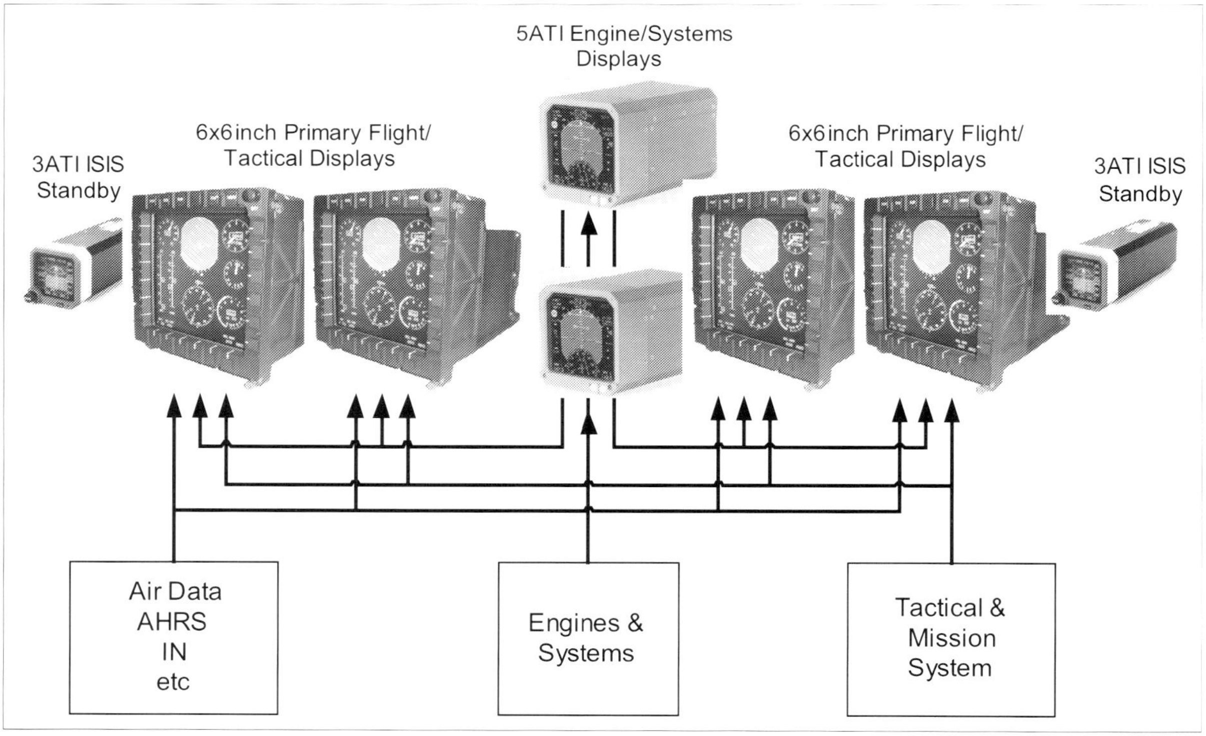

429 serial data buses. Information from the tactical and mission system is transferred via MIL-STD-1553B, and sensor video is relayed in analogue video form. The standby displays are 3ATI form factor Integrated Instruments (ISIS) which provide a combined attitude, speed, and height format.

Regulatory requirements

The civil airworthiness regulations are expressed in the Federal Airworthiness Requirements (FARs) and Joint Airworthiness Requirements (JARs) published by the US and European Aviation Certification Authorities respectively. There is a broadly similar but more fragmented set of regulations or standards for military air vehicles. The principles and processes established for civil airworthiness encapsulate best design practice for any airborne system, military or civil. Though not usually contractually imposed for military display systems, the application of these practices will result in a good product design.

Apart from the Joint Airworthiness Requirements (JARs), the other documents are not mandated as being the only means of compliance to the JARs, but any designer would be well advised to follow the procedures and practices contained within these documents. They were developed by consensus of all the interested parties (regulatory bodies, aircraft manufacturers, airlines, and equipment manufacturers) and are therefore the expected means to demonstrate compliance.

The most significant FARs and JARs (**1**) are:

- 25.1303 Flight and navigation instruments.
- 25.1309 Equipment, systems, and installations.
- 25.1323 Airspeed indicating systems.
- 25.1333 Instrument systems.
- 25.1335 Flight director systems.

Other significant documents referenced include:

- FAA AC 25-11 Advisory circular, transport category airplane electronic display systems (**2**).
- SAE AS 8034 Minimum performance standards for electronic displays (**3**).
- FAA TSOc113 Airborne multipurpose electronic displays (**4**).

JAR25 paragraph 1309 requires that:

- Equipment, systems, and installations must be designed to ensure that they perform their intended functions under any foreseeable operating conditions.
- The occurrence of any failure condition that would prevent the continued safe flight and landing of the aeroplane is extremely improbable.
- The occurrence of any other failure condition that would reduce the capability of the aeroplane or the ability of the crew to cope with adverse operating conditions is improbable.
- Warning information must be provided to alert the crew to unsafe system operating conditions and enable them to take appropriate actions.
- Systems, controls, and associated monitoring and warning means must be designed to minimize crew errors which could create additional hazards.

Compliance with the JAR requirements must be shown by analysis and by appropriate ground, flight, or simulator tests. The analysis must consider:

- Possible modes of failure including malfunctions and damage from external sources.
- The probability of multiple failure and undetected failures.
- The resulting effects on the aeroplane and occupants considering the stage of flight and operating conditions.
- The crew warning cues, corrective action required, and the capability of detecting faults.

The Certification Authorities have provided extensive guidance material to facilitate the design and analysis of systems that can demonstrate compliance with the JAR requirements. This material is to be found in AMJ 25.1309 (AMJ = Advisory Joint Material) (**5**).

The AMJ advocates a fail-safe design concept that considers the effects of failures in defining a safe design:

- In any system or subsystem, the failure of a single element during any one flight should be assumed, regardless of its probability. Such single failures should not prevent continued safe flight.
- Subsequent failures during the same flight, and combinations thereof, should be assumed.

The fail-safe design concept should use a combination of principles or techniques to ensure a safe design has:

- Designed-in integrity and quality.
- Redundancy or back-up systems.
- Isolation (spatial and physical separation) and independence of systems, components, and elements so that failure of one does not cause failure of the other.
- Proven reliability.
- Testability
- Failure containment to limit the safety impact of a failure.
- Failure warning and flight crew procedures for use after failure detection.
- Error tolerance and margins of safety.

The AMJ categorizes failure conditions and assigns rational probabilities of occurrence to them on the basis that an inverse relationship should exist between the probability of the loss of a function or its malfunction and the degree of hazard to the aeroplane and its occupants arising therefrom.

Failures are classified according to their severity as:

- *Minor*. Failure conditions that would not significantly reduce aeroplane safety and that involve crew actions that are well within their capability.
- *Major*. Failure conditions that would reduce the capability of the aeroplane or the ability of the crew to cope with adverse operating conditions to the extent that there would be a significant reduction in safety margins or functional capabilities and a significant increase in crew workload.

- *Hazardous*. Failure conditions that would reduce the capability of the aeroplane or the ability of the crew to cope with adverse operating conditions to the extent that there would be a large reduction in safety margins and physical distress or high workload such that the crew could not be relied upon to perform tasks accurately or completely.
- *Catastrophic*. Failure conditions that would prevent continued safe flight and landing.

The probability that a failure condition would occur may be assessed as being:

- *Probable*. Minor failure conditions may be probable.
- *Remote*. Major failure conditions must be no more frequent than remote.
- *Extremely remote*. Hazardous failure conditions must be no more probable than extremely remote.
- *Extremely improbable*. Catastrophic failure conditions must be extremely improbable.

Note that the earlier dated FAR does not distinguish between remote and extremely remote, it assesses both as improbable.
The AMJ provides guidance on acceptable techniques to substantiate the probability analysis. These include:

- Functional hazard assessment.
- Analysis of minor, major, hazardous, and catastrophic failure conditions.
- Operational and environmental conditions.
- Latent failures.
- Design assessment, which may include:
  - design methodology (validation and verification),
  - failure modes and effects analysis,
  - fault tree analysis,
  - probability analysis (based on component failure rates).

The AMJ defines the relationship between probability and severity of the failure condition as defined in Table 11.1.

Table 11.1 JAR 25 definitions of failure conditions and probability of occurrence

Severity	Probability	Analysis
Minor	Reasonably probable	No more frequent than 1 in 10^{-3} per flight hour
Major	Remote	No more frequent than 1 in 10^{-5} per flight hour
Hazardous	Extremely remote	No more frequent than 1 in 10^{-7} per flight hour
Catastrophic	Extremely improbable	No more frequent than 1 in 10^{-9} per flight hour

Certification guidelines

Representatives from all the major stakeholders in the civil aircraft business contributed to the preparation of guidelines that would capture best practice and provide consistency of approach to aid in the certification of complex airborne systems to satisfy the regulatory requirements. This guidance material is provided in the following documents:

- SAE ARP 4754 Certification considerations for highly integrated or complex aircraft systems (**6**).
- SAE ARP 4761 Safety assessment process guidelines and methods (**7**).
- RTCA DO 178B Software considerations in airborne systems and equipment certification (**8**).
- RTCA DO 254 Design assurance guidance for airborne electronic hardware (**9**).

The interrelationship between these documents is shown in Fig. 11.15. These documents provide specific guidelines for the design and production of systems, software, and hardware for airborne systems and equipment that perform their intended function with a level of confidence in safety that complies with the airworthiness requirements.

The guidelines are in the form of:

- Objectives for the design life cycle processes.
- Descriptions of activities and design considerations for achieving those objectives.
- Descriptions of the evidence that indicate the objectives have been satisfied.

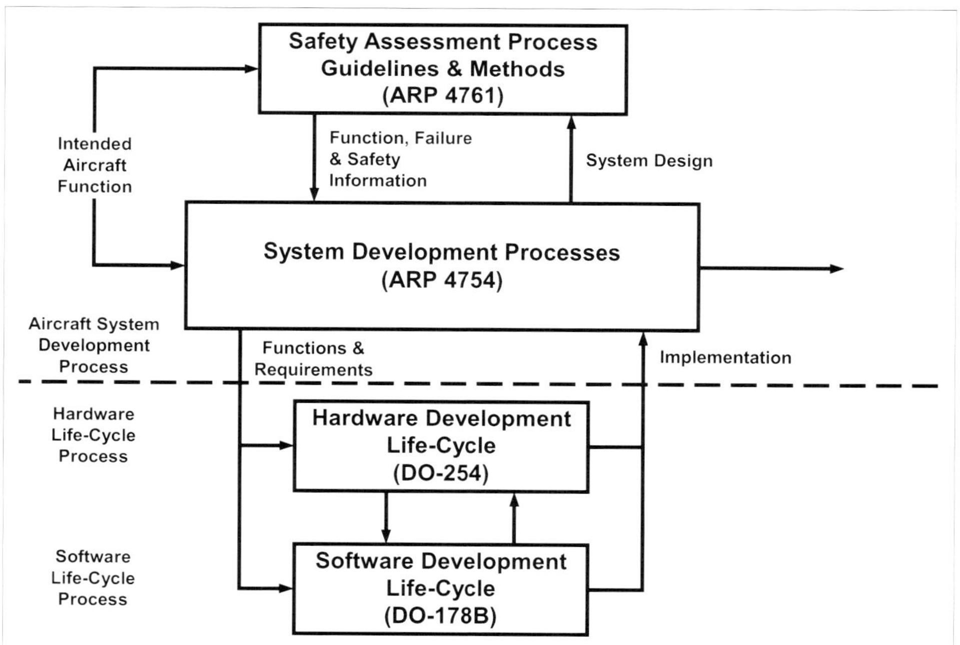

Fig. 11.15 *Interrelationship between guideline documents (Moir and Seabridge)*

The processes described embrace the whole design life cycle, including aspects of:

- The Planning process:
  - design methodology,
  - design standards,
  - design environment, tools, and procedures.
- The design process:
  - requirements capture,
  - conceptual design,
  - detailed design,
  - implementation.
- Supporting processes:
  - validation,
  - verification,
  - configuration management,
  - process assurance,
  - certification liaison,
  - data.

The three systems, software, and hardware guidelines interact. Analysis of the system considerations will result in a system safety assessment, which allocates systems requirements to functional areas in hardware and software. The software and hardware analyses feedback data into the systems analysis to complete the cycle.

The system safety assessment acknowledges architecture strategies for containment of anomalous behaviour such as:

- Serial – any component could produce a failure condition.
- Parallel – anomalous behaviour of more than one component is required to produce a failure condition.
- Partitioned – technique for providing isolation between functionally independent components.
- Multiversion dissimilar processing – to avoid some common mode error sources.
- Safety monitoring – protection against specific failure conditions in hardware or software or both.

The documents assign hardware and software design quality assurance levels in accordance with the safety categorizations in AMJ 25.1309 as defined in Table 11.2. The documents recognize the need both to validate and verify the design processes. They define:

- *Validation.* The process of determining that the requirements are correct requirements and that they are complete.
- *Verification.* The evaluation of the results of a process to ensure correctness and consistency with respect to the inputs and the standards applied to that process. Verification is not simply testing; testing cannot show the absence of errors. Verification is typically a combination of reviews, analyses, and tests.

Each step in the recommended design process has a set of objectives, and activities or guidance for the process step, moderated by the assigned quality level. Each

Table 11.2 Hardware and software quality assurance levels

Level	Definition
A	Design whose anomalous behaviour as shown by the system safety assessment would cause or contribute to a failure of system function resulting in a *catastrophic* failure condition for the aircraft
B	Design whose anomalous behaviour as shown by the system safety assessment would cause or contribute to a failure of system function resulting in a *hazardous* failure condition for the aircraft
C	Design whose anomalous behaviour as shown by the system safety assessment would cause or contribute to a failure of system function resulting in a *major* failure condition for the aircraft
D	Design whose anomalous behaviour as shown by the system safety assessment would cause or contribute to a failure of system function resulting in a *minor* failure condition for the aircraft
E	Design whose anomalous behaviour as shown by the system safety assessment would cause or contribute to a failure of system function resulting in a *no-effect* failure condition for the aircraft

process step is complete when its objectives and the objectives of the integral process associated with it are complete.

Traceability should be provided at each step of the process, between:

- System and hardware/software requirements.
- High-level and low-level design requirements.
- Implementation and verification.

Finally, the documents recommend a set of design life cycle data that include plans, standards, analyses, results, and reports for submission to the Certification Authorities as evidence of design assurance.

For obvious reasons, the level of criticality of software used in avionics systems determines the rigour applied to the development process, and hence rigours of software development in RTCA DO 178B are particularly strict. The software development process is generally of the form shown in Fig. 11.16, which shows the development activities evolving down the left of the diagram and the verification activities down the right. This shows how the activities eventually converge in the system validation test at the foot of the diagram, that is, the confluence of hardware and software design and development activities. Down the centre of the diagram the various software development stages are shown. It can be seen that there is considerable interaction between all the processes that represent the validation of the requirements and of the hardware and software design at each level. Any problems or issues discovered during the validation tests are fed back up the chain, if necessary back into the top level. Therefore, any minor deviations are reflected back into all the requirements stages to maintain a consistent documentation set and a consistent hardware and software design.

Whereas the earlier stages of software development and test might be hosted in a synthetic software environment, it is increasingly important as testing proceeds to undertake testing in a representative hardware environment. This testing

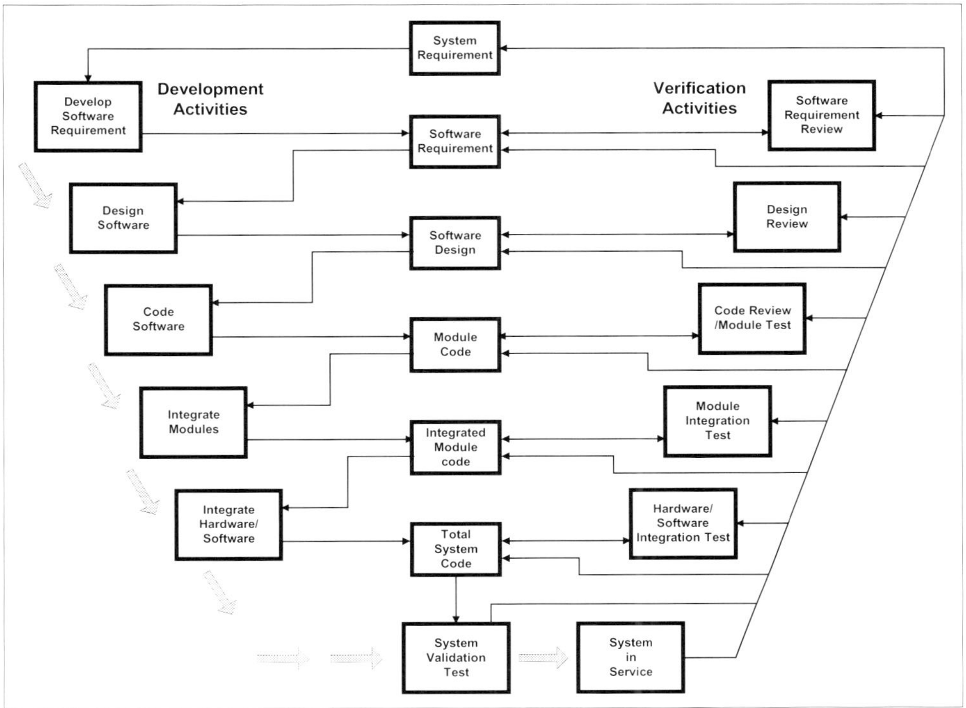

Fig. 11.16 *Software verification ('V') diagram (Moir and Seabridge)*

represents the culmination of functional testing of the LRU or system before proceeding to flight testing.

Environmental requirements

The environmental conditions and test procedures for civil airborne equipment are expressed in RTCA DO-160 (**10**). For military equipment the specifications tend to be aircraft specific, but with reference to MIL-STD-810 (**11**) or its equivalent DEF-STAN for European programmes. These documents define a minimum performance standard against a set of test conditions or categories. The onus lies with the equipment installer to assure himself that the standard and test conditions are appropriate for the application, seeking additional tests and analyses if required.

Electrical power supply transient immunity

During a power transient, an electromechanical instrument will temporarily freeze, and then continue to indicate, without glitch, once power is restored. Unlike an electromechanical instrument, electronic displays require power to produce an image. The image is lost once power is removed. Data processing may also be lost.

Recovery time after a loss of power can be a safety-related concern. The documents referenced earlier provide requirements and guidance about the performance of the display system under assorted transient conditions of the electrical supply system.

They advise:

- The design objective should be that the displays are insensitive to power transients.
- Bus transients caused by normal load switching (hydraulic pump actuation, ovens, generator paralleling, etc.) should cause no visible effect on the display.
- Expected abnormal bus transients (i.e. after generator failure not caused by engine failure) should not initiate a power-up initialization or cold start process.
- The display of attitude information should not be unusable or unstable for more than 1 s after the normally expected bus transients due to an engine failure, and should affect only displays on one side of the cockpit.
- For power interrupts of up to 200 ms in duration, the recovery time should not exceed 500 ms. Power interrupts of between 200 ms and 1 s in duration may cause disruptions in system operation for up to 2 s after power is restored. Recovery from power interruptions of up to 10 s duration should not cause any unsafe condition in system operation. In no case should power transients cause a frozen display.
- The large electrical loads required to restart some engine types should not affect more than one pilot's display.

Generally, these requirements have to be satisfied by a combination of display equipment performance and electrical power distribution design.

References

(1) JAR 25.1309, Joint Airworthiness Regulation: Equipment Systems and Installation, Civil Aviation Authority (UK).
(2) FAA AC 25-11, Advisory Circular: Transport Category Airplane Electronic Display Systems, Federal Aviation Administration.
(3) SAE AS 8034, Minimum Performance Standards for Airborne Multipurpose Electronic Displays, Society of Automotive Engineers.
(4) TSO-C113, Technical Standard Order: Airborne Multipurpose Electronic Displays, Federal Aviation Administration.
(5) AMJ 25.1309, Advisory Joint Material: System Design and Analysis, Federal Aviation Administration.
(6) SAE ARP 4754, Certification Considerations for Highly-Integrated or Complex Aircraft Systems, Society of Automotive Engineers.
(7) SAE ARP 4761, Guidelines and Methods for Conducting the Safety Assessment Process on Civil Airborne Systems.
(8) RTCA DO 178B, Software Consideration in Airborne Systems and Equipment Certification, Radio Technical Commission for Aeronautics.
(9) RTCA DO 254, Design Assurance Guidance for Airborne Electronic Hardware, Radio Technical Commission for Aeronautics.
(10) RTCA DO 160, Environmental Conditions and Test Procedures for Airborne Equipment, Radio Technical Commission for Aeronautics.
(11) MIL-STD-810, Environmental Conditions and Test Procedures for Airborne Equipment, US Department of Defense.

CHAPTER 12

Visual Performance Aspects

Introduction

As we enter the twenty-first century, we find ourselves in the unique position of being able readily to produce displays anywhere in the cockpit that are at the limit of the abilities of the human eye. Not to mention displays that are instantly reconfigurable and saturated with data. No longer will we have to ask for the very edge of technology and hope that we get close to the limits of the pilot and aircraft.

This chapter addresses some aspects of the human visual system, the presentation of flight information, and the ambient illumination environment to be found on the flight deck/crewstation of civil and military aircraft.

Human eye acuity

Resolution

Human factors research has shown that the human eye exhibits a limiting resolution under ideal conditions of about 1 min of arc (one-sixtieth of a degree) per optical line pair (**1**). For a head-down display viewed at 25 in (635 mm) this equates to a resolution of approximately 120 optical line pairs per inch (5 lp/mm). Interpreted another way, this means that the human eye can just perceive modulation depth in a series of equispaced black lines drawn at a pitch of 120 lines per inch (a line pair is the black line plus the 'white' line gap between it and the next line) when viewed at a distance of 25 in (see Fig. 12.1).

For a display medium such as an AMLCD, this would lead to the conclusion that the display should have a resolution of around 240 pixels per inch [or if a colour display, then 240 colour groups per inch (cgpi)] in order to replicate the black and the white line with a modulation depth of 100 per cent. However, recent empirical studies performed in flight environments indicate that this high a resolution is unnecessary – half this resolution is sufficient. Beyond roughly 120

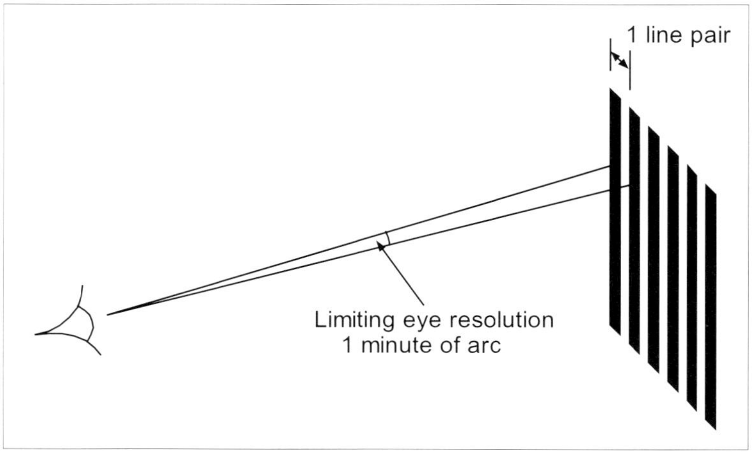

Fig. 12.1 Limiting eye resolution

colour groups per inch (cgpi), increased resolutions exhibit very limited increase in observable content. These studies support current design criteria and existing research which indicates that the normal eye can only discern meaningful information from spatial frequencies of 2 min of arc per line pair. Figure 12.2 shows the relationship between screen resolution and viewing distance.

There are numerous reasons to avoid resolutions higher than that needed. Increased resolution typically leads to escalating costs; e.g. doubling the resolution quadruples the number of pixels and quadruples the frequency to be supported by the associated electronics. Table 12.1 indicates typical display formats and the display dimensions each format will support at a viewing distance of 25 in (assuming 120 cgpi).

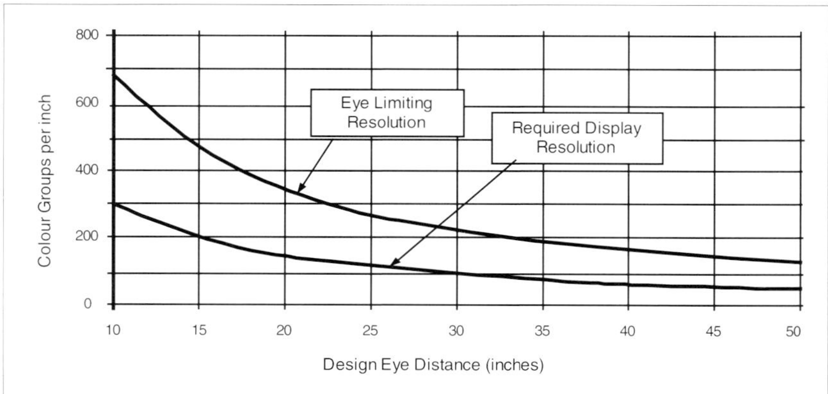

Fig. 12.2 Limiting eye resolution

Display scan times

As the data density increases, consideration must be given to the ability of the pilot actually to see the information yet alone understand and process it. For example, the current display suite in the F-22 consists of an HUD (20×24°), two UFDs (3×4 in), three SMFDs (6.25×6.25 in), and one PMFD (7.8×7.8 in). The typical time for the human eye to fix on an object is 0.3 s, and this provides clear vision over 5°. It has been reckoned that it would take the pilot about 18 s merely to scan the entire display space on the F-22. This assumes that the pilot only scans

Table 12.1 Display design characteristics for a 25 in design eye

Format	Resolution (cgpi)		Ideal size (in)		Typical avionics size
	Horizontal	Vertical	Horizontal	Vertical	
	400	400	3.33	3.33	3 ATI
VGA (square)	480	480	4.0	4.0	4 ATI and 5 ATI
	512	512	4.27	4.27	
VGA	640	480	5.33	4.0	ADI
SVGA (square)	600	600	5.0	5.0	5 × 5 in MFD
SVGA	800	600	6.67	5.0	HSI (portrait)
XGA (square)	768	768	6.4	6.4	6 × 6 in (nominal)
XGA	1024	768	8.53	6.4	8×6 in (nominal)
SXGA (square)	1024	1024	8.53	8.53	8×8 in square
SXGA	1280	1024	10.67	8.53	10×8 in (nom)
UXGA	1600	1200	13.33	10.0	
HDTV	1980	1080	16.5	9.0	

the displays and does not fix on any of the displayed information. Assuming a closure rate of 1700 nautical miles per hour, 18 s represents 8.5 nautical miles. It is therefore vital that the information is intuitive and attention grabbing when it needs to be so that the pilot can find what he wants when he wants it.

Aliasing

Unlike a monochrome CRT, the LCD image is discrete and sampled in both the horizontal and vertical directions, implying a spatial frequency degradation in the horizontal scan direction over the continuous beam of a CRT display.

Various methods of quantifying the resolution of discrete structured images and pixelation in displays have been employed over time. The Kell factor was originally derived for horizontally scanned CRTs where resolution was continuous in the horizontal direction, but in the vertical direction the imagery was discrete and sampled. The Kell factor was originally defined as the ratio between the effective and theoretically attainable vertical resolution for a given number of transmitted scan lines. The Kell factor value is less than unity for all video displays. The traditional Modulation Transfer Function (MTF) metric only applies to scanned CRTs in the horizontal continuous scan direction.

The resolution values for a pixelated display are based on the spacing to adjacent neighbours, typically, the nearest neighbouring pixels, to illustrate optimal performance. Actual performance for viewing arbitrarily oriented imagery is limited by pixels other than the nearest neighbours. The resultant degradation is represented by what is referred to as the Kell factor or the aliasing limit. The value of the Kell factor or aliasing limit of measured AMLCD displays varies from roughly 0.7 to 0.9 times the Nyquist limit, where the Nyquist limit is the cycles per picture width or one-half of the pixels per picture width.

Symbol size, shape, and colour

For a typical display an object needs to subtend at least 12 min of arc, and for an aircraft environment it is recommended that the object subtend at least 20 min of arc. This latter equates to a minimum character size of approximately 0.15 in (3.7 mm) when viewed on a display at 25 in (635 mm), or about 16 pixels at a screen resolution of 120 cgpi. Line width should be commensurate with font size.

A typical display format will use a range of font sizes to portray information, depending on the criticality and nature of the data. Large font sizes are used for critical data, enforced by line width and colour coding. The smaller font sizes are generally heavily used for textural information such as CNI and maintenance pages.

Display symbology has evolved over time, partly owing to the advancement of display technology and partly as a result of aircrew evaluations. Throughout the development process the philosophy of dual coding has been exploited by representing information in terms of both colour and shape. In some special cases, text readouts provide additional indications as well as numeric precision to shaped and coloured track file icons. This allows reinforced recognition of the data displayed, reducing errors. Initially, the dual coding involved basic colours and shapes (squares, triangles, diamonds, etc.), but today's graphics processors facilitate more complex shape coding that is icon-like in nature, providing easily recognizable graphic symbols resembling the actual object such as aircraft outlines for fighters, bombers, helicopters, etc., as well as radar dishes, SAM launchers, and so on, positioned to scale.

Display format guidelines

The advisory circular AC 25-11 issued by the US Federal Aviation Administration (FAA) encapsulates perceived wisdom for the design and certification of electronic display systems. Strictly, it refers to CRT-based displays, but it is equally applicable to AMLCD or any equivalent technology (**2**).

The document is advisory, it is not mandatory, but it does provide guidance and outlines methods to comply with the regulatory requirements encapsulated in the Federal Aviation Requirements (FARs) and Joint Aviation Requirements (JARs).

Other useful material is to be found in the following documents:

- SAE ARP 1068 Flight deck instrumentation and display design objectives for transport aircraft (**3**).
- SAE ARP 1874 Design objectives for CRT displays for Part 25 (transport) aircraft (**4**).
- SAE ARP 4256 Design objectives for liquid crystal displays for Part 25 (transport) aircraft (**5**).

Advisory circular AC 25-11 includes material on:

- Display function criticality and compliance considerations.
- Colour, symbology, coding, clutter, dimensionality, and attention getting requirements.
- Display visual characteristics.
- Failure modes.
- Information display and formatting.
- Specific integrated display and mode considerations including maps, propulsion parameters, warnings advisory checklists, procedures, and status displays.

The advisory circular provides specific requirement guidelines for electronic display systems integrity. It uses the FAR definition of 'improbable', which is taken to be 'extremely remote' in the JAR definition:

Attitude
- Display of attitude in the cockpit is a critical function. Loss of all attitude, including standby attitude, is a critical failure and must be extremely improbable.
- Display of hazardously misleading roll or pitch attitude simultaneously on the primary attitude displays for both pilots must be extremely improbable.
- Display of hazardously misleading roll or pitch attitude on any single primary attitude display must be improbable.

Airspeed
- Display of airspeed in the cockpit is a critical function. Loss of all airspeed display, including standby, must be extremely improbable.
- Loss of primary airspeed displays for both pilots must be improbable.
- Displaying hazardously misleading airspeed simultaneously on both pilots' displays, coupled with loss of stall warning or overspeed functions, must be extremely improbable.

Barometric altitude
- Display of altitude in the cockpit is a critical function. Loss of all altitude display, including standby, must be extremely improbable.
- Loss of primary altitude displays for both pilots must be improbable.
- Displaying hazardously misleading altitude simultaneously on both pilots' displays must be extremely improbable.

Vertical speed
- Display of vertical speed in the cockpit is an essential function. Loss of vertical speed display to both pilots must be improbable.

Rate-of-turn indication
- The rate-of-turn indication is a non-essential function.

Slip/skid indication
- The slip/skid or sideslip indication is an essential function. Loss of this function to both pilots must be improbable
- Simultaneous misleading slip/skid or sideslip to both pilots must be improbable.

Heading
- Display of stabilized heading in the cockpit is an essential function. Displaying hazardously misleading heading information on both pilots' primary displays must be improbable.
- Loss of stabilized heading in the cockpit must be improbable.
- Loss of all heading information in the cockpit must be extremely improbable.

Navigation
- Display of navigation information (excluding heading, airspeed, and clock data) in the cockpit is an essential function. Loss of all navigation information must be improbable.
- Displaying hazardously misleading navigational or positional information simultaneously on both pilots' displays must be improbable.

Propulsion system displays

- Powerplant instrument displays must be arranged and isolated from each other so that no failure that affects the propulsion system parameter for one engine can cause permanent loss of display or adversely affect the accuracy of any parameter for the remaining engines.
- No single fault or probable combinations of failures shall result in the permanent loss of display or the misleading display of more than one propulsion unit parameter for a single engine.
- Combinations of failures that would result in the permanent loss of required powerplant displays for more than one engine must be improbable.
- Combinations of failures that would result in hazardously misleading display of any parameter for more than one engine must be extremely improbable.
- No single fault or combination of failures not shown to be extremely improbable shall result in the loss of all propulsion system displays.
- Powerplant parameters not displayed continuously must be automatically displayed when the parameter exceeds an operating limit or threshold.
- Propulsion system parameters essential for determining health and operational status of the engines, including engine restart, shall be automatically displayed after loss of normal electrical power.
- Individual fuel tank quantity information must be displayed whenever the fuel system is not in the normal configuration for flight.

Crew alerting displays

- Crew alerting of selected parameters may be an essential function. Loss of crew alerting for essential functions must be improbable.
- Display of hazardously misleading crew alerting messages must be improbable.

Flight crew procedures

- The display of hazardously misleading flight crew procedures must be improbable.

Weather radar

- The display of weather radar in the cockpit is a non-essential function. However, presentation of hazardously misleading information must be improbable.

The above guidelines are summarized in Table 12.2.

The advisory circular recommends:

- The display should convey information in a simple uncluttered manner.
- Colour alone should not be used as a discriminator. Information content should be identified by at least two distinctive coding parameters (size, shape, colour, location, etc.).
- Symbols or messages should be logically and consistently positioned and collocated with associated information.
- Careful attention should be given to symbol priority to assure conspicuity and ease of interpreting 3D information on a 2D medium.
- Sensor warning flags should appear in the area where the data normally occur.

The advisory circular cautions about the compelling nature of the map display and notes that there have been incidents where gross map position errors have gone undetected or unbelieved because the flight crew have falsely relied on the map

Table 12.2 FAA AC 25–11 display data integrity guidelines

Parameter	Criticality	Probability of	
		Complete loss	Misleading data
Attitude	Critical	All=extremely improbable Primary=improbable	Extremely improbable
Airspeed	Critical	All=extremely improbable Primary=improbable	Extremely improbable
Barometric altitude	Critical	All=extremely improbable Primary=improbable	Extremely improbable
Vertical speed	Essential	Improbable	
Rate of turn	Non-essential		
Slip/skid	Essential	Improbable	Improbable
Heading	Essential	Extremely improbable	
Navigation	Essential	Improbable	Improbable
Powerplant	Critical	Extremely improbable	Extremely improbable
Crew alerting	Essential	Improbable	Improbable
Checklists			Improbable
Weather radar	Non-essential		Improbable

instead of correct raw data. It reinforces the need for operating procedures that require one crew member still to monitor raw navigation data:

Display elements

- Display elements should be natural, intuitive, and not dependent on training or adaptation for correct interpretation.
- The 'basic T' relationship should be followed except where deviations have been substantiated by satisfactory in-service experience.
- If side-by-side formats are used, then attitude, airspeed, altitude, and heading must still reside in the 'basic T' arrangement. Heading and attitude must be presented on the same display.
- Airspeed and altitude should be arranged so that the present value is located as close as possible to a horizontal line extending from the centre of the attitude indicator.
- Digital present value readouts should not totally obscure scale markings as they pass the present value box.
- Digital-only presentations of control parameters are to be avoided.
- An accurate, easy, quick-glance interpretation of aircraft attitude shall be possible for all expected but unusual attitude situations and command guidance display configurations.
- Compacted or combined formats are permissible in the event of failure of one of the display surfaces, but the format must follow the same style as the individual formats.
- Control knobs used to set digital data shall have adequate friction or tactile detents to avoid undue concentration.

Cautions and warnings

- A 'warning' should be generated when immediate recognition and corrective or compensatory action is required by the crew, the associated colour is red.

- A 'caution' should be generated when immediate crew awareness is required and subsequent crew action will be required, the associated colour is amber.
- An 'advisory' should be generated when crew awareness is required and subsequent crew action may be required, the associated colour should be unique, but not amber/yellow.

Colour coding

Standardized colour codings shown in Table 12.3 are advised to facilitate information discrimination, and those for weather precipitation and turbulence are shown in Table 12.4. The two option sets result from the evolutionary nature of the development of display formats and simply recognize different origins. Option set 1 is preferred for new designs.

Table 12.3 Standardized colour codings to facilitate information discrimination

<table>
<tr><th rowspan="2">Information</th><th colspan="2">Colour coding</th></tr>
<tr><th>Option set 1</th><th>Option set 2</th></tr>
<tr><td>Warnings</td><td colspan="2">Red</td></tr>
<tr><td>Flight envelope and system limits</td><td colspan="2">Red</td></tr>
<tr><td>Cautions, abnormal sources</td><td colspan="2">Amber</td></tr>
<tr><td>Earth</td><td colspan="2">Tan/brown</td></tr>
<tr><td>Sky</td><td colspan="2">Cyan/blue</td></tr>
<tr><td>Scales and associated figures</td><td colspan="2">White</td></tr>
<tr><td>Flight director</td><td colspan="2">Magenta/green</td></tr>
<tr><td>ILS deviation pointer</td><td colspan="2">Magenta</td></tr>
<tr><td>Autopilot engaged modes</td><td colspan="2">Green</td></tr>
<tr><td>Fixed reference symbols</td><td>White</td><td>Yellow*</td></tr>
<tr><td>Current data, values</td><td>White</td><td>Green</td></tr>
<tr><td>Autopilot armed modes</td><td>Cyan</td><td>White</td></tr>
<tr><td>Selected data, values</td><td>Green</td><td>Cyan</td></tr>
<tr><td>Selected heading</td><td>Magenta†</td><td>Cyan</td></tr>
<tr><td>Active route plan</td><td>Magenta</td><td>White</td></tr>
</table>

* Use of yellow other than for cautions is discouraged.
† Magenta used for 'fly to' or 'keep controlled' parameters.

Table 12.4 Standardized colour codings for weather precipitation and turbulence

Precipitation:	
0–1 mm/h	Black
1–4 mm/h	Green
4–12 mm/h	Amber/yellow
12–50 mm/h	Red
> 50 mm/h	Magenta
Turbulence	White or magenta

Visibility requirements

Civil transport aircraft requirements

The Federal Aviation Administration advisory circular FAA AC 25–11 (**2**) provides guidance on the visual performance requirements for civil display systems. Reference should also be made to:

- SAE ARP 1068 Flight deck instrumentation, display design objectives for transport aircraft (**3**).
- SAE ARP 1874 Design objectives for CRT displays for Part 25 (transport) aircraft (**4**).
- SAE ARP 4256 Design objectives for liquid crystal displays for Part 25 (transport) aircraft (**5**).
- SAE AS 8034 Minimum performance standards for airborne multipurpose displays (**6**).

These documents advise:

Chrominance and luminance

- Four lighting conditions should be considered in the display design:
  - direct sunlight falling on the display through a side window,
  - sunlight through a front window illuminating a white shirtfront which is reflected in the display,
  - sun above the forward horizon and above a cloud deck reflecting into the pilots' eyes (known as veiling glare),
  - night or dark environment in which the display brightness should be dimmable such that outside vision is not impaired.
- CRT and backlight phosphors are known to age and the start-of-life performance should take into account any degradation that will occur through the expected life of the device.
- Shade infill used as backgrounds (e.g. blue sky and brown earth for attitude) need not be so easily discriminated at high ambient light levels. The luminance of backgrounds must not distract from the legibility of information written over the background.
- The luminance of information underlays such as weather radar should be independently adjustable relative to the luminance of the instrumentation symbology.
- Automatic brightness adjustment systems can be employed to match display luminance to the prevailing ambient lighting conditions, but experience has shown that great care needs to be taken to position light sensors appropriately to adequately cater for all situations.
- Maximum display luminance should not be less than 75 fL (257 cd/m^2).
- Minimum contrast ratio should be greater than 3.0:1 at an ambient illumination of 8000 fc (86 000 lux).
- Dark ambient contrast ratio should be better than 20:1 at the design eye and better than 10:1 over the viewing envelope.
- The display should be dimmable to 0.1 fL (0.343 cd/m^2).
- Luminance uniformity should be better than 50 per cent.

Positional accuracy

- Absolute positional accuracy is not as significant as relative positional accuracy since most information is read relative to scales and features also written on the display surface.
- Similarly, linearity is only significant if distortion becomes apparent.
- However, jitter, i.e. rapid small spacial movement of a symbol, can cause mild fatigue and must be controlled to be imperceptible.

Jitter

- Rapid small spatial movement of a symbol can cause mild fatigue and should not be perceptible day or night either foveally or in peripheral vision.

Line width, symbol and character size

- Experience has shown that lines need to be of an adequate thickness to be easily assimilated.
- A colour display image is fabricated from three primary colours, generally as a series of adjacent dots or stripes. Care must be taken to ensure colour alignment (convergence) of the three images and that colour fringing does not occur.

Flicker

- Flicker is the undesired rapid temporal variation in display luminance of a symbol, symbol group, or a luminous field. Flicker can cause mild fatigue and should not be perceptible day or night either foveally or in peripheral vision.
- Refresh rate is a major determinant of flicker. To avoid flicker, CRT screen refresh rates should be in excess of 70 Hz.

Dynamics

- For those elements of the display that are normally in motion, any jitter, jerkiness, or ratcheting should be neither distracting nor objectionable. Conversely, smearing of moving symbols should be neither discernable nor objectionable.
- Screen update rates for analogue signals used in direct airplane control tasks should be equal to or better than 15 Hz.
- Any lag between input signal and display should be consistent with the airplane control task associated with that parameter. In particular, display system lag for attitude should not exceed 100 ms.

Military combat aircraft requirements

The military daylight legibility and readability requirements for electronic and electro-optical displays are far more prescriptive, as one might expect given the operational significance of the displays. Reference should be made to the following documents:

- JSSG-2010-3 — Joint service specification guide – crew systems – cockpit/crewstation/cabin handbook (**7**).
- JSSG-2010-5 — Joint service specification guide – crew systems – aircraft lighting handbook (**8**).
- MIL-HDBK-87213 — Electronically/optically generated airborne displays (**9**).
- MIL-STD-3009 — Lighting, aircraft, night vision imaging system (NVIS) compatible (**10**).

- EFJ-R-EFA-000-107 Cockpit lighting standard (**11**) (NATO restricted document).

Ambient illumination

The real-world natural ambient illumination is both too variable and too complex for every condition to be considered in the design of cockpit displays. However, it is useful to identify a few worst-case scenarios. For emissive Head-Down Displays (CRT or AMLCD) the cases to be considered should include high ambient day (sun rear and sun forward), dusk/dawn transition, and night. For transmissive Head-Up Displays and Helmet-Mounted Displays the cases should include high ambient sun forward and night. Useful conversion factors:

$$\text{Footlamberts(fL)} \times 3.426751 = \text{candela per square metre (cd/m}^2\text{) or NITS}$$

$$\text{Footcandles(fc)} \times 10.76391 = \text{lumens per square metre (lm/m}^2\text{) or lux}$$

Head-Down Display worst-case scenarios

High ambient – sun rear

This scenario (see Fig. 12.3) is the high ambient illumination condition that degrades the visibility of emissive displays by 'washing out' the presented information. The condition is experienced when:

- The aircraft is flying straight and level at 30 000 ft above 8/8 cloud.
- The solar disc is low (30° elevation) and to the rear of the aircraft.
- The display is bathed in direct sunlight – around 100 000 lux.
- Display specular reflections of the general cockpit area predominate.
- The forward ambient scene is diffused – around 15 000 lux.
- The aircrew helmet tinted visor is down.

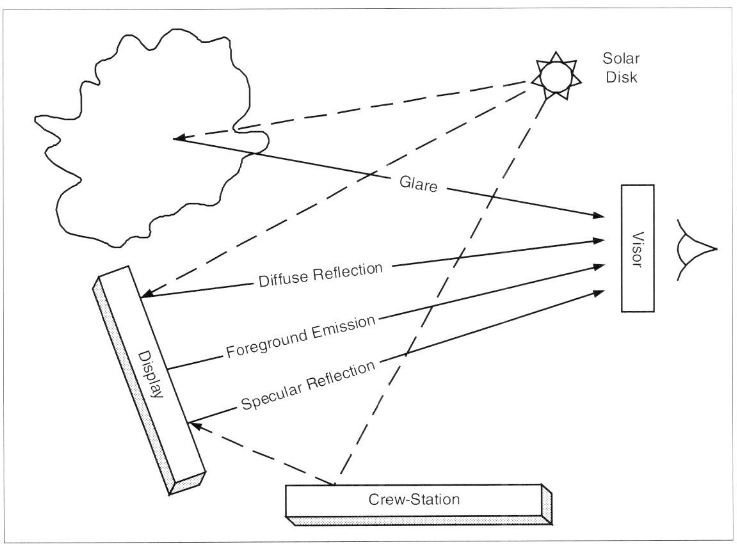

Fig. 12.3 HDD high ambient – sun rear scenario

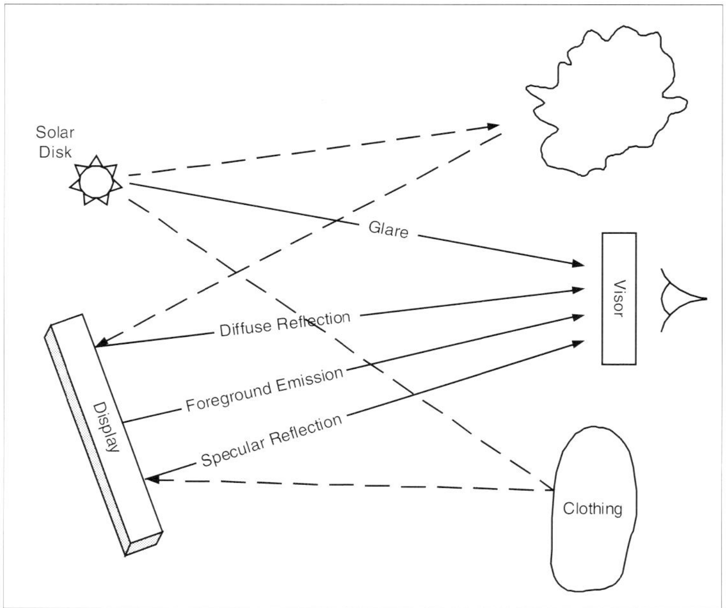

Fig. 12.4 HDD high ambient – sun forward scenario

High ambient – sun forward

This scenario (see Fig. 12.4) is less familiar but equally severe on display presentations, because the principal effect of the ambient illumination is to produce 'solar glare' which degrades the pilot's perceptual capability. The condition is experienced when:

- The aircraft is flying straight and level at 30 000 ft above 8/8 cloud.
- The solar disc is very low (15° elevation) and forward of the aircraft.
- The display is in shadow and illuminated by diffused skylight – around 8000 lux.
- Display specular reflections of the pilot's flying suit predominate.
- The aircrew helmet ND visor is down and the forward scene is dominated by the solar disc – around 110 000 lux.

Low ambient – dusk/dawn transition

The third scenario is the low dusk/dawn transient illumination condition, when the sun is forward of the aircraft and close to the horizon. The clear air conditions result in the cockpit being in hard shadow, while the pilot is subject to solar glare. Reflective displays are most affected, because the limited solar glare is sufficient to degrade the pilot's perception of the display low reflected luminance:

- The aircraft is flying straight and level at 30 000 ft in clear air.
- The solar disc is forward of the aircraft close to the horizon line – around 1500 lux forward ambient.
- The sky hemisphere is clear and of low luminosity – around 200 lux.
- The cockpit lighting control 'night' is selected and set to maximum.
- The aircrew helmet ND visor is up.

Night

In practice, night ambient illumination is not a single condition but the summation of multiple and complex radiation sets based on lunar reflected sunlight, various sky-glow derivatives, and individual starlight spectra:

- Night illumination can be assumed to consist of a diffused illumination over the range 1.0–0.0001 lux.
- The aircrew may be flying night adapted without any vision aids.
- The crew may be using night vision goggles to aid perception of the external scene.

Aircrew perception

The ability of the crew to detect information presented on a cockpit head-down display will depend on the visual difference or contrast between the foreground image and its background. In practice there are three display components, two resultant from the ambient illumination plus a component of veiling glare resultant at the eye, all modified by the visor transmission, if down:

- **Display emission.** The first component is the emission of the display itself, both as a foreground emission and a background emission. The foreground emission is the information content; the background is unwanted display emissions. For a stroke-written CRT the background emission from the black area is zero. This is not true of an AMLCD in which the backlight illuminates the whole screen area. Light is still transmitted by the black 'off' pixels.
- **Diffuse reflection.** The second background component is the result of an ambient illumination spectrum being modified (reflected) by the diffusing surface of the display. The ambient may be direct illumination emanating from the solar disc or a diffused component emanating from the clear sky or cloud bank. The display diffused reflecting surface may be the white phosphor coating on a CRT or the diffusing surface of the matt black matrix and colour filters of an AMLCD.
- **Specular reflection.** The third background component is the result of a further ambient illumination spectrum being modified (reflected) by the specular reflective component of the display front surface. The ambient spectrum will invariably be the result of a secondary reflection of the solar disc, sky, or cloud, modified by the immediate cockpit environment (i.e. flying suit and/or cockpit painted surfaces). Primary specular reflections of the solar disc, in the display front surface, along the display eye line, are normally prevented by good cockpit design and/or by tilting the display front plane downwards to ensure that only secondary lower-order cockpit reflections occur.
- **Veiling glare.** A fourth component exists which, although independent of the display physical characteristics, is dependent on the display location and immediate cockpit geometry. Any luminous point source in the visual field surrounding the display will induce glare and influence the observer's ability to perceive the foreground to background difference by introducing the effect of a veiling luminance superimposed over the complete task. The higher the luminance of that source, or the larger its solid angle area, or the smaller the angle it makes with respect to the task, then the greater will be the effect of the veiling glare component. In terms of the combat aircraft cockpit environment,

the primary source of veiling glare is the direct solar disc when it is in the forward visual hemisphere, and/or its diffused reflective component on the forward cloud face when it is in the rear hemisphere.

Head-Up Display and Helmet-Mounted Display worst-case scenarios

Practical flight experience indicates that the worst-case ambient illumination condition for the HUD transmissive display visibility is when the display symbology is presented against a cloud face, with a clear high-luminance solar disc producing additional glare just outside the display field of view. This scenario (shown in Fig. 12.5) represents:

- The aircraft is flying straight and level at 30 000 ft above 8/8 cloud.
- The solar disc is within 15° of the display central vision line.
- The display field of view is dominated by an illuminated cloud face – around 25 000 cd/m^2.
- The aircrew helmet ND visor is down.

The ability of the crew to detect information presented on the HUD or HMD will again depend on the visual difference or contrast between the foreground image and the background. In this case the foreground image is the HUD symbology reflected off the display combiner. The background is forward ambient scene illumination modified by the combiner optics. As before, there is an element of forward ambient veiling glare which reduces the crew's perceptual capability through the HUD/HMD. All of the above modified by the visor, if down.

Fig. 12.5 HUD/HMD high ambient – sun forward scenario

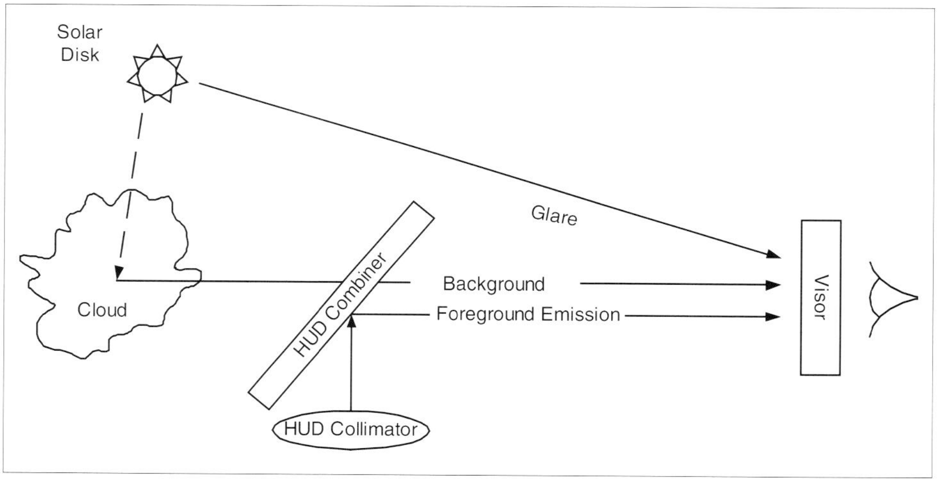

US DoD definitions and requirements

The requirements identified in the above-referenced US DoD documents embrace the legibility and readability requirements for electronic and electro-optical display contrast and luminance in a single combined environment of both diffuse light [producing 108 000 lux (10 000 fc) on the display face] and the specular reflection of a glare source [with a luminance of 6800 cd/m^2 (2000 fL)]. There is also a 'minimum luminance difference' requirement, which basically requires displays to have a high luminance in addition to achieving adequate contrast to combat veiling glare.

The combined diffuse and specular environment is intended accurately to simulate the lighting conditions in a fighter crewstation in direct sunshine. The 108 000 lux (10 000 fc) diffuse requirement represents sunlight of passing through an aircraft canopy (typically 80–90 per cent transmission) and striking a display somewhat off-axis. The 6800 cd/m^2 (2000 fL) glare source represents a reflection of the sun from interior parts of the crewstation or from the pilot's flight suit or helmet. The visibility requirements, under the prescribed conditions, are shown in Table 12.5:

Table 12.5 MIL-STD-85762 Table II

Information type	Luminance difference (minimum)	Contrast (minimum)
Numeric only	100 fL (343 cd/m^2)	1.5 : 1
Alphanumeric	100 fL (343 cd/m^2)	2.0 : 1
Graphics plus alphanumerics	100 fL (343 cd/m^2)	3.0 : 1
Video (high ambient) (six grey shades)	160 fL (550 cd/m^2)	4.66 : 1
Video (dark ambient) (eight grey shades)	160 fL (550 cd/m^2)	10.3 : 1

- Minimum difference luminance. The minimum luminance difference is the difference between the display foreground and the background luminances. The requirement is intended to overcome the eye adaptation and veiling glare problems that a pilot may experience when looking at a bright source of light outside the crewstation. Eye adaptation can limit display performance when a pilot must look quickly from an outside scene [a 34 000 cd/m^2 (10 000 fL) cloud, for example] to a dim display.
- Minimum contrast. The minimum contrast requirements are based on a wide variety of studies, existing military specifications, and hardware specifications of existing successful equipment. They represent the monochrome contrast needed to assure rapid, accurate reading of the various kinds of information in a wide variety of fighting, stress, and vibration environments. Displays with lower contrast can be used but will tend to look 'washed out' in some situations and may not provide the reading speed and accuracy required in a fighter crewstation. The video high-ambient figures represent the requirements to make visible an image with at least six shades of grey, with a ratio of $\sqrt{2}$ between each grey shade. The video dark ambient figures make eight shades of grey visible. In the special case of a hybrid CRT (one displaying both raster video and stroke symbols), the alphanumeric and symbology contrasts should be interpreted as relative to white video since the stroke symbols will generally be overlaid on video. Generally, this requirement is contained in specifications as a higher contrast requirement for stroke symbols (relative to black) rather than for video. These contrast requirements are based on the requirements and capabilities of monochrome devices and may not be achievable for colour displays, but they may still be adequate in some situations because they are

partially protected by a roof or a glare shield. Colour displays can also provide the advantage of colour contrast in addition to luminance contrast.

- Compensation multipliers. 'Compensation multipliers' (not shown in Table 12.5) define a proportionate increase or decrease in the contrast requirement for larger or smaller characters, or those with narrower stroke widths, relative to the nominal ranges stated. These multipliers are also the result of extensive studies and experience. They account for the fact that a display device that is not particularly bright or high contrast can be made more legible by making the characters larger and using a thicker stroke width, a 'trick' that is often used in real systems.

European (Eurofighter Typhoon) definitions and requirements

The requirements identified in the above referenced Eurofighter Typhoon document describe the legibility and readability requirements for electronic and electro-optical display contrast and luminance in a comprehensive but complex manner using the concept of Perceived Just Noticeable Differences (PJNDs). The concept is based on the premise that the ability of the crew to detect information presented on a cockpit display will depend on the visual difference between the foreground image and its background.

If both foreground and background are of the same colour, that is, they exhibit identical luminance and chrominance characteristics, then differentiation is impossible because the colours match. Two images only exhibit the same 'colour' when they have identical luminance and chrominance specifications.

Where an isochromatic condition exists, that is, when the foreground image and background image have the same chromatic signature, visual detection can only occur when the observer's threshold level of luminance contrast is exceeded. This luminance just detectable contrast level (ratio of foreground luminance to background luminance) represents one Luminance Just Noticeable Difference (LJND) between the foreground and background. Similarly, where an isoluminance condition exists, detection will occur only when the observer's chromatic threshold is exceeded and one Chromatic Just Noticeable Difference (CJND) occurs.

The visual difference between the foreground and background will invariably be a combination of both luminance and chrominance contrasts. The result can be quantified in terms of LJND and CJND combinations, as Perceived Just Noticeable Differences (PJND), where

$$\text{PJND} = \sqrt{\text{LJND}^2 + \text{CJND}^2}$$

The luminance threshold value, or one LJND, is specified as a logarithmic function of the luminance contrast ratio formed between the foreground and background that is necessary to achieve a 50 per cent detection probability of a target with a 10 000 cd/m^2 task luminance. The threshold level has been set at 0.0051. Thus

$$\text{Display LJND value} = \log_{10}(\text{contrast ratio})/0.0051$$

The chrominance threshold value, or one CJND, is specified as the difference in chrominance between the foreground and background (in the CIE 1976 L,u′, v′

UCS colour space) that is necessary to achieve a 50 per cent detection probability. The threshold level has been set at 0.0042. Thus

Display CJND value = colour shift/0.0042

The PJND values are computed for a particular display device from three sets of data:

- The display measured performance characteristics.
- The 'worst-case' ambient lighting conditions applicable to that display.
- A set of perception equations that represent a 'standard pilot's eye'.

Each of these datasets is itself a set of spectral characteristics which are then manipulated mathematically to derive two power spectra, display foreground and display background, relative to the pilot eye position. These two power spectra are the respective sums of all contributing power spectra (emissive and reflective) relative to two points on the display surface that represent:

- The active display symbol (foreground).
- Its surrounding or background.

when subject to the appropriate and relevant 'worst-case' ambient lighting conditions.

Figure 12.6 indicates the manipulation process and the spectral sets involved.

The airframe manufacturer provides the spectral datasets associated with the crewstation ambient illumination environment for the worst-case scenarios to be analysed, namely:

- Diffused ambient mission component.
- Specular ambient emission component.

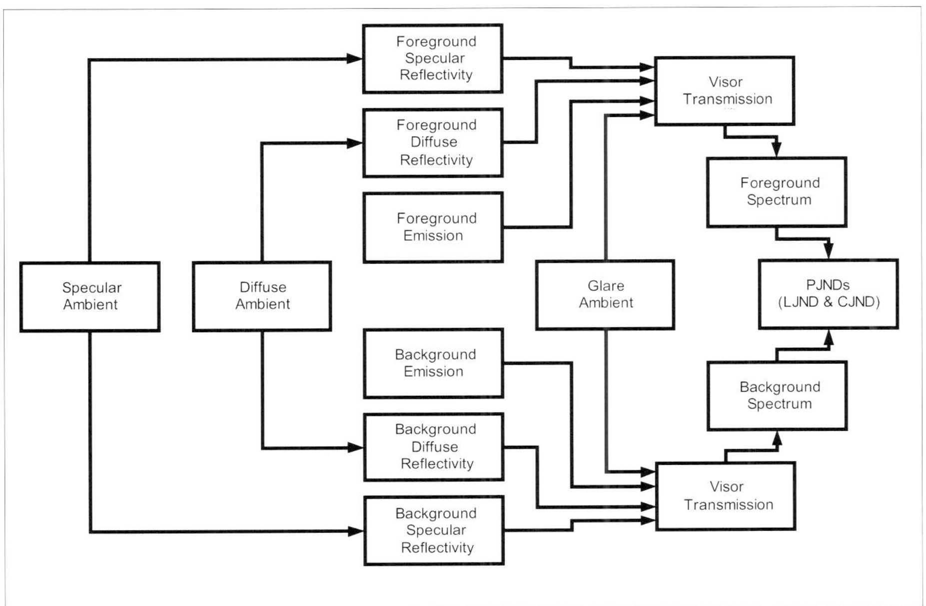

Fig. 12.6 Spectral transform process

- Glare ambient emission component.
- Helmet visor transmission.

The display equipment manufacturer provides the spectral datasets that define the operational performance of the product for both the display foreground and the background area, namely:

- Emissive power spectra.
- Diffused reflectivity spectra.
- Specular reflectivity spectra.

The resultant PJND (and the inherent LJND and CJND) numbers can be regarded as a figure of merit for the display device. Higher numbers reflect greater ease of perception by the crew of the information content of the display. The PJND value is compared against the predetermined required PJND for the display device.

Though technically elegant, and a viable analysis of product performance during product formal qualification testing, it is impractical to perform this level of testing on a 100 per cent basis on series-production articles. Routine testing has to revert to the more traditional tests of display luminance, specular and diffuse reflectance, and contrast ratio under the much simpler test regimes described in MIL-STD-85762A, or equivalent.

Viewability examples

The following examples illustrate the performance of typical head-down and head-up products against some of the worst-case scenarios identified above. This analysis is much simplified for illustrative purposes.

Colour CRT Head-Down Display

The simplified worst-case scenario is shown in Fig. 12.7. The colour CRT display performance cardinal points are as follows:

CRT active area	5 × 5 in square format
Bare CRT raster luminance	1000 fL white (3400 cd/m^2) (red = 180 fL, green = 800 fL, blue = 60 fL)
Bare CRT stroke luminance	2250 fL white (7700 cd/m^2) (red = 480 fL, green = 1600 fL, blue = 200 fL)
Filter transmission	25% primary colours, 18% daylight
CRT diffuse reflectance	25%
Filter specular reflectance	1%
Ambient illumination	10 000 fc diffuse sunlight (108 000 lux), 2000 fL point sources (6800 cd/m^2)

The display background is made up of a diffuse component of sunlight reflected off the CRT phosphor (R:G:B phosphor dots surrounded by a black matrix) which passes twice through the spectrally selective contrast enhancement filter, plus a specular component which is reflected off the front surface of the filter. The display foreground is the CRT phosphor emission attenuated by a single pass through the spectrally selective contrast enhancement filter which has a

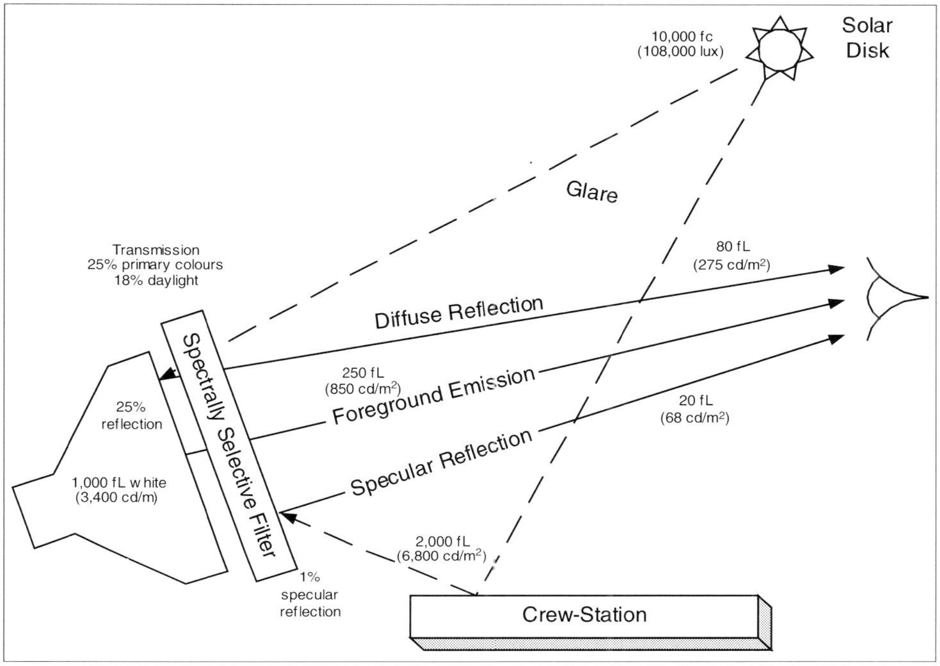

Fig. 12.7 Colour CRT HDD visual performance

higher transmission to the CRT primary colours than to the broadband daylight white light spectrum:

Background luminance = $(10\,000 \times 18\%^2 \times 25\%) + (2000 \times 1\%) = 100$ fL
Display luminance = $(1000 \times 25\%) = 250$ fL white raster (850 cd/^{2})
= $(2250 \times 35\%) = 560$ fL white stroke (1930 cd/m^2)
Contrast ratio = foreground/background
= $(250 + 100)/100 = 3.5{:}1$ white raster: black
= $(560 + 100)/100 = 6.6{:}1$ white stroke: black
= $(560 + 250 + 100)/(250 + 100) = 2.6{:}1$ white stroke: raster

The foreground luminance is the display + background since the ambient illumination will be equally reflected off all parts of the screen irrespective of image content. The luminance and contrast ratio of the primary colours in stroke and raster mode can be computed similarly.

AMLCD Head-Down Display

The simplified worst-case scenario is shown in Fig. 12.8. The AMLCD display performance cardinal points are as follows:

AMLCD active area	6 ×in square format
Backlight luminance	10 000 fL white (34 000 cd/m^2)
AMLCD transmission	4%
AMLCD contrast ratio	40:1
AMLCD diffuse reflectance	1%
AMLCD specular reflectance	1%
Ambient illumination	10 000 fc diffuse sunlight (108 000 lux), 2000 fL point source (6800 cd/m^2)

Fig. 12.8 *AMLCD HDD visual performance*

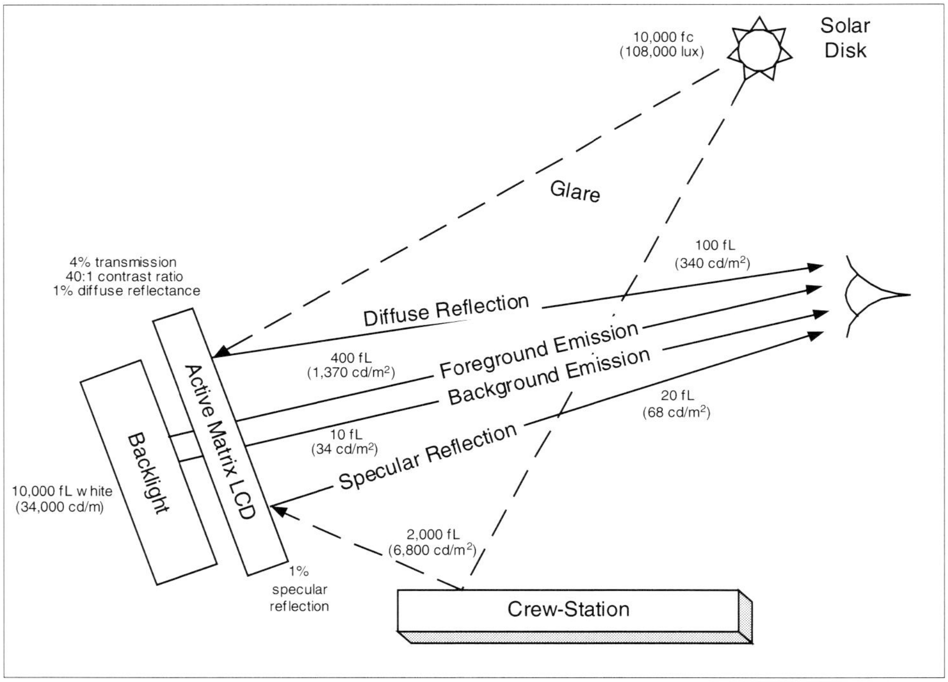

The display background is made up of a diffuse component of sunlight reflected off the AMLCD black matrix, plus a specular component which is reflected off the front surface of the AMLCD, plus the transmission of the 'off' (black) pixels. The display foreground is the AMLCD transmission of the 'on' (red, green, and blue) pixels:

$$\begin{aligned}
\text{Background luminance} &= (10\,000 \times 1\%) + (2000 \times 1\%) + (10\,000 \times 4\%)/40 \\
&= 130\,\text{fL}(450\,\text{cd/m}^2) \\
\text{Display luminance} &= (10\,000 \times 4\%) = 400\,\text{fL white}(1370\,\text{cd/m}^2) \\
\text{Contrast Ratio} &= \text{Foreground/background} \\
&= (400 + 130)/130 = 4.1 : 1
\end{aligned}$$

Head-Up Display

The simplified worst-case scenario is shown in Fig. 12.9. The HUD display performance cardinal points are as follows:

Bare CRT stroke luminance	10 000 fL (34 000 cd/m^2)
Relay optics efficiency	70%
Combiner reflectance	25% tuned to CRT phosphor wavelength
Combiner daylight transmission	90%
Ambient illumination	7300 fL (25 000 cd/m^2) illuminated cloud face

The display background is the illuminated cloud component attenuated by the combiner transmission. The display foreground is the CRT emission attenuated by the relay optics transmission and the HUD combiner reflectance at the wavelength

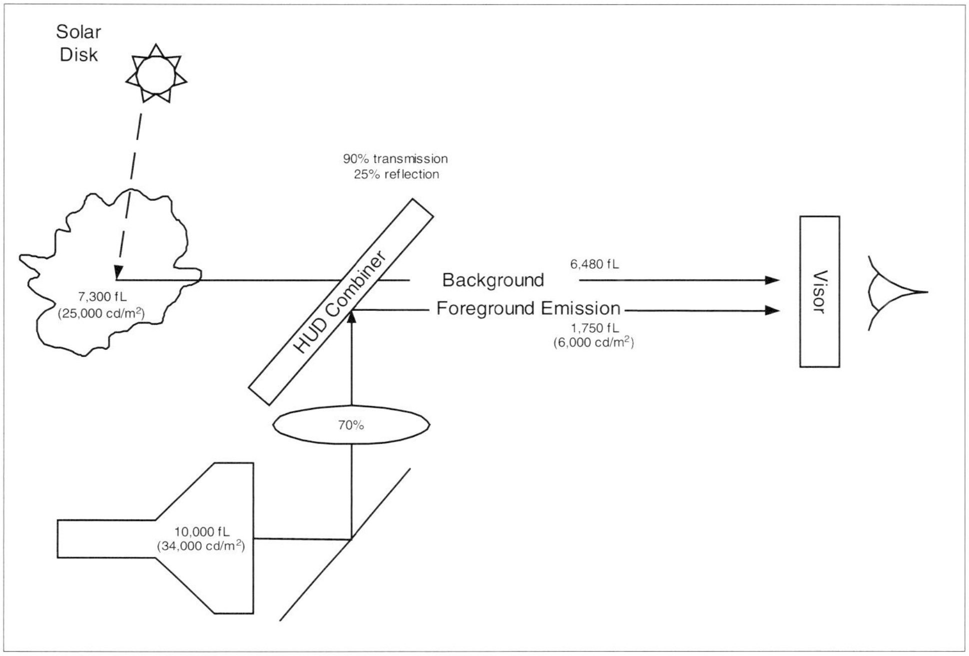

Fig. 12.9 HUD visual performance

of the CRT phosphor:

$$\text{Background luminance} = (7200 \times 90\%) = 6480\,\text{fL}$$

$$\text{Display luminance} = (10\,000 \times 70\% \times 25\%) = 1750\,\text{fL}(6000\,\text{cd/m}^2)$$

$$\text{Contrast Ratio} = \text{Foreground/background}$$

$$= (6480 + 1750)/6480 = 1.27 : 1$$

Night vision imaging system compatibility

The earliest known NVGs were demonstrated in prototype during 1934 and were manufactured by Holst and DeBoer. Their earliest use was in 1941, when ground troops were equipped with image-intensifying devices known as starlight sighting scopes on rifles. These devices are now classified as Generation Zero (GEN 0).

Night vision goggles (NVGs) were first used in a true combat situation in the late 1960s by US ground forces during the Vietnam conflict. These GEN I devices used the then new photocathode technology. Next came the first production NVGs, with the added benefit of Bright Source Protection (BSP) and Automatic Brightness Control (ABC), which ironed out the problem of persistent streaks across the image area associated with bright object exposure, and protected the tubes from damage. These were commonly known as GEN II. The most impressive step forward came with GEN III devices, which are used in almost every aviation sector today. Here, the improvements were the ability to function at very low light levels, achieved by the use of a gallium arsenide coating on the photocathode.

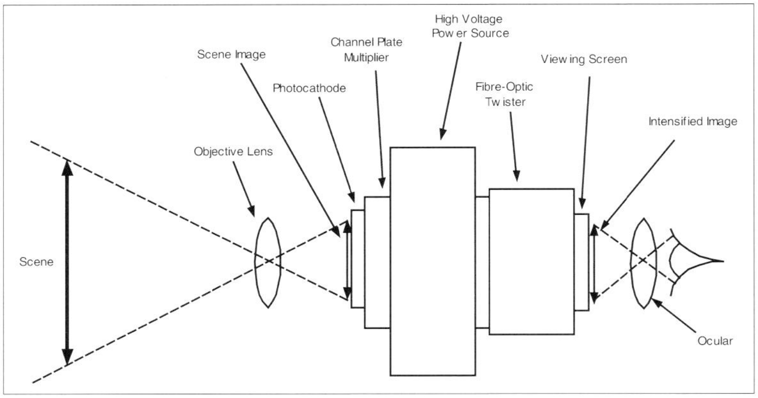

Fig. 12.10 *Diagram of an image intensifier*

Night vision goggles, also called Night Vision Imaging Systems (NVISs), are passive, helmet-mounted, binocular image intensification devices. NVISs operate by converting photons of the outside night scene into electrons using a gallium arsenide photocathode (see Fig. 12.10). The photocathode releases one electron for every photon it receives, thereby converting the light energy to electrical energy. The electrons are multiplied by passing through a wafer-thin microchannel plate, which is coated to cause secondary electron emissions which are accelerated by the electric field and finally collide with a phosphor screen. The phosphor screen then converts the electrons back into photons, displaying the image, which now is amplified around 2000 times. The image is shown in green as the human eye can differentiate more shades of green than any other colour. They must not be switched on in daylight, as there are possibilities of overload which may result in damage being caused to the intensifying tube (**12**).

The majority of today's NVGs intensify the light at both the higher end of the visual spectrum at 615 nm (red) and infrared beyond the visual spectrum (665 nm) and up to 900 nm. Modern NVGs, primarily the ones in use with military applications, have a built-in filter that cuts off the light in the lower visual waveband, which in turn makes the NVGs blind to corresponding (blue and green) colours.

For aviation use, the goggles are mounted on the front portion of the helmet and hang down in front of the pilot's eyes. The viewing eyepiece sits about 20 mm in front of the eyes, which enables the pilot to view and scan the outside world by looking at the image, and also enables the pilot to see all the cockpit instruments and displays by looking underneath the goggles. One of the major limitations of using NVGs is that the pilot only sees a 40° field of view (FoV), whereas under normal circumstances the pilot is used to 220° peripheral vision.

Because NVISs magnify the intensity of light sources, bright lights emitting energy in that portion of the electromagnetic spectrum in which the NVISs are sensitive produce severe veiling glare which can obscure the overall image. To prevent this, and to protect the image intensifier assembly from permanent phosphor burns, NVISs are equipped with an Automatic Gain Control (AGC). Whenever the NVISs are exposed to crewstation lighting that is too intense (from approximately 600 to 900 nm in wavelength), the AGC is activated, the NVISs become less sensitive to the radiance of the outside scene, and the pilot can no

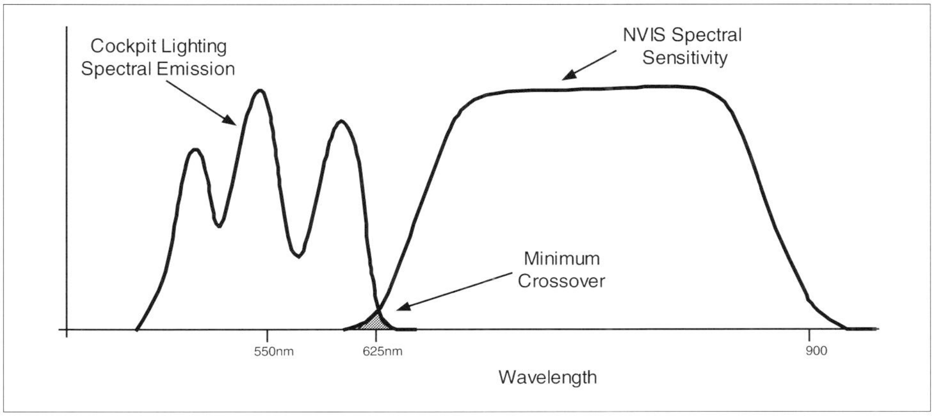

Fig. 12.11 *Conceptual diagram of the spectral distribution of NVIS-compatible lighting*

longer see objects outside the crewstation. To achieve compatibility of crewstation lighting with NVISs, the lighting should have a spectral radiance with little or no overlap into the spectral response of the NVIS image intensifier tubes. Figure 12.11 illustrates the requirements for NVIS-compatible crewstation lighting.

The brightness and spectral characteristics of conventional aircraft lighting cause the AGC of NVISs to activate. Thus, NVISs cannot be used with most existing crewstation lighting. To alleviate this problem and provide compatibility with NVIS operation, the crewstation lighting must be modified or redesigned to reduce the spectral radiance output in the NVIS-sensitive portion of the electromagnetic spectrum.

The rationale and criteria for achieving compatibility of crewstation instruments, displays, and lighting with night vision goggles is encapsulated in the US DoD document MIL-STD-3009 – Lighting, aircraft, night vision imaging systems (NVIS) compatible (**11**). MIL-STD-3009 is derived from the earlier MIL-L-85762A.

Many different methods are used to achieve NVIS-compatible aircraft lighting. For instrument lighting, successful methods include floodlighting as well as external bezel and integral lighting using filtered incandescent lamps and filtered ElectroLuminescent (EL) lamps. For light plates, filtered incandescent lighting, filtered EL lighting, and filtered Light-Emitting Diodes (LEDs) have been successfully used. For emissive displays, a combination of choice of primary colours and filters with special attention to the spectral emission of a backlighting source is required.

In defining NVIS-compatible aircraft interior lighting, a number of interactions must be considered. When a pilot uses an NVIS in an aircraft, energy from both the outside world and the crewstation lighting enter the NVIS. In addition to viewing the intensified image, the pilot's eyes also receive unintensified energy from the crewstation lighting. Figure 12.12 depicts these interactions. During the design of an NVIS-compatible crewstation, the system designer must consider the balancing of the energy the eye receives from the NVIS and the unintensified aircraft lighting.

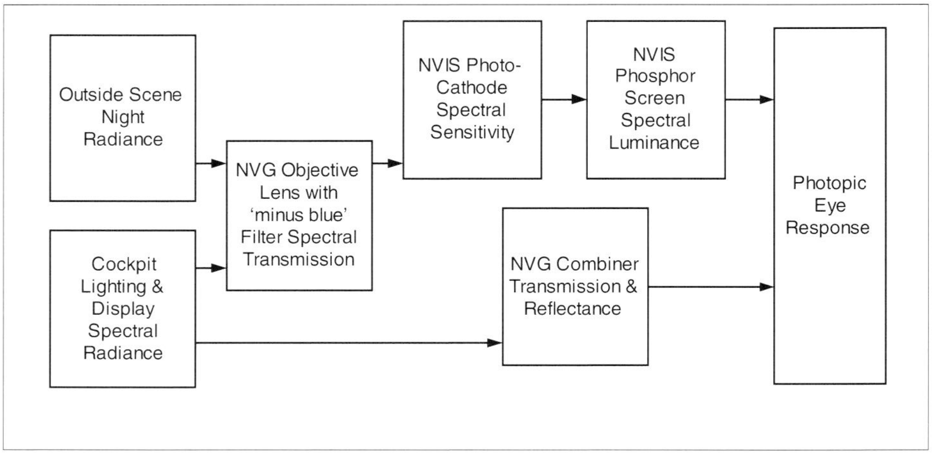

Fig. 12.12 *NVIS interfaces*

Classification

NVIS lighting requirements are classified into the following types and classes:

- *Type I.* Type I lighting components are those lighting components that are compatible with *direct-view* image NVISs. Direct-view image NVISs are defined as any NVIS using Generation III image intensifier tubes which display the intensified image on a phosphor screen in the user's direct line of sight.
- *Type II.* Type II lighting components are those lighting components that are compatible with *projected-image* NVISs. Projected-image NVISs are defined as any NVIS using Generation III image intensifier tubes which project the intensified image on a see-through medium that reflects the image into the user's direct line of sight – such as the Cat's Eyes (see Fig. 12.13) – or integrated with an image source (e.g. CRT) into a day/night Helmet-Mounted Display (HMD).

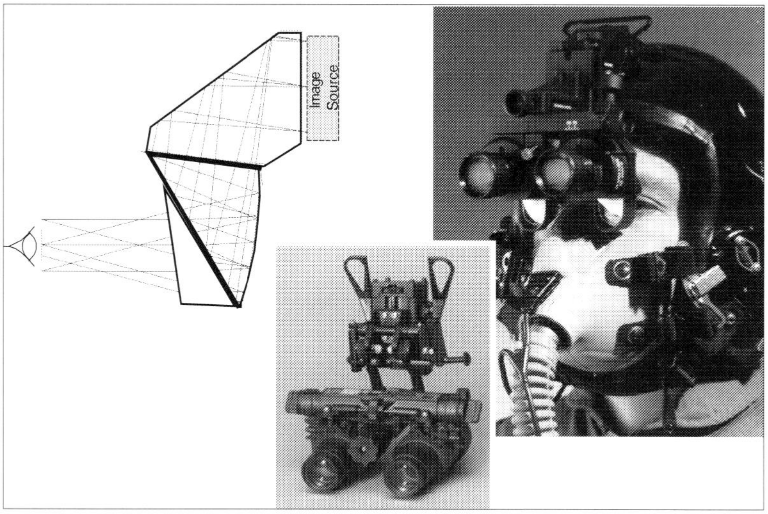

Fig. 12.13 *Cat's Eyes NVIS*

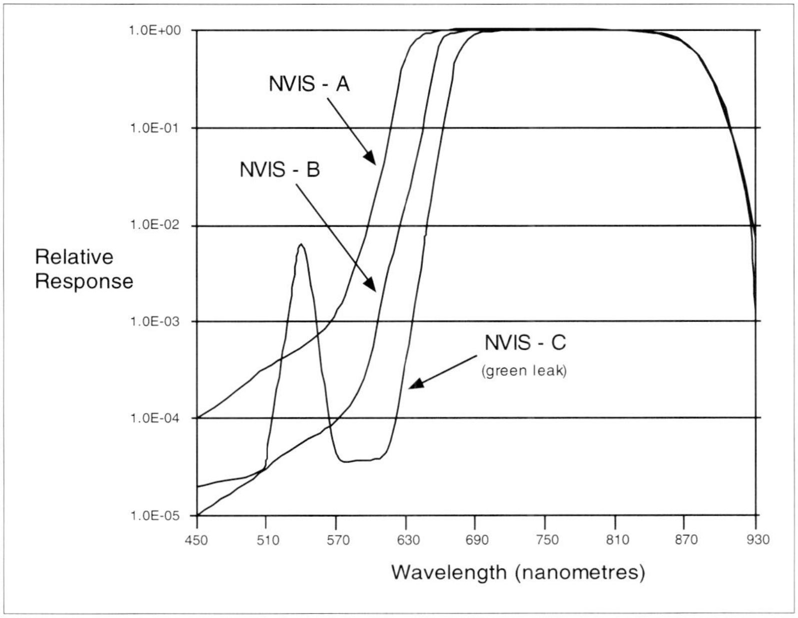

Fig. 12.14 *Spectral transmission of NVIS classes A, B and C*

- *Class A*. Class A lighting components are those lighting components that are compatible with NVISs using a 625 nm minus-blue objective lens filter which results in an NVIS sensitivity as shown in Fig. 12.14.
- *Class B*. Class B lighting components are those lighting components that are compatible with NVISs using a 665 nm minus-blue objective lens filter meeting the requirements specified in Fig. 12.14.
- *Class C*. Class C lighting components are those lighting components that are compatible with NVISs using a minus-blue objective lens filter with a 'green leak' meeting the requirements specified in Fig. 12.14.

The required 'minus-blue' filter (i.e. a filter that blocks blue-green light and transmits red light) is built into the NVIS objective lens during the manufacturing process.

The class A filter was developed by the US Army for use in the ANVIS (Army NVIS). This filter only allows blue, green, and yellow lights to be used in the crewstation. Red cannot be used in the crewstation because class A NVISs are extremely sensitive to colours with radiance at wavelengths longer than 600 nm.

The class B filter was developed by the US Navy for use in the NVISs being procured for fixed-wing tactical aircraft. The class B filter was developed primarily to allow three-colour CRTs to be used with NVISs. However, the class B filter also allows the use of a limited amount of red for warning lights. The class B filter, however, reduces the sensitivity of the NVISs when compared with a class A filter because it cuts off more of the image intensifier spectral sensitivity than a class A filter. Note that the red acceptable for use with class B is not a 'deep red' as might be expected in a full colour display, but more an orange red as a performance compromise.

The class C filter was introduced more recently to allow the NVISs to have higher sensitivity in the green, to make HUD symbology more visible. Some aircraft have HUDs that use a hologram as the reflective element in the combining glass. Holograms typically work with only a narrow wavelength of light. This feature can be used to improve the efficiency and see-through clarity of the HUD, but it means the light coming from the HUD is concentrated at one wavelength. Since this wavelength is in the green part of the spectrum and is blocked by the minus-blue filter in the NVISs, it is nearly impossible to see a holographic HUD with class A or B NVISs. Consequently, modified NVISs have been built and tested that have a 'notch' or 'leak' in the green part of the spectrum.

NVIS radiance

The accepted metric to demonstrate NVIS compatibility is that the image of the crewstation lighting/instruments/displays, when viewed through an NVIS, should be no brighter than the outside scene. Operational experience has shown that, because of its low reflectivity, a defoliated tree is the terrain feature that is the most difficult and important object to see at night. The luminance levels of the crewstation instruments, displays, and lighting should be set to be around $0.343\,cd/m^2$ (0.1 fL). US Air Force tests have shown these luminance levels, or lower, are typically used by pilots when wearing NVISs.

Mathematically, the NVIS is modelled in a similar manner to that used to model the response of the human eye. NVIS radiance is the amount of energy emitted by a light source that is visible through the NVIS. NVIS radiance is calculated as the integral of the curve generated by multiplying the spectral radiance of a light source by the relative spectral response of the NVIS. The units are NVIS radiance (NR), just as photopic luminance units (fL, cd/m^2) represent the amount of energy that is visible to a 'standard' human eyeball.

The NVIS radiance of a defoliated tree illuminated by starlight was calculated by multiplying the spectral radiance of starlight by the reflectivity of tree bark (see Fig. 12.15). The calculation yields a value of 1.7×10^{-10} NR_A for class A and

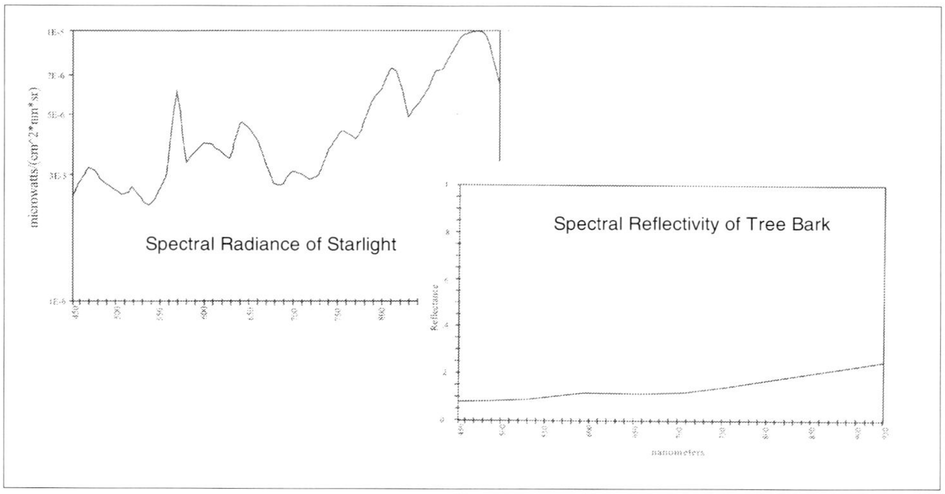

Fig. 12.15 *Spectral radiance of starlight and spectral reflectivity of tree bark*

Table 12.6 NVIS radiance requirements

Lighting components		Class A		Class B		Luminance	
		Not less than (NR_A)	Not greater than (NR_A)	Not less than (NR_B)	Not greater than (NR_B)	cd/m^2	fL
Illuminated controls			1.7×10^{-10}		1.7×10^{-10}	0.343	0.1
Monochrome Head-Down Displays (imaging)			1.7×10^{-10}		1.6×10^{-10}	1.71	0.5
Colour Head-Down Displays (imaging):	White		2.3×10^{-9}		2.2×10^{-9}	1.71	0.5
	Maximum		1.2×10^{-8}		1.1×10^{-8}		
Head-Up Displays		1.7×10^{-10}*	5.1×10^{-9}*	1.6×10^{-10}	4.7×10^{-9}	17.1	5.0

* For type I NVIS only.

1.6×10^{-10} NR_B for class B NVISs. Therefore, to keep the crewstation lighting, instruments, and displays dimmer than the outside scene, the NVIS radiance of those items (when illuminated to produce an acceptable level of luminance required for unaided eye viewing – typically less than 0.1 fL) should not exceed these values. The NVIS radiance requirements for displays are shown in Table 12.6.

Illuminated controls

Most of today's crewstations use incandescent sources for controls, which have spectral emissions reaching into the infrared portion of the electro-magnetic spectrum. The majority of these lighting sources can be made compatible by using a filter that cuts out the IR portion of the emission spectrum so as not to interfere with the NVGs.

MIL-STD-3009 specifies NVIS-compatible colours for illuminated information (alphanumeric and symbolic) on instruments, controls, and control panels. These lighting colours and tolerance limits are shown on the chromaticity diagram in Fig. 12.16 and are designated as NVIS GREEN A, NVIS GREEN B, NVIS YELLOW, NVIS RED, and NVIS WHITE in the CIE colour coordinate system.

Monochromatic Head-Down Displays

The NVIS spectral radiance requirements for monochromatic displays are the same as those for general crewstation lighting with the display luminance set to 0.1 fL (0.343 cd/m^2) except when the display is required to display shades of grey sensor video. An imaging display should be NVIS compatible at 0.5 fL (1.72 cd/m^2) in order for the crew to perceive eight shades of grey, with the lowest grey level at 0.04 fL (0.137 cd/m^2).

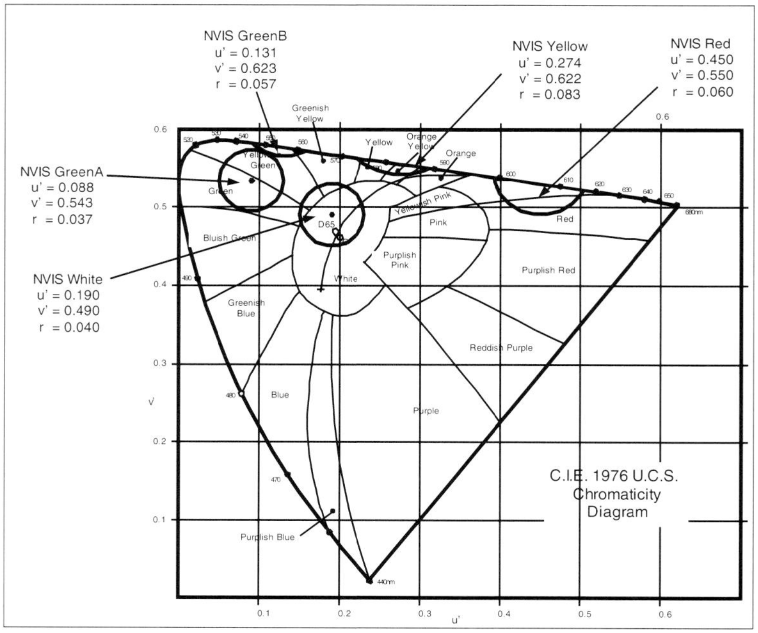

Fig. 12.16 NVIS lighting colour limits

Multicolour Head-Down Displays

Colour displays are in general not able to meet the NVIS radiance requirements for general crewstation lighting. Therefore, to allow the use of colour displays, the NVIS radiance requirement had to be relaxed. This trade-off was made for two basic reasons. First, the NVIS radiance requirement for general crewstation lighting assumed that most of the crewstation would be illuminated, but in the case of colour displays the display graphics were assumed to occupy a smaller percentage of the total crewstation area. Second, most colour displays are located in an area that will not be in the NVIS FoV when looking outside the crewstation.

The standard has two NVIS radiance limits for multicolour displays, one for 'white' light and the other for the 'worst-case' colour that produces the greatest NVIS radiance.

The original MIL-STD-85762A recognized that testing displays at low levels is extremely difficult owing to the signal-to-noise ratio in the measuring instruments at these low levels and allowed tests to be done at somewhat higher levels and then scaled down to the specified level. More recent tests on CRTs and Liquid Crystal Displays (LCDs) have shown that the scaling technique is not always valid. In a CRT, as well as in the fluorescent tubes used to backlight LCDs, the hot filament produces relatively constant IR emissions, while the phosphor produces emissions proportional to luminance. Since the IR emissions do not necessarily scale with luminance, this can allow a display with excessive emissions to pass the test. Therefore, MIL-STD-3009 requires any scaling to be validated for the particular technology being tested.

The assumption that the radiating area of a colour display will be small is not particularly valid with newer AMLCDs either. AMLCDs in use now are larger

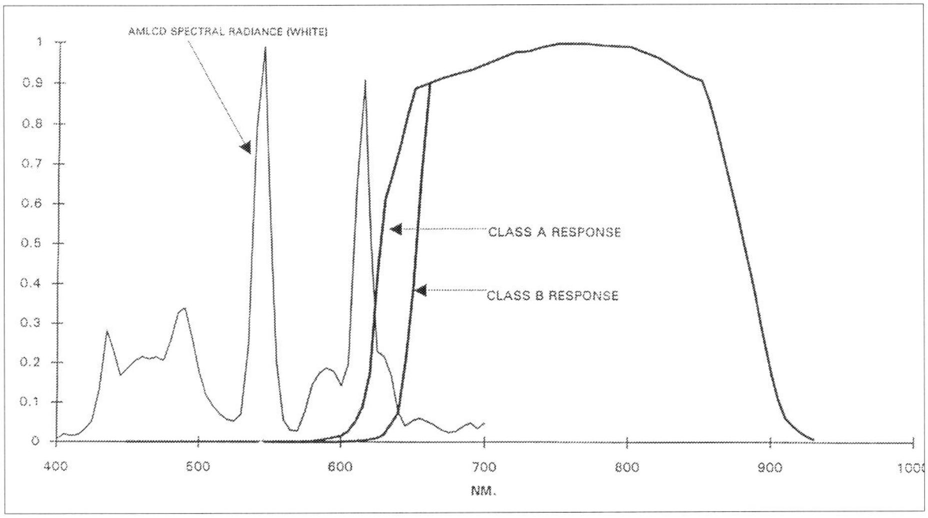

Fig. 12.17 *Unfiltered spectral emission of a typical AMLCD HDD and NVIS responses*

than anticipated when MIL-L-85762A was developed, and they tend to radiate IR from dark areas as well as light areas, increasing the total radiated energy. On LCDs, the IR emission is nearly constant regardless of display content; that is, the image is all 'white' to an NVG. This is because the colour filters used on an AMLCD generally only block radiance in the visual part of the spectrum to achieve red, green, and blue, but are almost totally transmissive in the IR part of the spectrum.

Figure 12.17 shows the typical emission spectra for an AMLCD display. The peak emissions at the wavelengths of red, green, and blue are easily distinguishable. Overlaid on that figure are the class A and class B NVIS responses, from which it can be seen that it is not possible to make a display with a red component compatible with a class A NVIS.

Figure 12.18 shows the same data as Fig. 12.17 but plotted on a logarithmic scale from which it is clear that the colour filters are transparent to IR. Without filtration the AMLCD display is not compatible with the NVIS.

To achieve compatibility it is necessary to add a red cut-off filter to the display backlight to suppress energy emitted in the IR part of the spectrum. The filter must have a sharp cut-off to pass visual red, which has a peak at about 625 nm, but strongly suppress emissions in the NVIS pass band at 650 nm (class B). Typically, a multilayer interference filter is used which will have a transmission better than 90 per cent in the visual part of the spectrum. Figure 2.17 also shows the display spectral emission with the filter incorporated. Using this approach, it is possible to achieve full colour compatibility with class B NVIS, and near compatibility with class A NVIS in green-only mode (red and blue pixels set to black).

An alternative approach is to use two backlights, one for day use to provide conventional high-brightness white illumination to the AMLCD, the other providing a low-brightness, NVIS-compatible, narrow-band, monochromatic (green) illumination to the AMLCD.

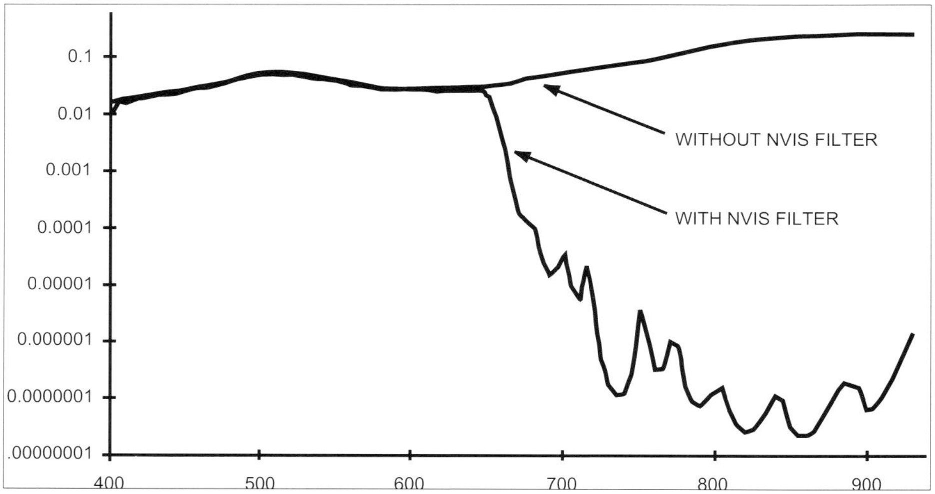

Fig. 12.18 *Unfiltered and filtered display spectral characteristics*

Head-Up Display (HUD) System

With type I (direct-view) NVISs, the HUD must be viewed through the NVIS. Therefore, there is both a lower and an upper limit on the NVIS radiance. The lower limit was selected on the basis of practical testing with a specific HUD/filter solution, which allows the HUD image to be just visible through the type I NVIS when set to 5.0 fL. The upper limit is 3 times the lower limit. There is an obvious loss in HUD image quality when viewed through the NVIS owing to the resolution (and graininess) of the NVIS, but image quality is improving as NVIS technology advances.

The 'leaky green' characteristic is a newer, alternative approach to viewing a holographic HUD with NVIS. It has been found to be nearly impossible to meet the minimum NR requirements discussed above with holographic HUDs. HUDs that use holographic techniques are now in use in several aircraft. The very wavelength selective nature of the holographic HUD combiner prevents HUD emissions being within the NVIS spectrum. The 'green leak' allows the NVIS to 'see' the HUD image.

With type II (projected) NVISs, the HUD is viewed with the unaided eye. There should be no intensified image seen through the image intensifier tubes, or the pilot will be presented with a double image (the unaided eye image overlaid with the intensified image). Therefore, only a maximum NVIS radiance has been established for type II NVISs.

There is an added complication with the operational aspects of type II NVIS if the HUD is used to present an FLIR image. When looking through the HUD, the pilot sees not only the FLIR sensor image of the outside world but also the intensified NVIS image of the outside world. These two images are in different portions of the electromagnetic spectrum (one in the near IR, one in the far IR). They are different and may at best be confusing and at worst disorienting. The solution is to suppress the NVIS image when the pilot looks through the HUD. The NVIS image is reinstated if the pilot looks away from the HUD. The on–off sequencing of the NVIS needs careful attention to avoid being a distracting irritant.

Canopy reflections

Reflections from the canopy, windshields, and windows should be minimized. Irrespective of NVISs, reflections that affect outside vision should not be permitted. Any such reflections can be extremely disconcerting at night, and potentially hazardous during landing.

Compatible external lighting

In the mid-1990s the issue of compatible external aircraft lighting was recognized. The problem came about when aircraft using NVGs had to turn off their external lighting in certain situations. When aircraft were flying close to trees and buildings, reflections from anticollision lights momentarily 'bloomed' the goggles and caused spatial disorientation. Fast jets could not be flown in formation without the anticollision lights being turned off, and the navigation and formation lights bloomed the goggles within 1 nautical mile. It therefore became necessary to modify the existing external lighting to remove a portion of the near-infrared emissions while maintaining the required lighting levels and distribution angles required for safe flight. The term 'NVG friendly' was adopted for such technology and is enabled by adding filters to the existing lighting to cut out a large portion of the near-infrared emissions.

References

(1) **Greeley, K.** and **Schwartz, R.** *F22 Cockpit Avionics A Systems Integration Success Story*, Lockheed Martin, Marietta, Georgia.

(2) FAA AC 25-11, Advisory Circular: Transport Category Airplane Electronic Display Systems, Federal Aviation Administration.

(3) SAE ARP 1068, Flight Deck Instrumentation and Display Design Objectives for Transport Aircraft, Society of Automotive Engineers.

(4) SAE ARP 1874, Design objectives for CRT Displays for Part 25 (Transport) Aircraft, Society of Automotive Engineers.

(5) SAE ARP 4256, Design Objectives for Liquid Crystal Displays for Part 25 (Transport) Aircraft, Society of Automotive Engineers.

(6) SAE AS 8034, Minimum Performance Standards for Airborne Multipurpose Electronic Displays, Society of Automotive Engineers.

(7) JSSG-2010-3, Joint Service Specification Guide – Crew Systems – Cockpit/Crew Station/Cabin Handbook, US Department of Defense.

(8) JSSG-2010-5, Joint Service Specification Guide – Crew Systems – Aircraft Lighting Handbook, US Department of Defense.

(9) MIL-HDBK-87213, Electronically/Optically Generated Airborne Displays, US Department of Defense.

(10) MIL-STD-3009, Lighting, Aircraft, Night Vision Imaging System (NVIS) Compatible, US Department of Defense.

(11) EFJ-R-EFA-000-107, Cockpit Lighting Standard, Eurofighter.

(12) **Wilson, A.** (2000). All right on the night, *Flight Deck International*, October.

Glossary of Terms

Advection: Advection differs from convection only on scale. It is the change in property of a moving air parcel because the parcel moves to a region where the property has a different value e.g. when warm air moves into a cool region. Convection is air transported by thermally induced currents, advection is air transported by steady (generally vertical) currents.

Anisotropic: The quality of exhibiting different properties in different axes. In the context of adhesives as used in Chapter 5, an anisotropic conducting adhesive is one which is electrically conductive in one axis, but which does not conduct in the axis perpendicular to the first.

Anthropomorphic: Anthropology means the study of man. Specifically and in the context of Helmet-Mounted Displays, applying to the study of the physical and diverse features of the human head shape.

Aspheric lens: Normal lenses have spherical surfaces. Some lenses in complex optical arrangements may have lenses whose centre of curvature is not the physical centre of the lens. An aspheric lens is a lens whose surface is not spherical, nether may the centre of the asphere be the physical centre of the lens.

Bi-level: A display capable of representing (or being driven into) only two states – black and white.

Chromaticity: Colour discrimination is possible because the human eye contains three classes of photoreceptors (cones) that differ in the photopigments they contain. R cones are maximally sensitive at a wavelength of 570 nm, G at 540 nm and B at 440 nm. The perception of colour depends on the excitation ratios of the cones – the ***tri-stimulus*** values. It is independent of luminance. No two wavelengths of light can produce the same exitations in all three cones, however a suitable additive combination of long (red), medium (green) and short (blue) wavelength lights can produce the same ratios as light containing energy at all wavelengths. Thus from three colours suitably mixed it is possible to perceive all colours. This principle of ***chromatic metamerism*** is used in CRT and AMLCD displays. The trivariant nature of colour vision permits colours to be specified in terms of three values. Ideally these would be the tri-stimulus ratios of the R, G, and B cones, but because too little was known about the cone spectra in 1931 the International ***Commission on Illumination (CIE)*** adopted a different, but related, system based on the prediction of metamers. The ***CIE chromaticity diagram*** plots colours on a two-dimensional chart which shows the relationship between tri-stimulus values, independent of luminance. The first diagram published in 1931 used an x, y co-ordinate system in which $x = X/(X+Y+Z)$ and $y = Y/(X+Y+Z)$ (where X, Y, and Z are the supposed tri-stimulus values). [These two coordinates carry all the information, the value of $z = Z/(X+Y+Z) = 1-(x+y)$ carries no additional information.] Subsequently other chromaticity diagrams have been produced as linear transforms of the 1931 chart. The most widely accepted transform is the 1976 chromaticity diagram which annotates its axes as u', v', see Chapter 5, Fig. 5.4. The continuous locus of spectrum colours (i.e. colours produced by a single wavelength) is represented by the outermost contour. All non-spectral colours are contained within an area defined by this boundary and the line joining red to violet. Any colour can, therefore, be described by its two u', v' co-ordinates.

Colour gamut: The colour gamut of a display device is the range of colours that can be displayed by it, most conveniently described on the CIE chromaticity diagram as the triangle of inscribed colours defined by the display device primary colours. (see Chapter 5, Fig. 5.4).

Contrast: Contrast is a measure of the ability to perceive the information content of an image. It is a measure of the luminance contrast only. Specifically it is the relative luminance of the foreground (the data) against the background luminance, expressed as a ratio relative to unity.

Convergence: Convergence is the requirement to fuse or register primary images of a colour display so that they are perceived as a single full colour image. Mis-convergence is the measure of any convergence error and should be less than the ability of the eye to perceive it at the viewing distance of the display. (see Chapter 12, Fig. 12.2).

Diffuse reflection: Diffuse reflection is the reflection that would be obtained from a surface having a diffuse matt-finished of microscopic roughness, e.g. from minute crystals. An ideal diffusing surface reflects source illumination with a spatial distribution which contains a cosine distribution about the display normal, irrespective of source angle. The observer sees no image of the source. An ideal diffusing surface is also known as a Lambertian surface.

Display normal: The normal to a display is the angle perpendicular to the display surface.

Greylevel (also greyscale): Greylevel is the ability of a display to represent image luminance between the states of black and white. In a digital (or digitally driven) display, the number of greylevels per colour will determine the number of colours available. The relationship between greylevels is usually non-linear, and should be matched to the human eye Relative Contrast Response (RCS) function. In this context, the display response is usually termed its greyscale response or greyscale transmission (for a T-LCD).

Greyshade: The ability of a display to faithfully represent video scene images is often described as the number of greyshades discernable. In this context the test image usually contains a set of greylevel bars ranging in shade from black to white with a relative luminous intensity of $\sqrt{2}$ between each shade.

Illumination: Illumination (or illuminance) is the density of radiant energy incident upon a receiving surface evaluated throughout the visible spectrum in terms of the spectral luminous efficiency (the human eye response). Illumination is the density of luminous flux per unit of surface area. The SI unit of illumination is the lux = 1 lumen/m^2. The imperial unit is the footcandle (fc) = lumen/ft^2.
1 fc × 10.7639 = 1 lux.

Luminance Many sources of luminous flux are not a point source, consequently another concept called luminance or photometric brightness is convenient. The SI unit if measurement is the candela per square meter (cd/m2), also known as the nit. The imperial unit is the footlambert (fL) = cd/ π ft^2.
1 fL × 3.426751 = 1 cd/m^2 = 1 nit.

Luminous intensity: The luminous intensity of a point source is the radiant energy of that source evaluated throughout the visible spectrum in terms of the spectral luminous efficiency (human eye response). The SI unit of measurement is the lumen (l). The luminous intensity of a point source when viewed in a particular direction is related to the luminous flux emitted over a solid angle. The SI unit is the candela (cd),
1 candela = 1 lumen/steradian.
SI defined the candela as 1/683 watts/steradian.

Occipital condyle: The occipital is the bone forming the back and back part of the base of the cranium, the part of the skull that encloses the brain. The occipital condyle is the opening through which the medulla oblongata passes, linking the spinal cord and the brain.

Pixel (and sub-pixel): Shortened form of 'picture element' defines the smallest element of a digitally addressed or quantised display. A colour display typically has a number of sub-pixels in each primary colour that together form the display element, pixel or colour group. (see Chapter 5, Fig. 5.10).

Planar doublet: An optical doublet is an optical surface made of two elements optically cemented together. A planar doublet is two optically flat pieces of glass cemented together, usually having different refractive indices or having an optical film, such as a hologram, sandwiched between them.

Purity: Purity is a measure of the saturation of a displayed colour. An absolutely pure colour may be approximately assigned to its spectral wavelength. As other colours are added, the original colour becomes less vivid, less pure, less saturated and tends to white. In the context of a shadow mask CRT, impure colour can result if electrons from one electron gun 'illuminate' phosphors of another colour.

Raster: A means to transfer an image, usually a pictorial (video) image. Developed originally for broadcast television the image is scanned sequentially from top left to bottom right in horizontal swathes. Each element of the image is compiled sequentially. The whole image is refreshed at a rate sufficient for the observer to be unaware of the sequential nature of the image construction (see Chapter 5, Fig. 5.5).

Shadow mask CRT: The most common type of cathode ray tube used to display a full colour image. The image is produced by three electron guns, one for each primary colour. The shadow mask is a thin metal sheet perforated in a uniform series of holes placed between the electron gun and close to the phosphor screen to ensure only electrons from the appropriate electron gun (say red) land on the correct colour (in this case red) phosphor dot. The other phosphor dots (say green and blue) are in the shadow formed by the mask (see Chapter 5, Fig. 5.2).

Spatial frequency: The frequency of image components expressed in relation to their physical or spatial dimensions or angle of subtense at the point of observation.

Spectral response (also spectra): The spectral response is the emission, transmission or reflection characteristic over the frequency of interest. For example the spectral response of the human eye to radiation at normal (photopic) illuminance levels (see Chapter 6 Fig. 6.7) is commonly known as the spectral luminous efficiency.

Specular reflection: Specular reflection is the classic reflection that would be obtained from a perfect mirror. That is the observer sees a direct (but mirrored) image of the source, with the angle of incidence equal to the angle of reflection.

Stroke (also cursive): A means to construct a graphical (usually a line) image. The image is constructed as a series of pen strokes as one might construct a line drawing with a pen on paper, hence its other names, cursive, calligraphic (see Chapter 5, Fig. 5.6).

Twisted nematic: Nematic liquid crystals are cigar shaped molecules oriented with their long axes parallel. They behave like toothpicks in a box, maintaining their orientation but free to move in any direction. Nematic substances can be aligned by electric and magnetic fields.

Visible spectrum: The human eye is sensitive to all wavelengths in the electromagnetic spectrum from 380 (blue) to 760 (red) nanometres.

Index

Bits & B

A-Z of Pudendal Neuralgia

Thank You.

Marion Jones

Hello

I'm Marion and I'm going to be your lecturer on all things Pudendal Neuralgia but first a little about myself.

I am a retired bank clerk and worked in admin at my local hospital. I have lived in Herefordshire for 41 years and led a very active life which included many long haul holidays enjoying the culture of every country I went to. I loved going to the cinema and to see West End shows and going out until I was admitted into hospital one day 11 years ago for what I thought was a routine operation.

The following 11 years have left me in agony with suicidal thoughts but my aim in writing this book has never diminished.

Acknowledgements:

This book is especially dedicated to all GPs, Consultants in Obstetrics and Gynaecology, Midwives and Nurses in the hope that they may learn something from all the information given by sufferers.

Also all my fellow sufferers of Pudendal Neuralgia in the Facebook support group. Without your wonderful input this book may never have materialised.

To Chloe Nichole Designs for some of the drawings. They are perfect and I hope this will bring you recognition and more work in the future.

Finally to Katie Shearer from Orphans Press who has had to guide me through every step of the way and I thank you for your patience.

Firstly, let me introduce myself. My name is Marion Jones, a 67 year old woman, born in West Sussex. As my parents both suffered with cancer, I left school early and was employed as a typist by the Council in the Public Health Department. After getting married, I spent seven years in Germany but have spent most of my working life in Herefordshire, formally working in a high street bank and latterly in the Human Resources department at my local hospital.

It was whilst working at the hospital I discovered that I needed a rectocele prolapse repair (Posterior Vaginal Prolapse) which is a fairly "routine "operation and one that is carried out every day in a NHS hospital somewhere in the country. However, as you start to read through my story and the A-Z of Pudendal Neuralgia, you will discover this was not a "normal" operation for me.

I sadly never returned to my job in Human Resources and for the past ten years I have been attending hospitals, GP and Chiropractor appointments up and down the country.

In late 2019 I decided that the medical profession needed to know how a sufferer with Pudendal Neuralgia coped with this debilitating condition on a day to day basis. I opened up an Instagram and Twitter account to start raising awareness for more research and training. This is ongoing.

The Facebook group for Pudendal Neuralgia and Pelvic Pain was created on 15th September 2015. I became friendly with another sufferer, Jane Lewis, who suggested that we set up our own Facebook group here in the UK, for those suffering from Pudendal Neuralgia and Pelvic Pain. This we did and the numbers today are exceeding 1200 sufferers.

It's not only women who suffer from this condition, men suffer too but they are slightly embarrassed to talk about it, whereas the ladies will go into great depth explaining their symptoms and they don't hold back which is a good thing in this day and age. All the members of the Facebook group suffer from the same condition, so we all understand.

When I had my initial appointment to discuss the operation, I was told by the Gynaecologist that normally a hysterectomy (leaving the ovaries in place) would be carried out at the same time for women of my age who needed a posterior vaginal prolapse repair. (I was 56) and it was all going to be carried out through the vagina. Who was I to question that? He was the Lead Consultant in Women's Health after all.

The day of the operation arrived. I signed all the paperwork including the consent form and walked down to the operating theatre. I had been looking forward to returning to the ward with the prolapse repair back in place but minus my uterus.

Following the operation, I was in a tremendous amount of pain in both the vagina and anus. I was told by the registrar that they had also done a repair on the anterior wall otherwise known as a Cystocele. This happens when the bladder drops from its normal position in the pelvis. I had shown no symptoms that this needed operating on, but when I questioned the pain, I was told it was early days and it would settle down in a week or two. Needless to say it didn't.

This is where my journey began to find out what was wrong with me. Why was I in so much pain following what I assumed to be a routine operation? I couldn't sit down or wear underwear, leggings or trousers because it hurt too much. There was constant throbbing, stabbing and burning pain in both areas. It was debilitating having to live with this after leading such an active life with travelling, meeting friends for meals, and going to concerts and cinema etc. All these activities came to a standstill.

I asked my GP to arrange for a second opinion on this debilitating pain. This turned out to be at the Birmingham City Hospital where I was operated on within two weeks for a Vestibulectomy which is a surgical procedure that can be used to treat vulva pain, specifically in cases of provoked vestibulodynia. (I also suffer from Vulvodynia). I had Botox injected into the anus and finally to treat the Bartholyn Glands which had also become infected.

Unfortunately, none of these helped with the pain. The consultant then referred me to the in-house vulva specialist who then diagnosed me with Pudendal Neuralgia. I was over the moon that I had at last got a name to the agony I was experiencing every minute of every waking hour. He then informed me that he was sure this was caused by the first operation and that the Pudendal nerve in the pelvic area had been damaged. I was under the Professor's care for a few years trying, all the nerve blocks and unpleasant medication until it got to the point when he said he couldn't help me anymore. I was very despondent when I left that hospital, but he did refer me to the Pain Management Department at the Queen Elizabeth Hospital in Birmingham. I knew that they treated members of the armed forces there who have been injured in a war zone so I thought my problem would be a minor issue for them and I could be treated.

I arrived at the QE Hospital Birmingham feeling confident that a miracle was about to happen, and I would be cured in a couple of appointments. I had previously emailed the story so far, so the Pain Management Specialist who I was booked in to see would have time to read and digest it. As soon as I entered the room, he told me that everything possible had been tried on me and there was nothing they could do. However, he told me to go back to the waiting area and take a seat (by that time my rectum and vagina were screaming at me). Within a few minutes a physiotherapist called me into her office. She could see I was upset about being told there was nothing else I could try. She told me that The University College London Hospital had a good reputation for dealing with nerve pain, so I grabbed the opportunity of going there as my last resort.

Whilst waiting for my first appointment in London, I discovered that an old biopsy site had grown bigger and occasionally it would bleed. I was referred back to my local hospital for investigation by the GUM Clinic (Genito-Urinary Medicine Clinic) as I was unable to get an appointment at my GP.

I had an appointment at the hospital within the two weeks but when I got into the room and up on the bed, I told the lady consultant that however gentle she was, could she please use the smallest speculum because it would hurt me. Eventually she agreed that I was in too much pain for an investigation to go ahead that day but advised me that she would have me in as a day case patient and examine me under anaesthetic. I was happy to go ahead with her advice. At least I wouldn't know or feel any pain.

Unfortunately, when I arrived on the day, it wasn't the same consultant carrying out my procedure.

I was having blood pressure checks and putting on the lovely theatre socks when the nurse announced that the consultant wanted to speak to me before I was put under, I sighed, then walked to theatre. The consultant greeted me with, "I just want to give you an examination before we put you out". I immediately protested and said that it would cause me great pain, plus it wasn't what was agreed at the outpatient's appointment. I then went on to inform her that I suffered from Pudendal Neuralgia, but she didn't seem to take any notice of what I was saying.

The consultant came back with "It will be fine, it won't hurt at all". I saw she had the speculum in her hand which she proceeded to push into my vagina.

I grabbed hold of the sides of the theatre bed and screamed...

She looked at me and held up the speculum and said, "It can't possibly hurt; I've used the smallest one".

By this time rage was building up inside me and I was shaking with fear. The nurses were trying to calm me down and I could tell they weren't happy the way I was being dealt with. Then came the final straw, she came up beside me and said...

"I think all this pain is psychological"

It was at that moment after being told, "All the pain was Psychological" I decided I was going to make a formal complaint against the consultant. I had suffered 8 years of pain from Pudendal Neuralgia, and I felt appalled that she suggested it was all in my head.

As soon as I left the hospital and I was back being looked after by my sister-in-law, I picked the IPad up and started typing. There was no stopping me. I was well and truly wound up with what I had been through. I made my complaint and heard back from PALS at my local hospital to say that my complaint would be investigated thoroughly, and I would hear within 28 days. Unfortunately, they gave my complaint to the head of Obstetrics & Gynaecology and it was he that had damaged the nerves in the first place, so I asked for a different person to look into my complaint.

Eventually, I had a letter of apology from the consultant, and she admitted she had used the wrong type of words and she should have explained "It's all in the head" a bit better. I fully understood that the brain is connected to all the different parts of the body but the way she said it cut very deep. I then began thinking perhaps everyone thought I was making the pain up. I asked relatives if that's what they thought....

There was a resounding NO. They all knew how much pain I suffered day in day out.

I asked for a meeting with the investigator and that was arranged for a few weeks later. The Clinical Lead and the Service Manager for Women's Health were both in attendance. They went through my complaint thoroughly and explained and apologised for the upset this had caused me. (I took a counselling course whilst I was waiting for the complaint to be finalised. Before the meeting was wrapped up, I asked the Clinical Lead whether he knew about the condition Pudendal Neuralgia. He replied, "I know very little about it". I then asked the Service Manager the same question. She replied, "To be honest with you, I have never heard of it".

As I was passed through the front doors of the hospital, I mentioned to my friend who came with me that I had managed to educate one hospital about Pudendal Neuralgia. Now for the rest of the country but where do I start?

I started by opening a Twitter Account in the name of @PNawarenessUK and an Instagram Account @pnawareness_uk_.

My first appointment at the University College London Hospital arrived quickly and I set off on the train with a friend who was much needed (I can no longer see the timetable board as I am partially sighted). I was nervous about going, probably because I knew it was my last chance of getting a cure. I saw a lovely sympathetic consultant who didn't want to examine me due to the pain I was in but really needed to. I told her that I had a four-hour journey on the train back to Hereford which would be really uncomfortable if she performed an internal. I watched as she held the speculum under the warm tap and I asked for lubricant, but she told me they don't use it now, just tap water to warm it up.

She could see that I was nervous and took her time examining me until she told me to get dressed. I had already told her what medication I was on and what I had already tried, but she booked me in for more nerve blocks (This meant more journeys up to London on packed trains and waiting around at stations when the trains were cancelled at the last minute).

The first block eased the pain for a couple of days but the second one made the Pudendal nerve pain worse. I was in a flare for about ten days after the block.

When I had my next appointment, she told me she was going to try me on the drug Tapentadol. The potency of the drug is somewhere between that of Tramadol and Morphine. It is generally regarded as a moderately strong opioid. It is used for the treatment of moderate to severe pain for acute (following injury, surgery, etc and chronic musculoskeletal) pain. Before I could start on the drug, I needed to come off the antidepressant Duloxetine.

They couldn't prescribe me the Tapentadol at the University College London Hospital but needed to make an appointment with my doctor who would be able to help me. I booked in to see my doctor and she told me that it wasn't necessary to reduce the Duloxetine gradually, just stop it and start taking the Tapentadol. I didn't question her on her advice, but I soon found out that this was incorrect. I started to suffer horrendous side effects again like a drug addict waiting for the next fix. Shaking, my head jerking but my brain couldn't keep up. I was aching from head to toe and feeling dreadful. But I kept off the drug. I started taking the Tapentadol.

Now this drug also has side effects (as all do). This time I itched all over and because of this I had no sleep for four nights. I got myself to A & E and the doctor there prescribed Antihistamines, I tried those for a week and had no change in the itching. Then I made an appointment to see my GP again and she decided that it was a good idea if I stopped taking them as they obviously didn't suit me. A few days later I sat with nearly 300 pills and a drink in front of me ready to take.

I had no reserves left in my body to cope with the pain anymore. I wanted out.... gone.... Whenever I hear of people taking their own lives I always think, how could they do that to the friends and relatives that they leave behind? That was the last thing on my mind. I couldn't care less who I hurt. I couldn't be bothered to write any notes to relatives on why and wherefore. I had left the front door unlocked in the hope of one of the neighbours finding me eventually. All I was interested in was getting rid of the pain in the easiest way possible.

I sat on my bed for about two hours, thinking it all through, and then a message arrived on my phone from a neighbour. She wanted to know if I was ok. This was unusual for a Sunday evening, but I replied with just one word… No.

She immediately dropped everything and came to the house; I was sitting on the living room floor by this time. She took control and said the medical profession should see me at my lowest ebb. Maybe then I would receive help. I told her about the pills, and she called 111. The responder was very sympathetic but of course she didn't have a clue what Pudendal Neuralgia was. I explained that I was in so much pain I didn't want to live anymore. She told me an ambulance would be sent, but it may be a couple of hours before it was my turn.

My neighbour stayed with me the whole time and I remember her telling the two paramedics that I would need a bed when I arrived as it was impossible for me to sit. They told her not to worry; I will be in good hands and be looked after.

On arrival at about 9.00pm I noticed that A & E was exceptionally busy, but the paramedics found me a seat and said that I would be seen shortly. I was up and down off that chair so many times and pacing the floor because it was too painful to sit. At midnight, I went up to the service desk and asked how much longer I would have to wait. The nurse explained that there were another three patients to be seen before it was my turn. I went back to my seat and waited… and waited. I spent my time messaging my sister-in-law. She wanted to travel up from Sussex immediately, but her husband persuaded her to get some sleep and to maybe drive up in the morning.

At 5.00am. I walked up to the service desk again. This time I was told there were another three people to be seen before it was my turn. I lost my rag at that point. I told her that I had been brought into the hospital because I wanted to take my own life. Those feelings had been heightened in the time I had been waiting to see someone. I told her I was going to leave to which she replied "You will need someone to let you out because the door is locked, and if you do leave, we will call the police".

I had reached the age of 65 without a police record and I didn't want one now, so I went back to my seat yet again. By this time, I had been in A & E for nearly nine hours without seeing any doctor or nurse.

Eventually, a lady from the Crisis Unit appeared beside me. We went into a room where she said she wanted to just talk to me about why I wanted to end my life. She was incredibly sympathetic and could see that I was in a mess, not only from being awake for 5 nights but because I was suffering intense pain from sitting. Even now as I am typing this, my mind drifts back to that night -a very difficult time in my life.

I got a taxi back home from the hospital and all I wanted to do was go to sleep. I lay down on the sofa, drank a cup of tea, and the next thing I knew my sister-in-law was beside me.

She gave me a hug and I started to cry. She tried to cheer me up by saying, "Come on, we are going on a cruise to see Alfie Boe in three weeks". Now I must confess, I did have a soft spot for Alfie but all I said back to her was, "I couldn't give a s..t about Alfie". She knew then that I was ill.

The Crisis team arrived that evening to chat to me and take away all the pills I had in the house. The way they did it was incredibly kind. They went through each drug and asked if I was still taking it then they went through the rest and put them all in a bag. They handed me a box containing 2 pills which I had to take that night for me to get some proper sleep.

I did sleep that night, in fact my sister-in-law had to wake me at 1.00pm because the Crisis Team would be back to talk to me at 2.00pm. I felt like a zombie, but at least I had slept.

They continued to visit me daily, they took control of what medication I took and promptly put me back on Duloxetine which the doctor took me off. I started to feel better once they were in the system. The Crisis Team were appalled at the treatment I had received.

Each day, they appeared on my doorstep to check I was ok and just talk to me. They suggested I needed to go on a counselling course so that I didn't have those thoughts again.

Looking back, I realise I was mentally ill at that particular time but failing to care about the people I would leave behind was so unlike me. My sister-in-law travelled up from Sussex a couple of times afterwards but no one else in the family knew. I didn't want my lovely mum-in-law to be worrying herself sick about me.

MY MISSION

This is personal to me now. I have had an awareness campaign in my head ever since I was told "It's all psychological". I vowed on the day of the procedure that I would spend my time raising awareness for this debilitating condition and not let anyone else go through what I had.

The employees of the NHS and possibly consultants that work in the private sector need to be taught about the nerves in the pelvis. When they are dealing with patients who are suffering a lot of pelvic pain from Pudendal Neuralgia, Vulvodynia or any other pelvic condition, the patient will be believed and understood. This isn't happening at the moment, and it needs to be. Sufferers are sent away still in pain or are being pushed from one department to another.

The second part of my mission is to obtain more research into the nerves in the pelvic area. The Pudendal nerve is the main nerve in the pelvis. Unfortunately, very little is known about that area of the body which I can never understand. Everyone has had a pelvis since Adam and Eve.

The final objective for me is to ensure that whilst medical students are at college, more time is devoted to learning about the Pelvic area. It seems to me as though they are saying that nothing medical ever goes wrong in that area, so why learn about it? I can guarantee there is a lot that can go wrong.

To sum up briefly, I will ensure every GPs surgery and hospital in the UK is aware of Pudendal Neuralgia. That includes Obs & Gynae, Maternity Units and Nurses in charge of cervical screening.

More research into the Pudendal Nerve.

Medical Students taught about the Pelvis in depth.

Finally, I have produced a Pudendal Neuralgia Information Leaflet which lists a lot of the causes and symptoms for both male and female within it.

A IS FOR ASHAMED

No, I'm not ashamed but the word "Pudendal" comes from a Latin word meaning "To be ashamed".

It is interesting that this term is an archaic reference to the external genitalia... However, the term is still used to identify the artery, vein and Pudendal nerve. Pudendal is also a term which relates to a medical procedure used during the delivery of babies. A Pudendal block is used to anaesthetise the area surrounding the vagina before passage of the baby's head.

Men and women should never be ashamed that they have this condition. We didn't ask to be inflicted with this debilitating condition but why isn't the Pelvic area taught at length at medical school? After all, everyone has a Pelvis!!

B

B IS FOR BIRTH TRAUMA, BITS & BUTTS AND BRANCHES

BIRTH TRAUMA

I asked the MASIC Foundation if I could include a story that a group member had told them about her birth trauma. This was agreed and I'm so grateful for that.

The MASIC Foundation (MASIC) exists to support women, challenge stigma and drive change.

The MASIC Foundation is the only multi-disciplinary UK charity to support women who have suffered severe injuries during childbirth known as OASI (Obstetric, Anal Sphincter Injuries). The charity is run by injured women and healthcare professionals who are committed to better detection and prevention of injury during childbirth.

"My son's forceps delivery, and subsequent 3b OASI tear, happened eight years ago. It's hard to remember now, several years later, how my body and life felt before the day my baby was born. I remember dreaming about new motherhood, and how wonderful it would be to hold my new baby for the first time – to look forward to first walks with the pram, all the beautiful moments women are told to expect with their new arrival. My reality was being so traumatised and unwell after his delivery that I couldn't get out of bed for weeks, couldn't pick my tiny baby up, breast feed him, or lift the weight of his new car seat or pram.

I sustained permanent Pudendal Nerve damage due to being in the pushing stage for three hours, followed by two failed attempts to deliver my baby with the ventouse, which then necessitated a forceps delivery where an episiotomy tore further into the anal sphincter and caused deep vaginal tearing. I suffered a postpartum haemorrhage and lost 1.5L of blood.

Short-term, this delivery caused a great deal of trauma. My body went into shock, and even though I was repaired with emergency surgery following the delivery, I suffered PTSD and severe postnatal depression. I couldn't bond with my baby, and I felt shame and a failure that giving birth had left me with these life changing injuries.

If I tried to speak about how I was feeling to the health visitor or GP, or even to my family, I felt like I was being judged on my ability to parent. I was told my injury was 'all in my head' on numerous occasions and another health professional suggested that maybe it was because 'deep down I didn't want my baby'. The ignorance and judgement I faced only worked to compound my isolation and distress and I feared I was a bad mother because of the things I was being told every single day. I was given a mental health diagnosis you could throw at somebody rather than anybody acknowledging that the brutal injuries I had sustained had affected me profoundly as a woman and had greatly impacted my quality of life.

Long-term, I have lost normal bowel function. I have daily urgency and an impaired ability to empty the bowel fully. I cannot fully flatus and have a lot of scar tissue on the back wall of the vagina next to the bowel which causes ongoing problems. I suffered for several years with debilitating Pudendal Nerve pain before finally finding a specialist physiotherapist who was able to work the scar tissue internally to offer me some relief and improved function. Unsurprisingly, my sexual function has been greatly affected by the injuries I sustained. My muscles are weak and damaged, and I have very little sensation. I am constantly worried that my partner will leave me because of this. Nobody ever talks about the profound effect this has on a woman psychologically.

My ability to exercise and partake in the activities that I used to love has also been badly affected by my injuries. I was a keen runner prior to giving birth and I enjoyed going to the gym three or four times a week after work and keeping fit. Since the forceps delivery I have to be extremely careful about the activities I can safely do. I have slowly got back into walking and jogging on the treadmill, but it is only ever possible by using a pessary and tampon to hold everything in place. All spontaneity and joy have been taken from exercising. It now becomes something I do to try and keep the weight off, whilst hoping I'm not making prolapses and deteriorating bowel function more likely.

When I gave birth, I had no idea what had just happened to me, or what the years ahead would entail. Fighting to be heard and to get treatment. Countless visits to Gynaecologists, Urogynaecologists, Colorectal nurses, Women's Health Physiotherapists, and nerve specialists. Humiliating tests and countless internal examinations. Spending thousands of pounds on private appointments. Speaking to other women this has happened to. All of this has led me to understand what a forceps delivery did to my body, and it has led me to realise that I will never get back the quality of life I once took for granted."

I want to thank this lady so much for allowing me to include her traumatic story into my book. I am always banging on about forceps deliveries, how women walk into hospital to have their much-loved baby only to leave there damaged. Animals are treated better.

BITS & BUTTS

I shouldn't really call my Vagina and Anus 'Bits and Butt's, but I've had these words in my head for a couple of years as the title for my book.

Even some in the medical profession are too embarrassed or don't know the correct words for our anatomy. You might think I'm joking but I have heard many stories of names.

GPs and consultants have used... "Down There", "Down Below". Down where? Australia? It's about time we all were taught to say the proper names such as, Labia, Vulva, Clitoris, and Vagina. Tell your children as they are growing up the real names for parts of their body.

BRANCHES

My main area of pain with Pudendal Neuralgia is at the entrance of the vagina and inside the anus and rectum. Today as I am typing this, I am experiencing a high level of pain and I feel as though my anus is about to explode!! I'm sure if I told someone that they would burst out laughing, but it's far from funny. It feels as though I have a clock inside the anus and it's ticking away whilst vibrating. It's constant, it never stops. How am I supposed to live like this? It's because I have to. I have no other option.

What could possibly be worse than struggling with a painful condition and feeling too ashamed to discuss the problem due to the intimacy of its nature? Such is the case for over 1200 men and women suffering with Pudendal Neuralgia, a little-known condition that affects one of the most sensitive areas of the body. This area is innervated by the Pudendal Nerve which is located deep in the pelvis and follows a path that comes from the sacral area then later separates into three branches. One of these branches goes to the rectal area, one to the perineum and one to the penis or clitoris.

Because Pudendal Neuralgia is listed as being uncommon (a view with which I personally don't agree with), it is often misdiagnosed, leading some patients to undergo inappropriate and unnecessary surgery.

Early in the diagnosis process, it is crucially important to undergo an MRI of the lumbar-sacral and pelvic regions to determine that no tumours or cysts close to the nerve,

An accurate history is needed to assess whether there has been a trauma or an injury to the nerve from surgery.

C IS FOR CAUSES, CLITORIS AND CONSTIPATION

CAUSES

- Car Accident
- Coil Insertion
- Cycling
- Damage to Coccyx
- Endometriosis
- Excessive Exercise
- Fall
- Forceps Delivery
- Hysteroscopy
- Horse Riding
- Hysterectomy
- Mesh Implants
- Pelvic Surgery
- Prolapse
- Speculum Insertion
- Sexual Abuse
- Too much Sitting
- Vaginal Infections
- Weightlifting

CLITORIS

This part of a woman's anatomy is hardly spoken about. Only yesterday I read an article that described the clitoris and how its sole use is for women's pleasure only. It serves no other purpose. I hadn't realised that and I'm sure there are many people, including the medical profession, who didn't know that either.

Unfortunately, not every woman experiences pleasure. Far from it, judging by some of the accounts of the ladies in the Facebook group.

One lady went to see a top gynaecologist to enquire about freezing her eggs as she had a successful career at the time but wanted to make sure she could have children in the future.She told me that unbeknown to her, she suffered from vaginosis and the consultant hadn't realised either. She said:

"He was extremely rough when doing the examination but cared very little for me when I told him it hurt"

This has left her suffering from PGAD which is Persistent Genital Arousal Disorder. She also told me that the pain feels like barbed wire and needs to spend the majority of her life in bed. She has been prescribed Pregablin which only helps mildly. She has been damaged for life and is at the point of taking the consultant to court. I will go into how PGAD affects sufferers with another lady's story further on in the book.

CONSTIPATION

After talking about the rectum and anus, the subject of constipation follows on nicely.

It's unbearable at times to sit on the toilet and try not to push too hard in case you cause a prolapse, but you just want the stool to slide out with as little pain as possible. Many sufferers in the group just cry in agony. Personally, I often have a lie down on the bed on my side after the bowel movement, even if it's a reasonably soft stool. It just heightens the pain level. By being constipated, it only makes the problem far worse.

Symptoms of constipation include having a bowel movement fewer than 3 times in a week and finding it difficult to pass a stool.

A few members in the group are bed-bound so they don't have the choice of being active, but even if the sufferer can walk around the house and garden, it can get things moving.

SELF CARE

- Make changes to your diet to help make the stool softer and easier to pass, such as increasing fibre and eating fruits and vegetables.
- Add some wheat bran, oats or linseed to your diet.
- Drink plenty of fluids and avoid alcohol.
- Before I resort to medication, I eat a few prunes or drink prune juice on a regular basis.

If none of the dietary changes make any difference, invest in a Squatty Potty and please visit your local pharmacy or make an appointment to speak to your GP.

D IS FOR DEPRESSION AND DIGITAL RECTUM EXAMINATION

DEPRESSION

Many people who suffer from Pudendal Neuralgia also suffer with depression and other mental health difficulties. It is thought that this is because it is rarely diagnosed and often inappropriately treated. This causes significant impairment of quality of life. Some have issues with the basic principles of life such as being able to sit, stand, walk and be intimate. Often a sufferer can no longer take part in his or her hobbies. The simple tasks become near impossible due to pain and discomfort and it really does turn people's lives upside down.

Many sufferers in the Pudendal Neuralgia group are also taking antidepressants to help deal with the psychological impact. This is also very upsetting. The episodes and flare ups are often unpredictable and can last a day, week, month or a year.

Unfortunately, sufferers have attempted to or have taken their own lives due to Pudendal Neuralgia and its unmanageable symptoms.

There is help available if you ever feel this way, you can attend A&E, your GP or contact other mental health organisations. I will list telephone numbers at the back of this book.

"Tomorrow will be a better day"

DIGITAL RECTUM EXAMINATION

"A rectal examination is where a doctor or nurse uses their finger to check for any problems inside. It's usually very quick and you should not feel any pain"

This is a quote from the NHS website for DRE (Digital Rectum Examination) and yes it's probably very true for a large number of men and women but… not if you suffer from Pudendal Neuralgia.

If you suffer pain in the rectum as probably most of our Facebook members' group do, then this examination isn't going to be a walk in the park. I practically needed to be dragged off the ceiling when the finger was inserted until it was removed.

Just to give you an idea on what happens, I will go through what should be expected when the nurse starts the examination.

- He/She will ask you to lie down on your left side, with your knees lifted towards your chest. This is the easiest position for the nurse to examine the rectum.
- He/She will put on a pair of gloves and inspect the outside of your anus for any problems. Please don't feel embarrassed about this examination as they have seen it all before thousands of times.
- He/She will put some lubricating gel on a finger and gently slide it into the rectum. Try doing some deep breathing exercises whilst this is being done. It shouldn't be painful, but it may be a bit uncomfortable.
- He/She will sometimes ask you to squeeze around their finger so they can assess how well the muscles are working.
- If you are a male patient, the nurse may press on your prostate. This can make the patient feel the urge to wee but it shouldn't hurt.

Now, before the nurse gets to the first bullet point on this list, please please mention what condition you suffer from and warn them that it's going to cause you immense pain. Ask for more lubricant and local anaesthetic to be rubbed on the area to help numb it.

E

E IS FOR EXASPERATING. EMOTIONAL, EXHAUSTED AND SELF ESTEEM

EXASPERATING

Having this condition brings a rollercoaster of exasperating emotional feelings just trying to find the light at the end of the tunnel.

It's exhausting constantly pursuing answers as to why you have this condition and coming up time and time again against a brick wall.

It's exasperating trying to get the GP to take the issue seriously but when you mention what condition you suffer from, they don't know what to do apart from give you pills and nerve blocks which rarely help.

Living in constant pain is no joke. Ten years ago, this August and I'm still in the same amount of pain now as when I was the day after the operation. I realise that this doesn't do anything for the morale of new members who join the Facebook group, but there are some lucky ones out there somewhere but haven't told us what treatment has worked for them, or they have just left the group. I look forward to the day when I can do that.

The worst symptom of Pudendal Neuralgia is coping with the exhaustion of living with the pain as I've mentioned before. As women, we still have to cook, clean and do everyday chores plus there are some members who still hold down full and part time jobs. I take my hat off to them because I've no idea how they manage to do that.

There are a lot of men and women in the group who continue to go to work or work from home during this pandemic. No amount of rest or early nights takes that exhaustion away. At the moment, I am surviving on approximately three hours of sleep at night which isn't enough to keep the mind and body happy.

This condition also wrecks your self esteem. It doesn't help that you can't have full intercourse with the partner. You don't feel like going out, and you can't be bothered to put make-up on or keep the house spick and span. Why? Because our self-esteem has been shot to pieces just coping with the pain. We all know in the group that we can go onto Facebook day or night and there will be someone there that understands our pain. We are all going through this pain together.

F

F IS FOR FENTANYL PATCHES, FAMILY & FRIENDS

TRANSDERMAL FENTANYL PATCHES

These are safe and effective for relieving pain when used according to the instructions. However, the patches may cause harm to patients and others if the instructions are not followed.

Fentanyl is a strong pain killer which is often used in the management of severe pain. It is one of a group of drugs called Opioids.

The side effects of the patch are similar to other strong opioids; it causes constipation, feeling sick and drowsiness.

The patch lasts for 72 hours in total. After that a new patch is required. The only side effects of using these patches was that I got to the stage where I couldn't string a sentence together or think of words like kettle and microwave!! My mind was completely screwed up.

I decided that they weren't working for me anymore without increasing the dose, and I certainly didn't want to do that. I spoke to my GP and asked her if I could come off them. She agreed, but it was done very slowly as I experienced bad side effects which felt like the flu and constant twitching. In fact, I felt like a drug addict that hadn't had a fix. It was horrible, but I got there in the end.

Quite a few of our sufferers in the Facebook group use the patches and they get relief from the pain of Pudendal Neuralgia but after a while the dose that they started on which is normally 12mg will not be sufficient for the pain so the GP may need to increase the dosage. The patches release 12 micrograms to 100 micrograms of fentanyl every hour.

Remember to always read the leaflet that comes with the patches before you stick one on.

FAMILY AND FRIENDS

This is a difficult subject to write about because how our family and friends react to our medical condition varies between the group sufferers.

I have to say I have been extremely lucky with my extended family. Both my birth mum and dad passed away when I was still a teenager (both to cancer) and I had been going out with a lad who was in the army. His family lived just across the road from me and they took me in to live with them along with their own six children. That was in the early 1970s and the family have been with me every step of the way with this condition. They understand and sympathise when I'm in so much pain and always want to come and look after me. Every day mum asks me how I am, and I say, "I'm fine". Sometimes it's easier to say that than go into how I really feel. She never believes me though.

Friends have been a different story for me. I have lost a few along the way because I can no longer go out for meals, organise things with them or go to concerts and the cinema. I meet them in the street and they say, "You look well" Hmm. I'm polite and say thank you. They really have no idea how much pain I am in. Some even ask, "How is your back"?

We can't really expect everyone to understand Pudendal Neuralgia when there are GPs, Consultants, Midwives and Nurses who have never heard of it or understand it.

Therefore, I am actively trying to raise awareness for this condition which affects both men and women and there are currently over 1200 of us sufferers in a Facebook group and probably thousands who are still waiting to be diagnosed.

G IS FOR GANGLION IMPAR BLOCK & GRIEF

GANGLION IMPAR BLOCK

A Ganglion Impar block is an injection into a group of nerve cells located at the front of the sacrum/coccyx joint at the base of the back. The injections are used to treat pain around the tail bone in the lower back. It involves an injection of a local anaesthetic and steroid around the area.

The injection targets a group of nerves located in front of the tailbone (Coccyx) and is used to improve the pain caused by irritated or inflamed nerves in the lower back. Both the local anaesthetic and the steroid help to block the pain signals sent to the brain. The steroid may also reduce swelling and inflammation which may be irritating the nerves leading to some of the pain you are experiencing.

What are the benefits?

The injection may provide temporary pain relief; however it is important that you follow a regular gentle course ensuring that you stay within your comfort zone, otherwise the pain will return.

Are there any alternatives?

A Stellate Ganglion block is rarely a first line treatment, and you may consider trying other approaches such as simple painkillers, exercise, pelvic physiotherapy, Transcutaneous Nerve Stimulation (Tens Machine) or acupuncture.

- The injection fails to help with the pain.
- Worsening of the pain (Temporary or permanent).
- Bleeding or bruising to the injected area.
- Infections at the site of the injection.
- Fat dimple (a depression in the skin).
- Numbness (Temporary or permanently) in the injected area.
- Leg weakness.
- Allergic reaction.

What are the possible side effects for an X-Ray guided procedure?

Female patients who are or may be pregnant must inform a member of trained staff in advance.

One of our group warriors had this procedure done and unfortunately it failed to work. In her own words she describes how she feels about it.

"Thanks to this little 'pain relief' procedure in 2017, I have been left with constant sacral and lower back pain. Never again will I let a pain specialist armed with a needle anywhere near that delicate and highly malfunctioning area of my body''

GRIEF

These are the words of a fellow sufferer who led such an active life up until she was diagnosed with Pudendal Neuralgia:

"I grieve for the normal life that I was blessed with before this wretched condition changed every aspect of my life 13 years ago."

It's when members of the group mention how it affects them that I push on with my awareness campaign.

H IS FOR HELPLESS AND HOPE

HELPLESS

We in the Facebook group feel helpless with not having a straight-line cure for this condition.

We are pushed from pillar to post, from doctor to consultant, many of whom have never heard of our condition!! One of the many problems is that the pelvic area in Women's Health needs to be reviewed and that's not just for this condition.

Yes, some treatments and medication do work for a few people. The nerve blocks, the opioids, Pregablin, Gabapentin, Tramadol all certainly work for a time before an increase in the dose needs to be discussed with the consultant in Pain Management. Who wants to live with a mixed-up brain? I know I don't.

HOPE

To have hope is to want an outcome that makes one's life better in some way or other. It not only can help make a tough present situation more bearable, but also can eventually improve a life because envisioning a better future motivates a person to take steps to make it happen.

Every single day I feel the need to raise awareness. I have hope for not only a better outcome from doing this but also the hope that everyone in the Facebook group as well as my Instagram and Twitter followers notice that changes will be made within the NHS for all of us suffering from Pudendal Neuralgia and other pelvic conditions.

I IS FOR INTIMACY

Once upon a time, before they were diagnosed with Pudendal Neuralgia, the men and women in the Facebook group were leading healthy sexual relationships. Then a vast majority of their lives changed, and they can no longer bear to be even touched in the intimate area.

This has caused relationships between partners to be under pressure. The one suffering with Pudendal Neuralgia feels inadequate because they are unable to take part in penetrative sex. It's not only their partners they feel sorry for but for themselves too. Orgasms are a thing of the past because they cause flare-ups even if the sufferer instigates it.

There are, of course, other ways to keep a sexual relationship healthy by trying other ways to be intimate with kissing and cuddling. But sometimes that isn't enough. One of the members said that she misses the penetration so much that it makes her cry because it makes her feel totally inadequate as a partner and as a woman.

As a single woman, after my diagnosis I have always hoped that I might meet another man, so I purchased a set of dilators in order to keep the "runway open". I started off with the smallest size, I smothered it with lubricant and inserted it into my vagina. I was in control, and yes it made me cry with the pain, but I wasn't going to give up that easily. Sometimes it feels like I will never move off that first size, but I will keep trying.

My hopes of finding a partner who will be perfectly happy to just kiss, cuddle and be there for me are diminishing by the day. I mean, when do I bring this subject up? Not on the first date, nor can I leave it too long because one of these nights the man will think his luck is in and I will have to explain.

J IS FOR JEANS, JEGGINGS, JOGGING AND JOY

Most sufferers will tell you that wearing jeans and jeggings are most unbearable due to the thickness of the fabric around the Vulva and crotch area. This means it can be a bit nippy in the wintertime as I only wear skirts and dresses now.

The irritation from wearing jeans and trousers can possibly bring the sufferer out in a flare-up (more pain than the already heightened pain we all experience). It can also be the cause of thrush and a water infection. I care little for fashion now, it's all about being comfortable.

JOGGING

One member of the group mentioned that she is no longer able to do her favourite hobby, which is sad because we all know jogging is a way of keeping fit, but the constant friction and pain in the vagina and vulva is just too much.

JOY

"Happiness is not something ready made. It comes from our own actions." – Dalai Lama.

To be honest, there isn't a lot to be joyful about if you are a sufferer of Pudendal Neuralgia, but there are many things that do bring joy to our members' day.

I think partners bring a huge amount of joy to the sufferer by giving them support and understanding. It's difficult for anyone to understand how the sufferer is feeling because it is such a debilitating condition. Half the time a sufferer will say I'm fine, or don't worry. Children also figure in their joys.

Other joys that members have are from crafts, walking, the seaside, reading, watching wildlife, pets and music. Now mine are music, my support bubble neighbour and my Pudendal Neuralgia and Pelvic Pain Facebook group. It is these that keep my mind active and at times they make me laugh too. We are all ages, both men and women, who mostly all suffer the same pain. We are like our own family. We lift each other when someone is having a bad flare day and always there for one another.

K

K IS FOR KINDNESS, KNICKERS AND KNOWLEDGE

KINDNESS

We all know that our families and friends show their kindness to us, but what about the medical profession showing us some kindness?

One group member said that as soon as she mentioned Pudendal Neuralgia in an appointment, she was dismissed. Is this because there is such a lack of understanding, knowledge and awareness for this condition?

I was shown no kindness the day I went into the Women's Health Department as a day case. I knew by pushing a speculum into my vagina the consultant was going to hurt me. I had lived with the condition for long enough to know it wasn't all in my head.

If an Obstetric and Gynaecology consultant doesn't know about Pudendal Neuralgia, then my mission to raise more awareness, research and training will not be in vain.

KNICKERS

As soon as I came round after my Hysterectomy, Rectocele and Cystocele I suddenly realised all was not well with my pelvic area. Apart from putting too many stitches inside the vagina so that I could hardly get in the car, I felt so much pain, but the consultant had no idea why.

Almost immediately, I realised that it was more comfortable to go commando and not wear knickers, tights, jeans, trousers or jeggings. I started to wear just skirts and dresses. This is where I have a lot of sympathy for the male sufferers in the group. Several ladies have followed this idea and I will continue doing so for the rest of my lifetime.

KNOWLEDGE

"Knowledge in advance enables one to be prepared."
Once I was diagnosed, I started going onto Google which I know can be dangerous at times and see what I could find out about this condition from which the Professor said I was suffering. I would spend my days researching everywhere for information as to where I could go to get help, and what treatment was available. The NHS has a little information on their website about the causes and treatment, but not enough.

I try to encourage new members to the group to read up about Pudendal Neuralgia, but the Facebook group is very knowledgeable and friendly and will always help when they can. However, when sufferers have an appointment due, it is always a good idea to prepare in advance. I would advise everyone to write down the questions they wish to ask because once they get into that room with the consultant everything will go out of their head, and they will leave the appointment without an answer to their questions.

Forewarned is Forearmed.

L IS FOR LAXATIVES, LICODAINE AND LOVED

LAXATIVES

I am writing this from experience, and I know for sure that other members in the group have problems opening their bowels due to constipation and pain.

I remember after waking up from the operation, I felt 'full'. I told the nurse, but she told me that was a problem with operations as they pump you up full of air. I just felt as though I needed a good fart (sorry if that was too much information). She explained that a tube is inserted through the incision and carbon dioxide gas is pumped through the tube to inflate the abdomen. By doing that it allows the surgeon to see the organs more clearly and give them more room to work.

The nurse told me that I would have to remain in bed until the next day. They gave me a bed bath and I would be able to go to the toilet once all the packing was taken out of the vagina. What she didn't warn me about was opening my bowels.

The nurses on duty encouraged the patients to walk around to 'get things moving' and drink plenty of water. Well, I paced the floor of the Women's Health ward, I was bending over my bed to ease the pain, but nothing was moving. They then gave me Lactulose solution to drink which still didn't help and finally resorted to using a suppository. That did get things moving eventually but I must admit it was painful.

There are different types of laxatives and there is a list of them on the NHS website that is recommended. They come in various forms such as:

- Tablets or capsules to swallow.
- Sachets of powder to mix with water then drink.
- A capsule to be placed inside the rectum (bottom) where it will dissolve. This is a suppository.
- Liquids or gels that you place directly into the rectum.

If you are in any doubt, speak to your pharmacist for advice.

LIDOCAINE

This is a local anaesthetic similar to the one dentist uses to numb the gums. It works by calming nerves which are firing off when they shouldn't be. If successful, the infusion will reduce the pain.

The Lidocaine infusion is administered directly into the vein in the hand which then flows into the bloodstream. I had this procedure at Birmingham City Hospital as a day case. On arrival I was weighed, and this was recorded on their records as this intravenous drug is measured by weight. The infusion was continuous, and it took between 45-60 minutes to complete. I took an IPad with me to keep my mind occupied. I was observed the whole time I was there with nurses constantly checking the monitors and making sure I felt comfortable.

After the infusion had finished, I was still observed for a set period of time (up to an hour) then discharged.

Benefits of Lidocaine

Patients can experience pain relief and function better after the procedure. The drug can be effective in certain types of nerve related pain, for example Fibromyalgia. If there is no improvement after the first and second infusion, then a third is not required.

Risks

- Lidocaine is a relatively safe drug when given at a carefully calculated dose. (This is measured to whatever your weight is.)
- Unfortunately, no medical procedure is risk free.

Short term side effects

- Numbness around the mouth.
- Dizziness.
- Slurring of speech.
- Hallucinations
- Muscle twitching.

These effects will go away once the infusion is reduced or stopped.

I had three infusions in total. The first worked for 3 days. I felt incredible but then the pain of Pudendal Neuralgia returned. The second infusion gave me no relief at all, but they were willing to try once more, sadly that also failed.

LOVED

I just want to share with you a message that was received in the Facebook group recently. The lady is a fellow sufferer with Pudendal Neuralgia, and this is what she told us.

"Love… strange one, but I have never felt more loved than I do now dealing with this condition.

The love of my humans and my pets pull me through this pain every day. Love is worth fighting it for. It's precious even when it feels like a slog and nothing but pain filled.

I had to reach out for help, I had to show my vulnerability, and I had to rely on my family and friends, and I can honestly say we are all closer for it. Sadly, it gives and it takes but it has given me more love and more appreciation of those I love and the life I still get to live, even with so much pain and worry."

M

M IS FOR MESH MORPHINE SULPHATE, MARIJUANA AND MUSCLES

These are both powerful drugs and here are three sufferers' experiences with using them.

MORPHINE SULPHATE

"When in the thick of a PN nerve flare pain throughout the genital area, buttocks, lower back and often down the backs of my thighs, this is the only drug that helps me.

If I take this (Oromorph) and lie on my side with a pillow between my legs, I can just about get through it. I have been told by pain specialists on numerous occasions that opiates do not work for PN pain… to that I say… let's see how you cope when you can't even communicate and just want to cry because you are experiencing too high a level of pain… you would be reaching for it.

I probably experience this a couple of times a month and sometimes they can go on for a couple of days. It is difficult to work when I feel like this."

MARIJUANA

"After being diagnosed with Pudendal Neuralgia, I was put on prescription medication, but I didn't really cope with the side effects. They also do not take away all my pain so my doctor recommended Marijuana and even told me where to buy it. After being on a journey of trying tinctures to edibles to smoking, I found vaping worked the best for me. It is one of the quickest deliveries to help lower pain. It does raise the pain a bit until it kicks in, which is normally around 30 seconds.

The side effects are peace of mind and pain management. I didn't find them addictive as opposed to prescription medication."

This sufferer does not live in the United Kingdom.

MESH

I have included one lady's story of how she was damaged with Mesh:

"My journey with mesh started on 12th May 2003 when I had a Transvaginal mesh tape fitted for stress urinary incontinence.

In January 2011, I attended a few appointments with my GP due to urine infections which also had blood in them. I was given antibiotics which failed to clear the infection up. I was in so much pain and it was very difficult to sit down so my next plan was to ask to be referred to a urologist.

Following my appointment, I had surgery to remove the eroded tape from my Urethra and reconstructive surgery using my Labia Majora in a martius pedical flap for the repair. Sadly, I sustained Pudendal Nerve damage which caused burning in the Vagina and Urethra together with feelings of electric shocks. Sitting continued to be very difficult as was standing for long periods. The pain is better when laying flat on the bed.

On 19th July 2011 I saw a pain specialist to have Lignocaine injection and a steroid into my right and left pubic area to try and alleviate the nerve pain. Unfortunately, this failed to work.

I was admitted to hospital on 7th February 2012 where Miss Sohier Elneil removed the remaining tape. One piece had adhered to the neck on the bladder and needed to be cut out along with a Cystoscopy, (a procedure that lets the healthcare provider view the urinary tract). A Colposuspension (An operation to treat stress incontinence in women) and removal of fibrous tissue and adhesions from the pelvic area. But not all the mesh was removed and I still have 20cms in situ.

I had confirmation on the follow up appointment that I had sustained Pudendal Nerve damage. My GP prescribed Nortriptyline but they didn't decrease the pain. I also take Tramadol.

I continue to suffer to the present day with the Pudendal Nerve pain and have learnt to live with it. Sadly it is a debilitating disease with constant chronic pain and unfortunately will always be with me."

MUSCLES

It never dawned on me that muscles were such an important part of being diagnosed with PN. I was wide awake at 2.30am one morning as I was lying in the bath trying to calm down my knees. They were burning, literally the same symptoms as the Pudendal Neuralgia pain, but in the knees. They were extremely hot to touch.

What the hell was happening to me now, I wondered? To make things worse I hadn't been sleeping for a few weeks, maybe just two or three hours a night. Since being diagnosed with type 2 diabetes last October, I have tried hard at being careful about what I ate plus getting more exercise in. Anyone who suffers from PN will tell you that it isn't easy to walk, stride out or go very far.

I had to be truthful here because I had been really pushing myself to stop being prescribed Metformin for diabetics, but any walking tightens the muscles and puts knots into all of them. Maybe I wasn't walking properly but it's the best I could do having this condition.

REMEDY

I had a bath in the early hours of the morning with a couple of handfuls of Epsom Salts thrown in to try and release the tense muscles and stop the burning pain in my knees. It certainly helped, but I didn't function well the next day. It was a self care day: jobs can wait, lying on the sofa listening to music and feeling sorry for myself. But… it didn't last. I vowed to be back on form the next day and thinking up words that connect with Pudendal Neuralgia for letter N.

N IS FOR NEGATIVITY, NERVE BLOCKS, NEURALGIA AND NUMBNESS

NEGATIVITY

It's really hard not to be negative living with this condition but I can honestly say I have never felt alone. We are a group of men and women 1200+ (now, who said it was rare?). We have lost a few members because they could no longer cope with the pain, or they were simply not believed by their doctors and consultants. This must not happen again. If you are or know someone who is a sufferer of Pudendal Neuralgia, then please reach out to the group on Facebook. There is always someone around to talk to. Please remember this.

NERVE BLOCKS

Pudendal Nerve Block

This is a common regional technique to provide anaesthesia to relieve pain. It can be given through the vaginal wall and into the Pudendal Nerve in the pelvis. The injection called a Pudendal Block will cause the area between the Vagina and the Anus to become numb.

Bilateral Nerve Block – CT Guided

This procedure involves an injection into the joint or nerve. It consists of a mixture of local anaesthesia and a cortiscosteroid and should take between 30 to 40 minutes to complete and soon after the pain is expected to reduce for up to six months but unfortunately everybody reacts differently.

NEURALGIA

There are several types of Neuralgia. It means that there is an irritation, or the nerves are compressed or damaged. All of which are extremely painful. The Pudendal nerve is the main and largest nerve in the pelvis, in other words it has an important job to do.

One particular time I needed to call 111, the lady responder who answered my call told me she knew about Pudendal Neuralgia, and was even more surprised when she said she knew what the symptoms were. She then proceeded to tell me that the pain goes all the way down the face and in the head too. It was only then that I put her right. She told me at the end of the conversation that she was pleased to have learnt about the condition. Looking back, I was happy that I had been able to raise awareness of my condition even in my moment of distress.

NUMBNESS

As soon as a sufferer of Pudendal Neuralgia joins the group, the first things they want to know about are the symptoms. The new member will say that there is a feeling of numbness in the anus and rectum, and it throbs non-stop.

It's a horrible feeling because when you go to sit down, you can't feel the actual cushion, but in between the anus and rectum it's screaming "Get up, don't sit down". The buttocks and especially the Gluteus Maximus muscles become extremely tight which brings on more pain.

I know a few members of our group are avid cyclists, so I have been hammering on the doors of the British Cycling Association asking if any of their members suffer with the condition. I still haven't had a reply and I don't really expect one.

There is one positive point for the cyclists now that saddles are being made with a gap in the middle. I'm sure this will help a few to keep active in the sport they enjoy so much.

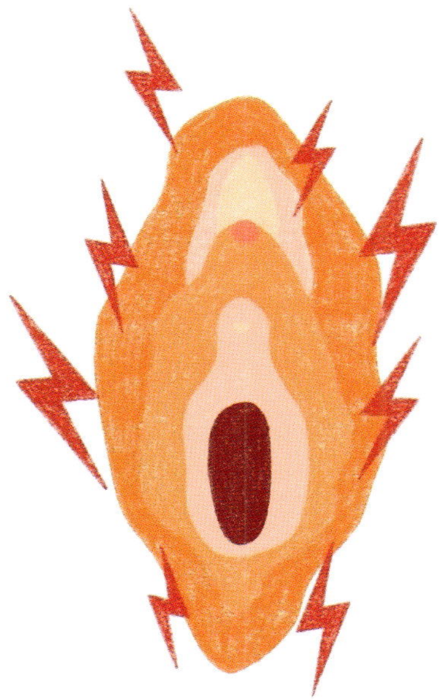

O IS FOR OATH AND ORGASM

"I will use my power to help the sick to the best of my ability and judgement."

"I will abstain from harming and wronging any man by it."

This is from the original version of the Hippocratic Oath, which orders doctors to do their best in their job and not use their skill or knowledge to harm or kill their patients.

While some medical schools ask their graduates to abide by the Hippocratic Oath, others use a different pledge or none at all. In fact, although "first, do no harm" is attributed to the ancient Greek physician Hippocrates, it isn't a part of the Hippocratic Oath at all.

The modern Hippocratic Oath is as follows:

"I swear to fulfil, to the best of my ability and judgement, this covenant: I will respect the hard-won scientific gains of those physicians in whose steps I walk, and gladly share such knowledge as is mine and with those who are to follow."

ORGASM

As I have mentioned before, Pudendal Neuralgia affects male or female, straight, gay or lesbian. It isn't particular about whom it chooses to target. There are numerous reasons why a sufferer is diagnosed with the condition. It may be that they enjoy cycling, had a smear test, an operation, a fall or gave birth where forceps were used or extreme exercise, to name but a few. One symptom that a lot of the members in the group tell me about is that they have been unable to have an orgasm.

If they can achieve one, they suffer for days or weeks afterwards with the Pudendal Nerve doing the Irish Riverdance inside their pelvis. So, they don't think it's worth the effort, and give up. It's terribly sad for both the sufferer and the partner. I am hearing that the partners have always been supportive to the sufferer and wouldn't want to hurt them at all. They find other ways to show their love. It isn't all about penetration.

P IS FOR PELVIC PHYSIOTHERAPY, PENIS, PERSISTENT GENITAL AROUSAL DISORDER (MEN AND WOMEN) AND POSITIVITY

PELVIC PHYSIOTHERAPY

This is something I had started having before Covid-19 blew in and all appointments were then cancelled. I did purchase a blow-up exercise ball that pregnant women relax on before giving birth. I found using it very therapeutic, rolling myself side to side and having the ball up against the wall and putting my back up against it and rolling side to side.

Seeing a Pelvic Physiotherapist is a must after having work done in the pelvic area.

The Pelvic Physiotherapist I saw was warm and friendly, and was incredibly patient. She explained to me that she wouldn't do an internal examination on me for a long time. Her theory was to work from the outside inwards. She told me at the very start she would not even think of giving me an internal until I was able to do the exercises and the muscles were slightly looser. The exercises were good and helped to stretch the pelvic area by holding on to one side of the bed and the pelvis hanging over the other side.

PENIS

I was very proud of one member in the group who was able to open up to me about the struggle he has with his Pudendal Neuralgia.

Firstly, he told me he feels very embarrassed about talking about it to his doctor who has been no help to him whatsoever. He has been prescribed Gabapentin which is supposed to help reduce nerve pain, but he has noticed very little reduction in the pain. We chatted for a while and gradually he was able to tell me that he suffered with dreadful nerve pain in both the penis and scrotum. This pain increases when erect and the sensitivity is unbearable. That sensitivity also brings with it an unwanted arousal. This understandably is embarrassing for the sufferer. He also told me that he had been circumcised as an adult and wondered whether this had anything to do with the diagnosis of Pudendal Neuralgia. I was unable to answer that question, but I didn't think it was the cause.

I'm so pleased that he was able to tell me his story, additionally I have learnt a great deal from him when he described what the pain is like for a man.

Thank you fellow sufferer and I hope this will help you talk to your doctor about this condition in the future.

PERSISTENT GENITAL AROUSAL DISORDER (PGAD) IN A FEMALE

What is PGAD?

Persistent Genital Arousal Disorder (PGAD is also called Persistent Sexual Arousal Syndrome (PSAS).

Women who suffer with this condition become sexually aroused without any sexual activity or stimulation even when there is no reason to feel sexually aroused. The sufferer may feel all the symptoms of sexual arousal such as a swelling of the vagina, and the stimulation of the clitoris but with PGAD the sufferer may feel like they are having orgasms constantly. In some cases, PGAD can last for hours, days or weeks at a time. It can be disruptive to their daily life.

PGAD is most commonly reported in women but it has also been reported in men (see below).

I have spoken to both women and men who suffer from this awful condition. Here is the story from a female member:

"I suffered from Pudendal Neuralgia for several years, but in a milder form. I didn't realise the numbness I had from sitting so long on my computer was anything but a nuisance. It was only when I had the rectal pain followed by a prolapse repair that it was far more than a nuisance. It was agonising pain, not just rectal but all around the genital/saddle area. Some days I also had a throbbing sensation from front to back and a stabbing feeling which made me jump. I could understand the rectal pain but not the perineal and the clitoral throbbing, agonising pain".

Another female story:

"After self-diagnosing as Pudendal Neuralgia, I had to see a colorectal surgeon and I told him my suspicions. He examined me and confirmed by pressing vaginally on the nerve, that it was Pudendal Neuralgia and he referred me to a gynaecologist, who in turn referred me to a dermatologist, who referred me then onto the pain clinic.

By this time, approximately two years after surgery, I had done tons of research and knew that the throbbing bussing clitoral pain was something called PGAD.

I went to bed at night in absolute misery, with a cushion between my knees watching TV to try and distract myself until four or five in the morning with constant waves of pulsating pain and sharp stabs that I couldn't stop. If I could get to sleep, I was fine but the severe pain made it impossible. This lasted for months then the PGAD more or less went for a while, though the perennial pain persists. Occasionally, if I do too much, it can be light housework or bend awkwardly, the PGAD returns. The surgery was five years ago but I will never be free of pain. My life is so restricted by it, it never goes, it's like a living hell.

PGAD on top of Pudendal Neuralgia makes it unbearable. I must live my life somehow with this and like others with a lot of other medical problems.

I thought having surgery for damaged discs was painful, but PN and PGAD are worse. There is no light at the end. There is no joy in life.

PERSISTENT GENITAL AROUSAL DISORDER IN A MALE

Now I want to give you an update on how our male member of the group is feeling after finding the courage to open up to me about his PGAD. He was dreadfully embarrassed that he had this condition and just didn't know what to do.

He told me that since our chat, he had been able to find a supportive, sympathetic doctor who listened to him and didn't just brush him off with saying "It's all in the head"

He has been in constant contact with his new GP, and she had prescribed Tramadol and was going to organise a Spinal CT scan.

There are so many sufferers who go to their GP in the hope of getting help, but they are constantly not being believed. This condition also impacts on the sufferers' mental health, and this should be taken into consideration by the medical profession.

If there are any males who feel too embarrassed to talk to their GP when they think they have the condition Persistent Genital Arousal Disorder, please pluck up the courage and talk to a friend about it. If necessary, change your GP to one that understands your embarrassment.

I have found that a few of the males in the group have found it easier talking to a female.

POSITIVITY

This is something we all need.

Positivity - that tomorrow is another day, and it will be better.

Positivity - that you can get through this.

It is difficult to remain positive when you are suffering from Pudendal Neuralgia. The constant pain can wear a sufferer down so much that they don't know how they are going to keep living with flares.

I always think to myself when I am having a really bad day, "Tomorrow will be a better day". We are all allowed off days when it's all too much and that is where the group plays its role. A sufferer never feels alone, they can rant and rave and there is always someone there to read the status and reassure the sufferer that their pain is quite normal and part of the symptoms.

I am proud of the group and have made everlasting friendships from talking about the condition.

Q IS FOR QUESTIONS AND QUALITY OF LIFE

The most important thing I have learnt from going from GPs to hospital then being transferred to another hospital or department is to keep all the paperwork I have received about my condition from the various medical professionals because it's so easy to look back if the consultant said one thing and I have a letter saying something else. I know one ends up with a whole file, but someone might thank me for this piece of advice one day.

QUESTIONS

Asking questions are more important than the first bit of advice above. Whenever you have an appointment with a medical professional, before you go, make a list of the questions you need answering as it's so easy to go into the consultants room and the mind goes blank or you are halfway home and think 'Oh I should have asked him that' or 'I forgot to ask him…'

So, before any appointment, be prepared, write your list of questions down, and never leave the consultant's room without an answer as to what they intend to do at the next appointment.

QUALITY OF LIFE

A sufferer will tell you that it is impossible to lead a "normal" lifestyle. It is all about never ending pain. Pain in the vulva, vagina, legs, penis, and scrotum all day everyday.

I have been talking to one member of the group in the last few days and I was told that they are anxious to have a life back without the constant pain.

There are mums and dads out there who want to be running around after their children, joining in playing football and taking them to their activities instead of being stuck at home in pain.

They do not feel they can do anything. All they can do is put a fake smile on their face and pretend everything is fine.

There is no quality of life for some members; a few sufferers I know of spend most of their time in bed because that's where it is most comfortable. It is too painful for them to be up and about doing 'stuff'.

I'm laying here in bed on my side typing this book on my phone. In the good old days, I would have been up, showered, dressed and with make-up on ready to enjoy the day.

For a member of the Pudendal Neuralgia Facebook group, it's a life of GP appointments, meetings with consultants about what they want to try next on the medication list. More often than not, the health professionals haven't got a clue what to do with us. Have you any idea why? It's because there are not many GPs, Consultants and Nurses know about Pudendal Neuralgia.

The group's numbers are increasing daily, and we are all searching for that one treatment that will cure us of the constant pain.

I am going to keep raising awareness for every member until such time that women's health is improved by more training and research.

It's not too much to ask someone to sit down or put underwear on but for us sufferers it's an impossible task. I want to make sure that I do everything in my power to improve the quality of not only my life, but the lives of the other members and for sufferers in the future.

R IS FOR REACH OUT, RELAD AND RESENTMENT

REACH OUT

If a sufferer has ever said that they cope well with the constant pain, I would be suspicious.

What with day-to-day life, with young children to support and look after, then it's just not possible. There comes a time when one morning they wake up after a fitful night's sleep and the Pudendal Nerve is stamping away in the anus and Vagina/Penis area, they don't know how they are going to get through the day let alone the week, month or year. I know I've been there!

But what I want to get across to everyone suffering any kind of pain is that there is always someone to talk to, whether it be within the Facebook group or by phoning the NHS mental health helpline.

I can hear you all saying talking doesn't help, but it does. I'm not ashamed to say I have been through several counselling sessions in the past 18 years for various reasons. Some have been face to face and some by telephone. Once that counsellor makes contact by saying hello, you can rant and rave and tell them everything you want to, and they will take it all in and listen. It helps, honestly it does.

Remember there is always someone that will listen:

www.nhs.uk/conditions/stress-anxiety-depression/mental-health-helplines/

RELAX

This condition never gives the sufferer a minute's peace. It is constant except when in a bad flare-up, that is when a sufferer must just admit defeat and rest up.

Early on, when I was first diagnosed with Pudendal Neuralgia, I attended a Pain Management course which basically taught me all there was to know about coping. I told them I started making various crafts and listening to music to take my mind away from the pain. At first, I thought this isn't going to be for me because I know how to live my life, however as the weekly course came to a finish, I realised how much I had actually learnt about relaxation.

The one big thing I learnt was to pace myself. I never did that before. It was all or nothing for me. If the grass needed cutting I would do it all but following the course I spread tasks like this over two days. I had to learn to pace myself. I can no longer cope with the mowing, but I still put that method to other tasks that need doing.

The Pain Management Course was very helpful to me, and I took away a great deal of information that I could work on at home.

RESENTMENT

I have spoken to many sufferers over the years and those who have been 'damaged' during an operation hold a lot of resentment.

They wish they had never had that surgery which was to change their life forever. It certainly changed mine. The retirement I thought I was going to enjoy is now out the window. Now I just live one day at a time like everyone else in the Facebook group.

Being angry is destroying to a person both mentally and physically and I didn't want anger in my life anymore.

S IS FOR SEXUAL ABUSE/SEXUAL ASSAULT AND SYMPTOMS

This is a subject that must be approached with care. Only those that have suffered this kind of abuse will truly know what it's like. I have taken these details from the NHS site.

If there are any men and women who have suffered this kind of abuse, please remember that you can reach out to the facebook group who will be very understanding. It could be that you have been diagnosed with Pudendal Neuralgia following an abuse/assault so the group is there to answer any questions that you may have regarding the condition.

If you have been sexually assaulted, there are services that can help. You don't have to report the assault to the police if you don't want to and you may need to think about what has happened. However, consider getting medical help as soon as possible, as you may be at risk of pregnancy or sexually transmitted infections (STIs). If you want the crime to be investigated, the sooner a forensic medical examination takes place the better.

Try not to wash or change your clothes immediately after an assault. This may destroy forensic evidence that could be important if you decide to report the assault.

Where you go for help will depend on what's available in your area and what you want to do. For specialist medical attention and sexual violence support, whether you decide to have a medical examination or not, your first point of call is a Sexual Assault Referral Centre (SARC).

- A doctor or practice nurse at your GP surgery.
- A voluntary organisation, such as Women's Aid, Victim Support, The Survivors Trust or Survivors UK (for male victims of sexual assault)
- The 24-hour freephone National Domestic Abuse Helpline run by Refuge on 0808 2000 247
- The Rape Crisis National freephone helpline on 0808 802 9999 (12.00 noon until 14.30 and 19.00pm until 21.30pm every day of the year)
- A hospital accident and emergency (A&E) department.
- A Genitourinary Medicine (GUM) or Sexual Health Clinic
- A contraceptive Clinic
- A Young People's Service.

If there are any men and women who have suffered this kind of abuse, please remember that you can reach out to the Facebook group who will be very understanding. It could be that you have been diagnosed with Pudendal Neuralgia following an abuse/assault, so the group is there to answer any questions that you may have regarding the condition.

SYMPTOMS

- Stabbing and burning pain in the rectum, anus and vagina.
- Throbbing
- Scratching in the Vagina
- Feeling of a 6 inch ruler in the anus sideways
- A knife in the anus and rectum.
- Rectal spasms
- Stinging
- Hypersensitivity in Gluteus Maximus

SUICIDE, WRITTEN BY A MEMBER OF THE PUDENDAL NEURALGIA FACEBOOK GROUP

"When I first felt the pain of Pudendal Neuralgia on my right side, I thought "I can't go on with this", not more illness as I had already spent countless years in pain from fibromyalgia and widespread myofascial syndrome, and I just felt trapped in my body.

I had Pudendal Neuralgia for about 3-4 years before I was diagnosed. I found there was no real help for it. I felt as helpless as I did when I was diagnosed with my other conditions.

I thought about it logically and decided I would stop the pain once and for all. There was only one option.

However, I was unsuccessful. I had tried with pills at the age of 35 when I first got the other conditions, however I am still here. Neither attempt worked.

I can't lie and say I am not disappointed because I don't have much quality of life, and I cannot say I would never try again; however I am still here and so life is still worth living.

I think one of the reasons is because it would be unfair to the one person who cares about me the most, my partner of 21 years."

T IS FOR TENSION, TRAMADOL, TRAVELLING AND TRAINING

TENSION

Where do you hold all your tension in your body? Mine is in my Gluteus Maximus muscles (my buttocks).

Every month I visit my chiropractor, I need acupuncture and the massage machine to loosen the muscles which I have been holding tight since the month before. He knows exactly where to place his finger for the trigger pointing therapy. He presses onto the muscle and waits for me to scream. I try to do this down the sides of the legs because I had to admit to the chiropractor only last month that I had been sitting with my legs up beside me in order to get comfortable. As you can imagine I had a good telling off, but unless you suffer from Vulvodynia or Pudendal Neuralgia no-one has any idea what it's like to sit down. Seeing a chiropractor has been the best decision I made with regards to treatments to help with the pain.

Another place I hold the tension from pain is in my teeth and jaw. I've read several articles where they say that if you unclench the jaw, the pain can become less. I've yet to prove this.

TRAMADOL

This is a strong painkiller. It is used to treat moderate to severe pain, for example after an operation or a serious injury. It's also used to treat long-term pain when weaker painkillers no longer work.

It is available only on prescription and comes in tablet, capsule and liquid drops that can be swallowed. It can also be given by injection, but this is usually only done whilst in hospital.

Common Side Effects:

- Headaches
- Feeling sleepy, tired, dizzy or "spaced out"
- Feeling or being sick (Vomiting)
- Constipation
- Dry Mouth
- Sweating
- Low Energy

Serious Side Effects

These are rare but call the doctor straight away if you experience any of the following:

- Difficulty with breathing or short shallow breathing
- Dizziness, fatigue and experiencing low energy levels – these can be a sign of low blood pressure
- Hallucinations (seeing or hearing things that are not there)
- Confusion
- Very sleepy
- Trouble urinating or you can't urinate at all
- Seizures (fits). If you experience a fit, go to A&E immediately.

This information has been taken from the NHS Website.

TRAVELLING

Whenever the word travel is mentioned to me my whole body freezes. Don't get me wrong, I used to love to travel all over the world with holidays in Australia, New Zealand and Kenya but that was all pre-Pudendal Neuralgia.

Now I cringe at the thought of just going into town on the bus to get food shopping. Why? Because having to travel is now so painful. With hard seats and encountering so many potholes, by the time I reach the supermarket, I am completely done in and so sore. I have to really think to myself when I'm in a flare, can the shopping wait for another day, or maybe I should order online.

The worst experience I had travelling was to an appointment at the University College London Hospital by train. It wasn't a pleasant journey into London with a packed out train which meant I had to stand, plus, my feet were trodden on every 5 minutes. I was glad to reach my destination.

The return journey was even worse. I got to the station to find that my train was cancelled which meant standing for over an hour in the main reception area. This did nothing for the Pudendal Neuralgia pain and eventually I got home just after midnight. It took a few days to get over that journey.

There are members in the group who are unable to travel at all, while some lie down on the back seat of a car to attend hospital appointments.

I always go prepared like the good girl guide that I was, by taking a cushion which I have made, a drink and something to occupy my mind on journeys.

TRAINING

I expect you all know by now why I'm working my way through the alphabet. It's simply to raise awareness for the condition and to help others that are suffering pelvic pain but as yet have not been confirmed to be Pudendal Neuralgia.

There are GPs, Consultants, Midwives and Nurses who know very little about Pudendal Neuralgia and I want to help change that. So many members of the group go to appointments and the health professionals say they have never heard of it.

U IS FOR UNDIAGNOSED, UNDERSANDING AND UNEMPLOYED

UNDIAGNOSED

It's not easy for a sufferer to get a diagnosis for Pudendal Neuralgia. After numerous trips to the GPs, all I was given were Antidepressants.

They weren't going to cure the horrendous stabbing pain that I had in the vagina and the constant throbbing pain in the anus/rectum. I know some anti depressants are muscle relaxers as well, but that part didn't work for me.

The GP did various tests to make sure I didn't have an STI (Sexually Transmitted Infection) but they all came back negative. I had noticed this terrible pain in the pelvis straight after I came round from the hysterectomy, front and back prolapse so it had to be something that happened during that operation.

After 8 months of going backwards and forwards to the GP, I came to the end of my tether and asked her for a second opinion.

I was referred to the City Hospital in Birmingham and had a further operation 2 weeks later. The Bartholin glands were infected, I had been stitched up incorrectly and I had a vestibulectomy on the vulva region. This is a procedure where surgery is performed to remove the affected skin. I was given Botox in my Gluteus Maximus muscles to see if that calmed down the throbbing in the anus/rectum.

It was at that hospital I was finally diagnosed with Pudendal Neuralgia by the Vulval Specialist who was also a professor.

I was pleased as punch to have a name for my debilitating pain which consumed me every minute of the day, but I still did not have a cure.

UNDERSTANDING

This is a massive problem for Pudendal Neuralgia sufferers. We are pushed from pillar to post with the condition being called CRPS (Complex Regional Pain Syndrome) which is basically an umbrella term for a lot of conditions that they don't know about or, "It's all psychological". Neither of which I accepted. My GP hadn't heard of Pudendal Neuralgia and neither had the Service Manager of the Obs and Gynae Department at my local hospital.

It was then I thought that if my GP and hospital didn't know about the condition, I had to start raising an awareness campaign. So here we are, I will keep going until the pelvic region of a man and woman are studied in depth at medical college and our present day medical profession are given more training.

UNEMPLOYED

This story has been written by a group member.

"This is me now, from 30th January 2021. It had been on the cards for some time, but it doesn't make it any easier to take. I loved my job but since last April, I have been off on long term sick because of Pudendal Neuralgia. I did grit my teeth and work through the pain at first, but it got worse and I could no longer manage the sitting. My cashier job involved a lot of this, and the Occupational Health Department was unable to make any adaptations to my working area. I was also not fit enough to move to a different department.

Over the past year, while off work they had treated me fairly. I had no complaints about that, and they were reluctant to see me go, as was I to leave.

I had gone from seeing hundreds of people on any one day to seeing no one, speaking to no one, feeling useless with no self worth or purpose.

Pudendal Neuralgia has robbed me of my independence, the ability to earn and to contribute. I miss my work colleagues, having conversations with lovely customers. I wasn't ready to retire; I am only 57 years of age.

Pudendal Neuralgia is a thief, stealing so much from sufferers. It is relentless.
I hope by increasing awareness and education within the medical profession, in time better treatment and a better quality of life for us all.

V

V IS FOR VOICE AND VULVODYNIA

VOICE

When I was growing up, I was a meek and mild girl born into a Catholic family who went to church every Sunday and every Feast Day.

I never answered anyone back in authority and I believed what they told me was true. I was also educated in a convent. So, what changed me? What gave me a voice that I'd never used before? It was the medical profession.

Yes, I agree that millions of people have "routine" operations every day and leave the hospital much better, but I didn't. Neither did the other 1200 plus other men and women in my Facebook group either.

I wasn't going to stand back and be told all my pain was psychological! That was the day I found my voice. As soon as I arrived home from having a biopsy taken from the inside of my vagina I started typing out my complaint of how I had been treated as a patient. From that day onwards I found my voice.

I have continued to use it in the hope that someone somewhere would read my A-Z of Pudendal Neuralgia on Facebook, Twitter or Instagram and understand how this condition can change a person's life dramatically.

VULVODYNIA

As well as being a sufferer of Pudendal Neuralgia, I also have Vulvodynia. For anyone who doesn't know anything about this condition, I will try and give you a brief outline of some of its causes and treatments.

When it was suggested, I include Vulvodynia I thought I would look to see how many awareness pages there were for the condition on Instagram. I was amazed at the number. It's another condition that is classed as taboo. Women are too embarrassed to talk about it, even to their own GP. So hopefully by me including Vulvodynia, it will help some sufferers feel they are not alone, this also includes the women who have yet to be diagnosed.

It wasn't until I was diagnosed with Pudendal Neuralgia by the Professor in Birmingham that he examined the vulva area and told me I also suffered from Vulvodynia (dynia meaning pain). It's extremely uncomfortable to say the least. It feels like the vulva is being torn to shreds, on fire and making sitting down uncomfortable.

I don't use any soap, shower gel and all I have when I do have a bath is just pain water. We don't need soaps smelling of roses and lilies on our vulvas and please never use bath bombs because they can do untold damage to that area. Just plain water is enough because our vulva is self-cleaning.

Here are some of the causes of this condition:

- Previous surgery
- Childbirth
- History of severe vaginal thrush
- Trapped nerves
- Smelly soaps and bubble bath
- Drop in Oestrogen levels
- Lichen Sclerosis and Lichen Planus
- Sexual abuse
- Sometimes no reason is known

Home Remedies:

- Wash and pat dry the area gently
- Don't wear tight jeans or leggings
- Underwear can aggravate the soreness
- Don't use smelling products
- Avoid horse riding and cycling
- Apply a cold compress to the area

Treatments:

- Medication
- Steroids
- Tricyclic antidepressants or anticonvulsants
- Pelvic Floor Therapy
- Talking Therapy
- Nerve Blocks
- Local Anaesthetics

This condition is very much like other medical conditions in the pelvic area. It needs more awareness, research and training (ART) for the medical profession.

W IS FOR WILLPOWER AND WITHDRAWN

WILLPOWER

We all use willpower consistently to get through the day with coping in pain and experiencing low energy. We often do not have much left for extra goals we want to achieve.

If you think you do not have any willpower, well let me tell you that you do. You have loads of it. It is what gets you through each day.

It's when people say to you, "I don't know how you do it". We think to ourselves we just must. But it is our willpower that is the force driving our limited energy through the day. We know from suffering chronic pain with Pudendal Neuralgia it's a constant struggle and some of the points listed below you will be able to resonate with.

- I use willpower to get out of bed on time instead of oversleeping
- I use willpower to not nap too much during the day
- I use willpower to do small amounts of housework when I am so tired.
- I use willpower to do my jobs because they make me feel good mentally.
- I use willpower to accomplish anything outside of the house when I am in pain and tired.
- Then I have used all my energy, done, that is all, finito.

Just by getting out of bed in the morning for many members takes tremendous willpower. It certainly does for me. I think to myself I will have another 5 minutes more, then 30 minutes have gone by. Or I am in too much pain to cook an evening meal, but I do it.

We all have the willpower to get up and fight our way through another day. We should all pat ourselves on the back for achieving this.

WITHDRAWN

If you have a friend or relative who suffers from Pudendal Neuralgia or Vulvodynia, then it's possible that the sufferer has withdrawn from the social life that they had before they were diagnosed.

I found myself turning offers down to go out for meals etc in the evenings because I couldn't sit for that long and enjoy a meal. I would cancel at the last minute to go on a car ride to visit friends for fear of being in so much pain I would spoil the fun for everyone else.

Bedtime for me is 6.30pm. By then I have had enough and all I want to do is lay on my side in bed and wait for the nerves to settle down. I'm a right party pooper (not that it mattered when being in lockdown) but on the odd occasion I have accepted an invitation, I have always paid for it with extra pain for the next few days.

This works both ways of course, friends you had, no longer invite you to social events, consequently after a while they drop away completely.

Pudendal Neuralgia can be a very lonely condition, but I am glad I have the social media groups of fellow sufferers who completely understand the pain we have all been dealt with.

X IS FOR X MARKS THE SPOT AND XEROSTOMIA

This letter has given me much grief since I worked on letter W. I wondered if I should pass it by but no, I wasn't going to give up that easily. So, X marks the spot it will be.

Just like when the surgeons mark the part of the body they are going to operate on with an X. They have been known to take the wrong leg off.

I am unable to show you any Xs on my operation site as it was all done through the vagina (hysterectomy, rectocele and cystocele). In fact, one news agency I won't mention the name of wanted photos of the operation wound taken whilst in hospital. Needless to say, I just shook my head and told them they really had no idea about Pudendal Neuralgia or where my pain was.

Here are just a few of the Xs that apply to us sufferers.

- X can also mean NO
- X to sitting
- X to mesh
- X to underwear
- X to trousers
- X to riding a bicycle without the new supported saddle
- X to using forceps during childbirth
- X to constipation
- X to travelling for long periods
- X to perfumed soap, shower gel and bubble bath

- X to medical professionals in the future not knowing about Pudendal Neuralgia

XEROSTOMIA

Xerostomia is defined as 'dry mouth' resulting from reduced or absent saliva flow. It is not a disease, but it may be a symptom of various medical conditions, a side effect of radiation for cancer treatment to the head and neck, or a side effect of a wide variety of medications.
This probably happens to a lot of people who are on a cocktail of drugs. It's that awful taste when you permanently want something to drink or a polo mint to suck on. I'm not keen on plain water so I need some kind of flavour in it. I've taken to the warm fruity teas and infusions that you can also add to cold water.

Y IS FOR YEARNING FOR CHANGE AND YOURSELF

By now you will all know that I have been yearning for change. ART (AWARENESS, RESEARCH AND TRAINING) for the GPs, consultants, midwives and nurses into Pudendal Neuralgia.

This condition affects far more people than we know about. Men and women are finding the Facebook group every day and they are so glad they have found us because they thought they were alone with their condition. I am just astounded that the medical profession knows so little about it, even in the Obs & Gynae Departments.

Pudendal Neuralgia for some sufferers turns into a mental health problem. I have touched on this previously, but the annoying part about it is, having to wait so long to get diagnosed, then try all the possible treatments available. Some find relief with nerve blocks or high dose medication but some like me just accept their pain as part of their everyday life.

I will continue to fight this cause for as long as it takes me to see these changes.

The words below were written by a fellow sufferer:

"I have already written about my yearning and my mission to bring more awareness, research and training for the medical profession, so they know how to treat us, talk to us, and touch us. The yearning I want to write about now in this section is slightly different. It's how Pudendal Neuralgia sufferers yearn for their old life back.

I had a member of the group message me one night suggesting I use this word because she yearns for the things she can no longer do. Her life is completely different from when the first day of Pudendal Neuralgia struck. She can no longer sit and relax, work or have sexual intercourse with her husband due to the horrendous pain. These are the main parts of her life that she finds difficult to cope with now.

I know from experience and from writing this book that it is impossible to sit down. Some days I have stood up and typed but that puts pressure on the back from being bent over. I literally can't wait to go and lay on my side on the bed at night and wait for the nerves to calm down in both the vagina and anus.

We all miss out on so many activities, things we used to do in our lives, going to concerts, watching a film, going on trips. The thought of going on a plane on holiday just causes PNS (Pudendal Neuralgia Stress) for me. Sometimes I think to myself, to hell with it, it's not going to go away, so I'm going to make the most of my life.

YOURSELF

You? Yes You. Do you look after yourself? If not, it's about time you did. It's not just a matter of taking all the medication every morning, noon and night, it's far more important than that.

Self Care

Why do I need to know about self care? It's basically all about the things that the medication can't do. For example, have you ever tried an Indian Head Massage? It's so relaxing; I have often gone to sleep during a session. It helps to relax the mind and releases the tension that has built up in the head, neck and shoulders. I will go into a couple of the therapies that you can do yourself at home. I took a video of myself having Reiki when the country was in between lockdowns. You will find this on my Instagram page @Pnawareness_ UK_.

How many of you have had a hand massage? That's something you can all try at the end of a stressful day. The hands are continually sore and rough from the constant washing but by rubbing some hand cream or oils into the hands, the benefits may include less anxiety and tension in the muscles of the hand. It also promotes better sleep and a better grip strength. Hand Massages are also beneficial to sufferers of arthritis, Fibromyalgia and MS.

YEARS

Obviously not all the sufferers took part in this survey, but I asked them how long they had been in the pain of Pudendal Neuralgia. Only 12 people answered so I dread to think what the number would have been if everyone had taken part.

The total came to 140 years and nearly all their stories to me say the same thing more or less.

- "My doctor has never heard of it"
- "The Consultant doesn't believe how much pain I am in"
- "Why can I not get rid of this pain?"
- "I am in agony everyday"
- "The Medical Profession know very little about it"
- "I can't go on much longer"

Z IS FOR ZOOM

I just want to talk about the first group Zoom chat we held at the end of January 2021. This event was the first time a few of us had got to see each other and talk. We didn't set an agenda but it just flowed with no awkward silences. It was a good way of finding out how we were all coping with our pain as well as the third lockdown.

Well, I have come to the end of the alphabet which was done twice because all the members seemed to enjoy doing it. I hope you have not only enjoyed reading this, but you have learnt a lot also about how it feels to suffer from this condition. I have found it very therapeutic writing it.

I'm hoping that this book will be read by sufferers, both men and women with other chronic pelvic pain but most of all by the medical profession.

I will continue to raise awareness for Pudendal Neuralgia until this condition is taught in depth on the student's medical college agenda and existing GPs, consultants, midwives and nurses take on extra training to include Pudendal Neuralgia and finally more research needs to be undertaken.

Our pain is not "All Psychological".